VOLUME SIX HUNDRED AND TEN

# METHODS IN ENZYMOLOGY

## Modern Approaches in Drug Discovery

# METHODS IN ENZYMOLOGY

VOLUME SIX HUNDRED AND TEN

# METHODS IN ENZYMOLOGY

## Modern Approaches in Drug Discovery

Edited by

**CHARLES A. LESBURG**
*Department of Chemistry*
*Merck & Co.*
*Boston, MA, United States*

ELSEVIER

ACADEMIC PRESS
An imprint of Elsevier

Academic Press is an imprint of Elsevier
50 Hampshire Street, 5th Floor, Cambridge, MA 02139, United States
525 B Street, Suite 1650, San Diego, CA 92101, United States
The Boulevard, Langford Lane, Kidlington, Oxford OX5 1GB, United Kingdom
125 London Wall, London, EC2Y 5AS, United Kingdom

First edition 2018



ISBN: 978-0-12-815383-3
ISSN: 0076-6879



*Publisher:* Fiona Pattison
*Acquisition Editor:* Ashlie Jackman
*Editorial Project Manager:* Shellie Bryant
*Production Project Manager:* Divya Krishna Kumar
*Cover Designer:* Christian J. Bilbow

Typeset by SPi Global, India

# CONTENTS

# CONTRIBUTORS

**Stephanie Kay Ashenden**
Department of Chemistry, Cambridge University; Discovery Sciences, IMed Biotech Unit, AstraZeneca R&D, Cambridge, United Kingdom

**Christian Bergsdorf**
Novartis Institutes for BioMedical Research, Basel, Switzerland

**Jose Brandao-Neto**
Diamond Light Source, Harwell Science and Innovation Campus, Didcot, United Kingdom

**Paul E. Brennan**
Nuffield Department of Medicine, Structural Genomics Consortium; Nuffield Department of Medicine, Target Discovery Institute; Nuffield Department of Medicine, Alzheimer's Research UK Oxford Drug Discovery Institute, University of Oxford, Oxford, United Kingdom

**Margaret T. Butko**
Biodesy, Inc., South San Francisco, CA, United States

**Juan J. Carrillo**
Lilly Biotechnology Center, San Diego, CA, United States

**Bason Clancy**
Biodesy, Inc., South San Francisco, CA, United States

**Patrick M. Collins**
Diamond Light Source, Harwell Science and Innovation Campus, Didcot, United Kingdom

**Jayme L. Dahlin**
Department of Pathology, Brigham and Women's Hospital, Boston; Broad Institute, Cambridge, MA, United States

**Ronald L. Davis**
Department of Neuroscience, The Scripps Research Institute Florida, Jupiter, FL, United States

**Alexandre Dias**
Diamond Light Source, Harwell Science and Innovation Campus, Didcot, United Kingdom

**Bradley C. Doak**
Medicinal Chemistry, Monash Institute of Pharmaceutical Sciences, Monash University, Parkville, VIC, Australia

**Alice Douangamath**
Diamond Light Source, Harwell Science and Innovation Campus, Didcot, United Kingdom

**Shelby K. Doyle**
David H. Koch Institute for Integrative Cancer Research; Center for Precision Cancer Medicine; Department of Biological Engineering, Massachusetts Institute of Technology; The Broad Institute of MIT and Harvard, Cambridge, MA, United States

**Helen L. Evans**
David H. Koch Institute for Integrative Cancer Research; Center for Precision Cancer Medicine, Massachusetts Institute of Technology; The Broad Institute of MIT and Harvard, Cambridge, MA, United States

**Gillian Farnie**
Nuffield Department of Medicine, Structural Genomics Consortium; Nuffield Department of Orthopedics, Rheumatology and Musculoskeletal Sciences, Botnar Research Centre, University of Oxford, Oxford, United Kingdom

**Suet Ling Felce**
Nuffield Department of Orthopedics, Rheumatology and Musculoskeletal Sciences, Botnar Research Centre, University of Oxford, Oxford, United Kingdom

**Mary Geck Do**
The University of Texas MD Anderson Cancer Center, Institute for Applied Cancer Science, Houston, TX, United States

**Tarun Gheyi**
Lilly Biotechnology Center, San Diego, CA, United States

**Kirk E. Hevener**
Department of Pharmaceutical Sciences, University of Tennessee Health Science Center, Memphis, TN, United States

**Huijue Jia**
BGI-Shenzhen; Shenzhen Key Laboratory of Human Commensal Microorganisms and Health Research; China National Genebank-Shenzhen, BGI-Shenzhen, Shenzhen, China

**Michael E. Johnson**
Center for Biomolecular Sciences; Department of Medicinal Chemistry & Pharmacognosy, University of Illinois at Chicago, Chicago, IL, United States

**Miklos Kepiro**
Department of Neuroscience, The Scripps Research Institute Florida, Jupiter, FL, United States

**Angela N. Koehler**
David H. Koch Institute for Integrative Cancer Research; Center for Precision Cancer Medicine; Department of Biological Engineering, Massachusetts Institute of Technology; The Broad Institute of MIT and Harvard, Cambridge, MA, United States

**Tobias Krojer**
Structural Genomics Consortium, Nuffield Department of Medicine, University of Oxford, Oxford, United Kingdom

**Hyun Lee**
Center for Biomolecular Sciences; Research Resources Center; Department of Medicinal Chemistry & Pharmacognosy, University of Illinois at Chicago, Chicago, IL, United States

**Becky S. Leifer**
David H. Koch Institute for Integrative Cancer Research; Center for Precision Cancer Medicine, Massachusetts Institute of Technology; The Broad Institute of MIT and Harvard, Cambridge, MA, United States

**Fiona E. McCann**
Nuffield Department of Orthopedics, Rheumatology and Musculoskeletal Sciences, Botnar Research Centre; Kennedy Institute of Rheumatology, University of Oxford, Oxford, United Kingdom

**Daniel R. McMasters**
Computational Chemistry, Vertex Pharmaceuticals, Boston, MA, United States

**Ben Moree**
Biodesy, Inc., South San Francisco, CA, United States

**Moses Moustakim**
Nuffield Department of Medicine, Structural Genomics Consortium; Nuffield Department of Medicine, Target Discovery Institute; Department of Chemistry, Chemistry Research Laboratory, University of Oxford, Oxford, United Kingdom

**Russell Pesavento**
Center for Biomolecular Sciences; College of Dentistry, University of Illinois at Chicago, Chicago, IL, United States

**Kiira Ratia**
Research Resources Center; Department of Medicinal Chemistry & Pharmacognosy, University of Illinois at Chicago, Chicago, IL, United States

**JinHong Ren**
Center for Biomolecular Sciences, University of Illinois at Chicago, Chicago, IL, United States

**André Richters**
David H. Koch Institute for Integrative Cancer Research; Center for Precision Cancer Medicine, Massachusetts Institute of Technology; The Broad Institute of MIT and Harvard, Cambridge, MA, United States

**Joshua Salafsky**
Biodesy, Inc., South San Francisco, CA, United States

**Martin J. Scanlon**
Medicinal Chemistry, Monash Institute of Pharmaceutical Sciences, Monash University, Parkville, VIC, Australia

**Gurpreet Singh**
Institute for Therapeutics Discovery and Development, University of Minnesota, Minneapolis, MN, United States

**John Strelow**
Lilly Research Laboratories, Eli Lilly and Company, Lilly Corporate Center, Indianapolis, IN, United States

**Romain Talon**
Diamond Light Source, Harwell Science and Innovation Campus, Didcot, United Kingdom

**Ali Tavassoli**
School of Chemistry, University of Southampton, Southampton, United Kingdom

**Ashley Taylor**
Medicinal Chemistry, Monash Institute of Pharmaceutical Sciences, Monash University, Parkville, VIC, Australia

**Jacob Valentine**
School of Chemistry, University of Southampton, Southampton, United Kingdom

**Boglarka H. Varkuti**
Department of Neuroscience, The Scripps Research Institute Florida, Jupiter, FL, United States

**Frank von Delft**
Diamond Light Source, Harwell Science and Innovation Campus, Didcot; Structural Genomics Consortium, Nuffield Department of Medicine, University of Oxford, Oxford, United Kingdom; Department of Biochemistry, University of Johannesburg, Aukland Park, South Africa

**Michael A. Walters**
Institute for Therapeutics Discovery and Development, University of Minnesota, Minneapolis, MN, United States

**Qi Wang**
BGI-Shenzhen; BGI Education Center, University of Chinese Academy of Sciences, Shenzhen, China

**S. Kirk Wright**
Novartis Institutes for BioMedical Research, Cambridge, MA, United States

**Fangming Yang**
BGI-Shenzhen; BGI Education Center, University of Chinese Academy of Sciences, Shenzhen, China

**Tracy A. Young**
Biodesy, Inc., South San Francisco, CA, United States

**Nancy Zaarour**
Nuffield Department of Orthopedics, Rheumatology and Musculoskeletal Sciences, Botnar Research Centre; Kennedy Institute of Rheumatology, University of Oxford, Oxford, United Kingdom

# PREFACE

Drug discovery is an aspirational term covering various activities to identify a chemical entity which will, at some future moment after successful clinical development, lead to a distributed therapeutic to address a medical need. That future moment is typically years—if not decades—after a decision is made to apply resources, both intellectual and financial. Once a chemical entity—for the sake of this volume generally limited to small molecules and peptides but not biologics—has been discovered, a development campaign is likely to cost between $1 and $3 billion (DiMasi, Grabowski, & Hansen, 2016) and will be fraught with unforeseen failures (Kola & Landis, 2004). With finite resources available it is therefore of utmost importance to maximize the success rate of drug discovery efforts. It remains a substantial challenge to know in advance that drug discovery efforts will actually result in a drug, or a better understanding of human biology, or perhaps nothing at all.

Drug discovery is a field both broad and deep practiced by academic, industrial, and government entities alike. It is a rich ecosystem populated by large global corporations, academic research groups large and small, public and private funding agencies and foundations, sources of capital large and small, intellectual property experts, focused biotechnology companies, technology incubators, independent consultants, and many others around the world.

The field has been reviewed, scrutinized, criticized, and refined at length, with innumerable reviews and also entire peer-reviewed publications dedicated to publishing all aspects of drug discovery. For the purposes of this volume, the scope of drug discovery is further limited to aspects of target identification, design and screening of compound libraries via biophysical, biochemical, and phenotypic techniques. Chapters do not cover any testing in animal models of disease nor preclinical toxicological evaluation. More specifically, this volume aims to address topics of utility which have recently emerged or have been better formalized.

These topics include not only methods to identify and assess compounds but also to understand the inherent risks present in this diverse field. To this point, this volume is organized first with a chapter on risk identification and management (Chapter 1). It is essential to contemplate a drug discovery effort from a perspective which will inform the application of resources.

The following chapters are grouped by traditional boundaries: target identification (Chapters 2 and 3), library design (Chapters 4–6), screening methods (Chapters 7–11), evolution of screening hits into lead molecules (Chapter 12), and the identification and assessment of off-target activities (Chapter 13). Each of these topics contains chapters addressing modern advances in their respective areas, each in an effort to improve the quality of the targets we choose to screen, to find high-quality molecules through different screening methods, and to condense of the soup of hits into high quality leads and ultimately into candidates for clinical development.

After all, the quality of the chemical entities emerging from drug discovery pipelines will have a dramatic impact on the ability of those molecules to emerge successfully from a drug development effort and improve human health.

Charles A. Lesburg
Department of Chemistry, Merck & Co., Boston, MA, United States

## REFERENCES

DiMasi, J., Grabowski, H., & Hansen, R. W. (2016). Innovation in the pharmaceutical industry: New estimates of R&D costs. *Journal of Health Economics*, *47*, 20–33.

Kola, I., & Landis, J. (2004). Can the pharmaceutical industry reduce attrition rates? *Nature Reviews Drug Discovery*, *3*, 711–716.

CHAPTER ONE

# Risk Management in Early Discovery Medicinal Chemistry

**Gurpreet Singh*, Jayme L. Dahlin[†,‡], Michael A. Walters*[,1]**
*Institute for Therapeutics Discovery and Development, University of Minnesota, Minneapolis, MN, United States
[†]Department of Pathology, Brigham and Women's Hospital, Boston, MA, United States
[‡]Broad Institute, Cambridge, MA, United States
[1]Corresponding author: e-mail address: mwalters@umn.edu

## Contents



## Abstract

Drug discovery is inherently very risky. The management of these risks can enable the effective use of limited human and monetary resources. A careful attention to risk management in early discovery is especially important given that what happens in the early phases of a project may dictate the course of a research program for months or years. Risk management in early discovery starts with high-level managerial concerns: careful project selection, sufficient staffing and funding, and access to the appropriate instrumentation and tools. Herein we describe the operational elements of risk management that range from the very broad to the extremely specific. These elements have as their base an embedded culture of risk management that extends down to the

*Methods in Enzymology*, Volume 610
ISSN 0076-6879
https://doi.org/10.1016/bs.mie.2018.09.012

experiment level, and project ownership in which all researchers anticipate risk, but are not paralyzed by it. In our model, on this base of culture stand the four pillars of early discovery risk management: right libraries, right assays, right series, and right structure–activity relationships. Appropriate attention to these considerations can decrease the risks inherent in early discovery medicinal chemistry, thereby potentially increasing the return on the investment of necessarily finite resources.

## 1. INTRODUCTION

Researchers in translational medicine increasingly need to navigate some of the risky landscape in early drug discovery and development. Any journey on this landscape is fraught with risks. However, we will explain how to engage these risks wisely, thereby offering the highest likelihood of success for those who have the will to venture into this complex terrain. Our focus here is primarily derisking the active compounds found in high-throughput screening (HTS) in an academic drug discovery and development center (Dosa et al., 2014). However, the advice we offer should be generally useful in any stage of drug discovery and development, be it in academia or industry.

## 2. RISK IN DRUG DISCOVERY

The discovery of drugs is an inherently risky enterprise whether it is tackled by academia or industry (small-biotech to large pharmaceutical companies) (Lombardino & Lowe, 2004). That being said, since this risk is heavily dependent on a balance of time, money, and will (or passion), academic endeavors aimed at the rather high bar of drug discovery will almost always face greater risks than similar work in established pharmaceutical companies ("big pharma"). Academia typically faces risk in the discovery of chemical matter and the early derisking of compounds up to IND (investigational new drug) status and Phase I clinical trials. Beyond this point, industry usually is employed as a licensing partner to help shepherd projects through the costly and time-consuming process of Phase I–III clinical trials that may ultimately lead to a new drug application.

## 3. MANAGEMENT OF RISKS IN DRUG DISCOVERY

The management of drug discovery projects in the academic setting has evolved to be similar but definitely not the same to that which is found in industry. Industrial drug discovery is baked into the fabric of the business and constitutes the only measurable output of the company. In industry, therefore, there is a typically a pyramidal organization structure with strategic portfolio management being conducted at a high level, while the tactical management of the pipeline (the collective of ongoing discovery and development projects) is managed by groups closer to the basic science. In academia, drug discovery is typically not even on the radar of collegiate administrators and many academic scientists may not even be aware that such work is ongoing at their institution. One of the principle purposes of the academy is to progress an extensive number of fields and to prepare educated alumni. However, translation of drug discovery projects in some institutions is already viewed as both serving a societal need and providing an alternative source of funds in times of decreasing influx of federal and state money. This being the case, many academic institutions have established drug discovery centers that facilitate early discovery. Some colleges and universities have set up a relatively informal management organization that, at the very least, is charged with managing the funding of discovery and development projects (Frye, Crosby, Edwards, & Juliano, 2011; Huryn, 2013; Slusher, Conn, Frye, Glicksman, & Arkin, 2013).

With this management organization comes the establishment of rules and strategies to manage risk. What projects should be worked on? How should they be funded? We suggest that it is best to manage risk by considering diversifying the goals of early discovery projects. We propose that four appropriate goals of early academic drug discovery can be (1) project termination, (2) enhanced funding and project continuance, (3) probe discovery, and (4) lead discovery and beyond. These goals are listed in ascending order of degree of difficulty. Each of these directly corresponds to similar milestones in industry. Admittedly, project termination does not sound too difficult and it certainly does not sound too much like success. Termination of a project can feel like failure to the principal investigator who may have spent an academic lifetime pursuing a particular target or a therapeutic regimen. However, if the decision to terminate a project is reached in a rational fashion, it can free up early discovery funding for other more tractable projects. The "pet" project can continue along the lines of fundamental research until it is perhaps again ready to enter the discovery pipeline. It is worthy of note that NIH funding contributed to every one of the new drugs approved by the FDA from 2010 to 2016 (Galkina Cleary, Beierlein, Khanuja, McNamee, & Ledley, 2018). We must point out that academic endeavors should most appropriately target the advancement of fundamental science and that it should be the rare case that a project becomes part of the "discovery" portfolio (Zoghbi, 2013). Nevertheless, projects that eventually are selected for "translation" must be subject to rigorous derisking to spare and focus limited discovery resources.

We follow Roy's recommendation that a probe mindset might be the most powerful way to derisk early discovery in academia (Roy, 2018). Probes are compounds that are potent, selective, have cellular activity, and are used to answer biochemical questions both in vitro and in vivo (Arrowsmith et al., 2015; Blagg & Workman, 2014; Huryn, Resnick, & Wipf, 2013; Skuta et al., 2017). Drugs for use in humans face much higher standards for safety, routes of administration, and commercial viability. Additionally, probes are not typically constrained by intellectual property concerns and therefore, studies of probes can be more readily published. This is an important consideration for academic groups where students might be unfairly hindered by the lack of publications. That being said, the institute to which two of us belong (G.S. and M.A.W.) has produced a drug that was discovered and developed solely by a graduate student and is currently in Phase II clinical trials (Patil et al., 2015) (ITDD, The Institute for Therapeutics Discovery and Development at the University of Minnesota). Two strategies may serve both probe and drug discovery, first, a strategic patenting and publication plan for either probes or drug discovery that is agreed to among all researchers on a project prior to or early on in the project; second, the discovery and development of probes with drug-like properties in mind.

## 4. GENERAL STRATEGIES FOR MANAGING RISKS

We and others have outlined risks that need to be managed up to IND and these need to be looked at through the lens of systems thinking, in which the endpoint of a process is always kept in mind. That is, even though risks such as patent expirations may only be of long-term concern for a select project, these risks must be considered even in the earliest phases of drug discovery. Our focus in this chapter is on the earliest stages of the discovery process in academia (or industry) since this is where we have the most experience and, most importantly, where the stage is often set for later success or failure. Several articles have been devoted to the outline of important risk-reduction considerations across the broad span of drug discovery and development activities (Dahlin, Inglese, & Walters, 2015). Ultimately it is important to have all of these risks in mind when embarking on the discovery of a new drug.

Roy (2018) presented an excellent overview of early probe and drug discovery in academia and highlighted the following short- and long-term concerns: (1) lack of correlation between target and disease, (2) synthetic difficulty, (3) potential target-related toxic effects, (4) inadequate endpoint and patient selection, (5) patent expiration and competition from generics, (6) drug discontinuation due to long-term safety issues, and (7) poor efficacies across larger, genetically diverse populations than those involved in the clinical trials.

## 4.1 Business Models for Managing Risk

Strategic guidelines for managing risk in business can be applied to the scientific enterprise at the project level. Consider the following rules for business risk management as applied to early discovery: (1) enhance risk strategy, (2) embed risk management, (3) optimize risk management functions, and (4) improve controls and processes (Ernst & Young Global Limited, 2013). A risk strategy should be a key component of any discovery project and helps the project team protect itself from misleading and irreproducible results. A culture of risk management should be embedded into each team member. This can be done by instilling a sense of ownership for the project at every scientific level. Reagents, methods, instrumentation, and conceptual models should be routinely validated (see Section 6). Risk management functions should be clearly spelled out in the team charter and revisited on a regular basis. Who is responsible for the validation of experiments, compounds, cells, antibodies, animal models, computational models, etc.? Finally, risk controls and processes should be viewed as a normal component of the discovery process. In the short term, they may appear to be burdensome, but in the long term, they will ensure more effective project operations and reduce the risk of "unforced" failure, i.e., failure that was (should have been) controllable.

Enterprise risk management (ERM) is another business model used to understand and deal with corporate risk that can be applied equally well to the discovery process. ERM continuous implementation typically involves four steps: (1) identification of risks, (2) assessment of these risks in terms of likelihood and magnitude, (3) determining a response to these risks, and (4) monitoring the progress of these responses (Fig. 1). For example, consider the risks associated with the keeping of experimental records in a paper-based laboratory notebook. There are risks associated with this practice: the notebooks may be lost, damaged, or stolen. These risks can be mitigated by the response of maintaining careful control of project notebooks and ensure they

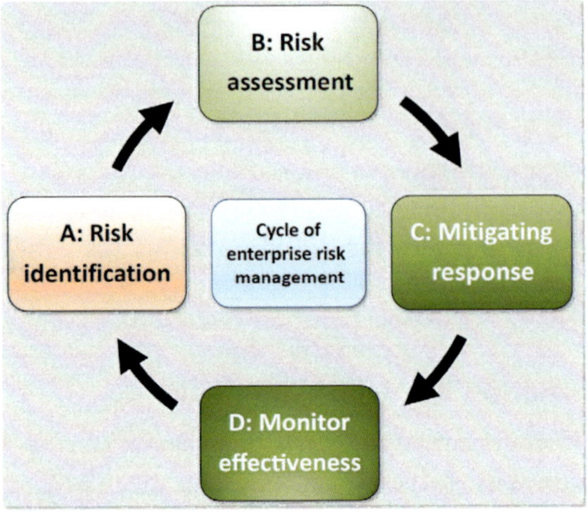


**Fig. 1** The cycle of enterprise risk management. (A) Risk identification, (B) risk assessment: what is the nature of the risk? How damaging could failure to manage the risk be to the success of the project? (C) Mitigating response: how will the project team deal with the risk? What procedures or protocols will be put in place? (D) Monitoring effectiveness: how well did the procedures that were enacted manage the risk? The cycle continues with the identification of other risks or refining of the mitigating responses on the initially identified risk.

are kept in locked, water-and-fireproof cabinets. In all likelihood, however, the greatest risk (most likely in academia) is that project experiments are not even recorded in the notebooks. In this case, the risks to the project are the lack of both the easy reproducibility of the missing experiments and the lack of evidence of the experiments having been executed. This, of course, could lead to the failure of the discovery project and is of high magnitude. The response to the risk could be frequent checking of notebooks and data annotation/organization to ensure they are being kept up to date. Another potential response would be to switch to electronic, cloud-based notebooks and institute a frequent review cycle of notebook pages. This response would ensure that both the experimental data were secured offsite to avoid the risks associated with loss and damage, and scheduled reviews would ensure record-keeping compliance. This risk analysis can be applied to minimize any number of risks that might face early discovery projects.

## 4.2 The Four Pillars

Pfizer scientists presented a Four Pillars concept for the development of probe compounds for target validation (Bunnage, Chekler, & Jones, 2013).

They observed that drug programs that achieved the "three pillars of survival" in cell-based assays were more likely to succeed than those that did not. These three pillars are (1) sufficient exposure at the site of action; (2) proof of target engagement; and (3) expression of functional pharmacological activity. They added a fourth pillar, proof of phenotype perturbation, to round out their recommendations for robust molecular target validation.

### 4.3 The Five Dimensions of Drug Discovery Productivity

Researchers at AstraZeneca reported a five-dimensional productivity strategy that improved the company's productivity, as measured by success rates, from 4% in 2005–10 to 19% in 2012–16 (Fig. 2). The five dimensions of this strategy are all about getting things "right," that is, risk reduction. The five dimensions are (1) right target, (2) right tissue, (3) right safety, (4) right patient, and (5) right commercial potential. This strategic mindset is heavily weighted toward the later stages of drug discovery (Morgan et al., 2018).

## 5. OUR FOUR PILLARS OF EARLY DISCOVERY DERISKING

We propose that early drug discovery projects can be more successful if the following four pillars are kept in mind: (1) right library; (2) right assays; (3) right compound(s); and (4) right series (Fig. 3). These pillars are in turn supported on the foundation of project ownership and an embedded culture of risk management. An embedded culture of risk management exists when the organization, from the administration down to the laboratory level, is knowledgeable in the potential risks associated with each step of the drug discovery and development process. At the laboratory level, this means that risk management is embedded into each experiment. This means that experiments should be designed such that one can learn something no matter the outcome (positive or negative). If the experiment fails, be sure to have included controls in it that let you narrow down where in the workflow something went wrong. If the outcome is a "true negative" have sufficient controls in place that you can be confident that you would have seen a positive result if your experimental hypothesis was true. For example, if no active compounds are found in a high-throughput screen (a "true negative" result), one should have confidence through controls, reference compounds, and optimization of assay performance that if an active compound existed in the screening deck, you would have found it. Project ownership supports risk management in that it becomes the responsibility of all members of a project team to use the right tools to get the true results. These results are

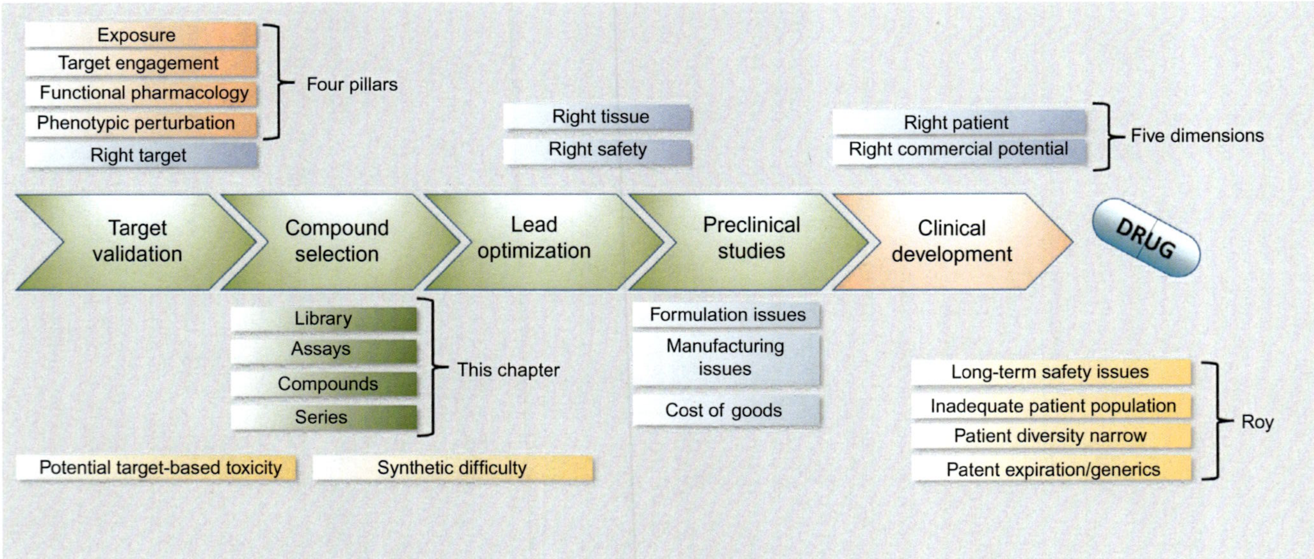


**Fig. 2** A summary of the fundamentals of risk management that have appeared in the literature and appear in this chapter. The process of drug discovery and development is shown with chevrons starting from target validation and continuing through clinical development. This chapter focuses on risk management in the compound selection stage, but many of the fundamentals of risk management discussed herein are applicable in any stage.

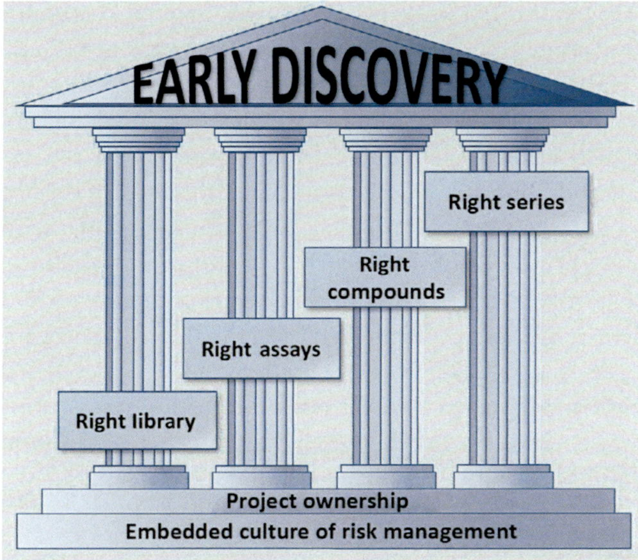


**Fig. 3** The success or failure of early discovery project stands on the scientific pillars of right library, right assay, right compound, and right series. The four pillars are supported by the organizational foundations of an embedded culture of risk management and project ownership.

not necessarily the results one may expect, but are results that can be trusted to be accurate. In early discovery this requires the use of validated libraries, validated assays featuring reference compounds and replicates, purified compounds, and structure–activity relationships (SARs) that can be used in the design of optimized compounds.

## 5.1 Right Library

HTS remains the conventional strategy for finding ligands for new targets (Macarron et al., 2011). Apart from in-house collections or synthesis efforts, millions of compounds are available for screening from various commercial vendors. In 2013, Baell conducted a study on a compound collection from commercial vendors representing 80% coverage of available chemistry space (Baell, 2013). By using filters based on physiochemical lead-likeness, functional groups, similarity, and PAINS, they designed a library that was representative of the available chemical space that included fewer than 350,000 compounds. A small-molecule library of an appropriate size and chemical diversity may result in a sizeable number of hits. However, appropriate steps must be taken so that screening effort results in a greater number of true hits (taking into account tractability), and fewer nuisance compounds that may lead to a predestined journey to failure at later stage (Lushington & Chaguturu, 2014). It requires due diligence and collective expertise of high-throughput technology, cheminformatics, and medicinal and analytical chemistry to inform the right choices pertaining to the selection of library compounds for primary screening. The properties of the library compounds that may interfere with the assay outcome can be of chemical (nonspecific reactivity) and/or physical (such as aggregation, solubility, inherent fluorescence, purity) nature. Though the field is not yet mature enough to precisely characterize every chemical structure in the context of above problems, however, a significant level of understanding has been developed based on the experience of various research groups across the world. In the following sections, we list the risk factors and the corresponding measures one should take while finalizing a library for screening.

### 5.1.1 Reactive Compounds

**a.** Highly reactive functional groups including, e.g., acid chlorides, acid anhydrides, sulfonyl chlorides, cyanates, iso(thio)cyanates, peroxides, acyl hydrazides, can be eliminated by using filters that include SMILES strings specific for such problematic functional groups. (For a comprehensive list of such reactive groups see Brenk et al., 2008.)

**b.** Redox cyclic compounds (RCCs) containing chemical moieties that can produce significant concentration of hydrogen peroxide under assay conditions and alter the assay outcome by oxidizing protein targets, assay components, or by interfering with cell permeability and signaling pathways (Soares et al., 2010). Several classes of compounds including certain pyrimidotriazinediones, *ortho*- and *para*-quinones, arylsulfonamides, tolylhydrazides, nitrophene-2-carboxylates, acetamido-dihydroisoxazolo-pyridine-diyl diacetates, and pyrrolo-quinoxalin-chlorobenzamides have been reported as RCCs (Dahlin & Walters, 2016; Johnston, 2011).

**c.** Several studies have highlighted the removal of assay interference compounds such as PAINS (Baell, 2013). These have provided a detailed list of compounds and associated cheminformatic filters one can employ to eliminate such classes of compounds from the pre-HTS collection. However, as the mechanistic basis for the reactivity of many such compounds is sometimes unclear and the reactivity is highly scaffold dependent, one may choose to only flag such compounds in the initial screening and employ secondary confirmation steps if these appear as hits in the results (Baell & Nissink, 2018).

### 5.1.2 Solubility and Physiochemical Properties

Compounds with poorly aqueous solubility or high lipophilicity are generally not suitable in a screening collection. Such compounds may form colloidal aggregates and appear as hits with limited optimization potential. Rapid elimination of swill (REOS), a cheminformatics filter that combines guidelines like Lipinski's rule of five with the reactive functional group filters, can be useful to eliminate compounds with reactive moieties and undesirable physiochemical properties (Walters, Stahl, & Murcko, 1998). Additionally, filters like the Abbott Physicochemical Tiering (APT) can be used to both help prepare better libraries and triage compounds after HTS (Cox, Gregg, & Vasudevan, 2012).

### 5.1.3 Purity and Structure Analysis of the Library Compounds

Once a library has been selected and acquired, quality control and analysis is required to ensure the chemical integrity and purity of the compounds. A detailed structural analysis is not feasible at this stage considering the large number of compounds. However, UV/MS/ELSD-directed LC-based purity analysis of a random subset of library compounds can inform the quality of the library. A cutoff of greater than 90% purity based on UV analysis is often preferred. Another important aspect is to consider the duration for which the plated libraries can be stored and be still viable for biological testing. Based on an assay-based QC of a library over several weeks, Auld and coworkers suggested a 4-month storage life before library daughter plates are replaced (McArthur et al., 2009).

## 5.2 Right Assays

Assays and screening trees must be well designed to select compounds that exhibit potent and selective modulation of the chosen target. The right assays should take into account the potential modes of compound interference.

### 5.2.1 Derisking Interferences Through Assay Design

Significant sources of risk related to bioassays include *irreproducible* bioactivity (false positives) as well as *reproducible* compound-mediated assay interferences. We recommend derisking compounds through (1) rigorous bioactivity confirmation, (2) key technical modifications to assay design, and (3) multiple counter-screens for generalized and technology-related assay interference and target selectivity (Dahlin, Inglese, et al., 2015; Dahlin & Walters, 2014).

### 5.2.2 Sources of Compound-Mediated Assay Interference

We broadly divide compound-mediated assay interference into two categories: generalized and technology-related interference (Fig. 4). Others and we have noted the importance of this distinction (Kenny, 2017; Xie et al., 2018). Generalized interferences represent on-target/on-many-targets due to poorly tractable mechanisms of action. We consider this "interference" in that the underlying chemistry of such bioactivity impedes the identification of ultimately useful, on-target chemical matter. As a result, generalized interferences often produce readout changes across multiple assay technologies, usually through nonspecific modulation of protein structure and dynamics (Baell & Nissink, 2018; Kenny, 2017). Generalized interference examples include aggregation, nonspecific reactivity, redox cycling, denaturation, and chelation (Thorne, Auld, & Inglese, 2010). Technology-related interferences interfere with specific steps in assay technologies. Examples include capture reagent disruptors (e.g., biotin mimetics in assays utilizing biotin–streptavidin tags),

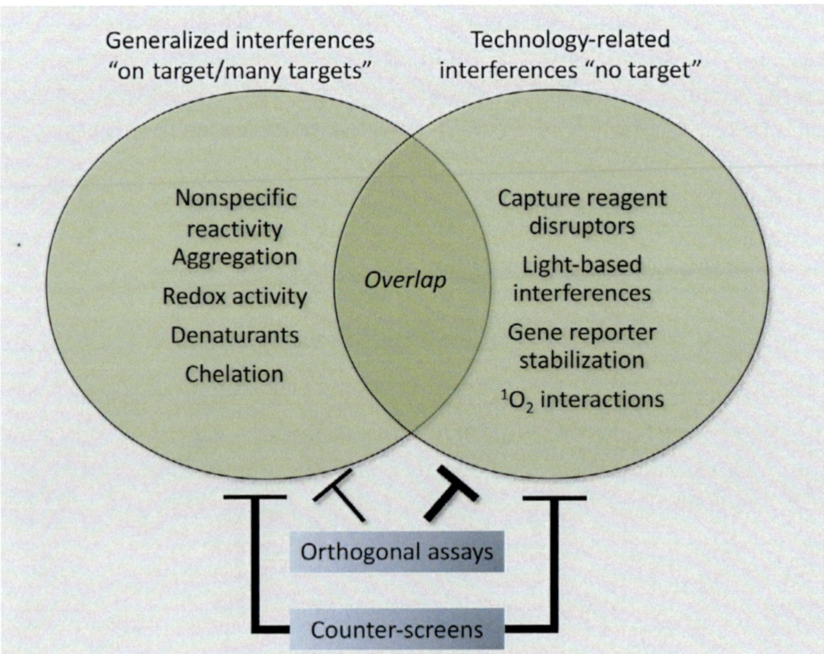


**Fig. 4** Compound-mediated assay interference schematic. Interference is broadly categorized as generalized- and technology-related interferences, though overlap exists when the interference source disrupts both the biological target and the assay technology. Counter-screens are useful to identify specific sources of interference, while orthogonal assays are more useful to identify technology-related interferences.

singlet oxygen quenchers and scavengers in AlphaScreen™, reporter gene stabilizers or inhibitors in luciferase assays, or fluorescence quenchers in fluorescence intensity or FRET assays (Auld et al., 2010; Brenke et al., 2016; Schorpp et al., 2014; Simeonov & Davis, 2015). A key concept is that generalized assay interferences often show highly reproducible bioactivity across orthogonal assay formats (though activity can vary based on certain technical factors such as protein concentration, buffer components, and reaction time). Technology-related interferences can often be identified when there are significant changes in bioactivity when truly orthogonal techniques are utilized. Note that there can be overlap between the two categories, such as when interference modulates the biological target(s) of interest and also interferes with the transmission of the assay signal. Examples can include (1) chelators, which can disrupt metal-dependent biological systems and interfere with certain capture reagents, and (2) nonspecific reactivity, whereby electrophiles can covalently modulate biological targets and also assay reagents (Dahlin & Walters, 2014).

### 5.2.3 Bioactivity Confirmation

A critical compound of risk mitigation as it relates to compound assays is to establish bioactivity across multiple testing formats, define bioactivity precision, and rigorously characterize the bioactive compound. To this end, we recommend the following general approach:

1. Confirm the bioactivity of the compound with multiple technical and biological replicates including technical replicates and concentration–response formats. This derisks the possibility of random error (false positive) and helps define the intra- and interassay precision (Inglese et al., 2006; Malo, Hanley, Cerquozzi, Pelletier, & Nadon, 2006).
2. Confirm the bioactivity with at least one orthogonal assay technology to derisk the possibility of technology-dependent activity. While promising, note that bioactivity confirmed by orthogonal assays does not necessarily represent *tractable* bioactivity (Dahlin & Walters, 2014).
3. Confirm the bioactivity of the compound with repurchased material (with and without purification) to derisk the possibility of material impurities (Dahlin, Baell, & Walters, 2015). See also Section 5.3.
4. Confirm the bioactivity of the compound with synthesized material to further derisk the possibility of material impurities and establish more control of compound production (Falk et al., 2011; Tarzia et al., 2007).
5. Perform rigorous chemical and purity characterization of top candidates.

### 5.2.4 Assay Design to Mitigate Compound-Mediated Assay Interferences

There are several strategies that can be employed in biochemical assays that should reduce the incidence of many common sources of compound-mediated assay interferences. We recommend the following:

1. Mitigate aggregation through the inclusion of nonionic detergents (e.g., Triton X-100, Pluronic) and decoy proteins (e.g., BSA) if assay performance permits (McGovern, Caselli, Grigorieff, & Shoichet, 2002; McGovern, Helfand, Feng, & Shoichet, 2003; Ryan, Gray, Lowe, & Chung, 2003).
2. Mitigate electrophiles through the inclusion of electrophilic scavenging reagents (e.g., dithiothreitol, DTT) and decoy proteins (e.g., BSA) if assay performance permits (Dahlin, Baell, et al., 2015; Dahlin, Nissink, et al., 2015).
3. Mitigate metal contaminants through the inclusion of chelation agents (e.g., EDTA) if assay performance permits (Hermann et al., 2013; Morreale et al., 2017).
4. Utilize longer-wavelength (red-shifted) channels for light-based assays, which reduces the likelihood of light interference by compounds (i.e., quenching, autofluorescence, inner-filter effects) by compounds (Simeonov et al., 2008; Turek-Etienne et al., 2003; Vedvik et al., 2004).
5. Include reference compounds whenever possible. Ideally, this includes both well-behaved controls (if available) and interference controls to gauge the sensitivity of the target and assay readout to common modes of assay interference (Dahlin et al., 2017).

### 5.2.5 Post-HTS Counter-Screens to Derisk Compound-Mediated Assay Interferences

Even if appropriate steps are taken to mitigate the incidence of compound-mediated assay interferences, there is still the possibility for poorly tractable compounds to pose as promising leads. In acknowledgement of this reality, we highly recommend multiple counter-screens for generalized- and technology-related interference.

1. Derisk nonspecific thiol reactivity by examining for direct evidence of compound-thiol adducts and also supporting mechanistic studies:
   a. Incubate compound with thiol-based probes and examine for evidence of adduct formation. Common thiol-based probes include glutathione (GSH), coenzyme A, cysteamine, which can be adapted to U/HPLC-MS and fluorometric readouts (Dahlin, Baell, et al., 2015). Users should also consider protein-based reporters such as ALARM

NMR, as some compounds may react with protein-based thiols but not with nonproteinaceous thiols like GSH (Huth et al., 2005).

   **b.** Assess for mechanistic hallmarks of covalent modulation, which include time dependence and reversibility (jump dilution, dialysis, SPR) (Dahlin, Baell, et al., 2015).

**2.** Derisk redox-active compounds by performing at least one counter-screen for $H_2O_2$ production by test compounds (Soares et al., 2010). Example assays include horseradish peroxidase-phenol red, resorufin, and TCEP NMR (Johnston et al., 2008; Lor et al., 2007; Tarnowski et al., 2018).

**3.** Derisk aggregation by performing concentration–response analyses, detergent titrations with the original target, and at least one counter-screen with an unrelated protein such as AmpC β-lactamase (Auld, Inglese, & Dahlin, 2017; Feng & Shoichet, 2006). Aggregate formation can also be confirmed with dynamic light scattering.

**4.** Derisk light-based interferences by (1) characterizing the absorbance and fluorescence properties of the bioactive compounds and (2) spiking compound to completed reaction mixtures (Dahlin et al., 2017; Simeonov et al., 2008).

**5.** Perform technology-specific counter-screens as dictated by the primary and orthogonal assay technologies. This is especially important if no viable orthogonal assays are available. For example, this could include changing the capture reagent system from biotin–streptavidin to GSH-GST or His-Ni to rule out tag mimetics (Falk et al., 2011).

**6.** Derisk nonspecific bioactivity by performing selectivity testing vs multiple related and unrelated targets, preferably at concentrations reflective of their use in biochemical and cellular assays. While initial selectivity may be promising, apparent selectivity may reflect intrinsic susceptibility of certain targets to interferences. Certain targets may be exquisitely sensitive to thiol-reactive compounds or aggregators, while others may be unperturbed (Dahlin et al., 2017).

## 5.3 Right Compounds

Of course, the "right compounds" should be potent and selective modulators of the chosen target. However, there are two further considerations when it comes to the "right compounds." The first is that researchers have recognized that some compound classes are highly promiscuous. This promiscuity can be caused by many factors including aggregation, fluorescence interference, metal chelating, electrophilic reactivity, and redox cycling. Identification of promiscuous compounds is aided by following the assay approach described in Section 5.2 (n.b., Promiscuous compounds are often confused with privileged compounds. Promiscuous compounds typically do not modulate targets by the formation of nonbonded interactions; their polypharmacology transcends known ligand–protein interactions. Privileged compounds have pharmacophoric substructures that are known to interact with common features on proteins). The second consideration is that the purity and identity of purchased compounds is often not as advertised. That is, the compound that has been tested is often not a pure compound and perhaps not even the compound whose structure is associated with the assay plate well that has been tested. Impurities often lead to irreproducible or intractable bioactivity. Incorrect structural assignments can lead to a waste of effort in synthetic follow-up, increasing the risk of loss of resources. We describe here five levels of compound characterization for finalizing the compounds for SAR evaluation (Fig. 5).

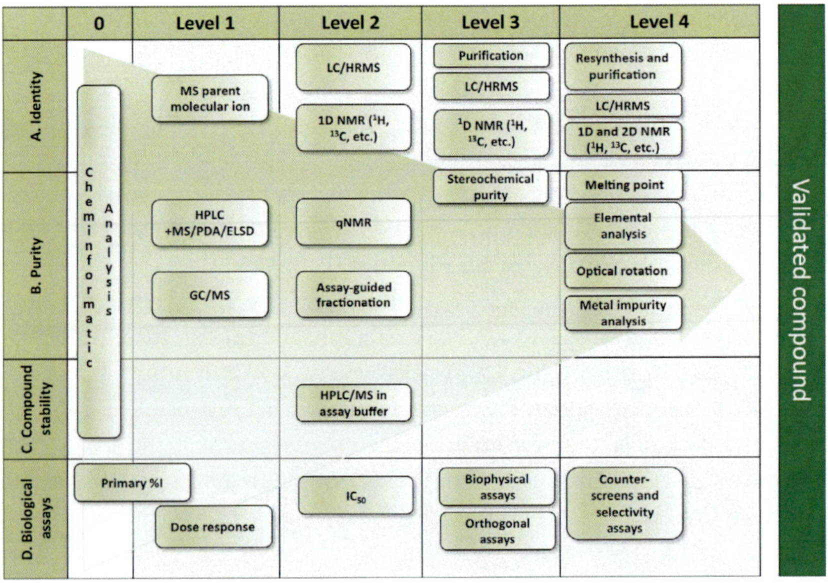


**Fig. 5** Chemical and biological validation of a compound from high-throughput screening. (A)–(D) The components of validation: identity, purity, compound stability, and biological assays. Levels 1–4 represent levels of confidence in the compound at each step of validation. For example, Stage 1 validation of compound identity is simply the observation of the molecular ion in mass spectrometry (sometimes used to QC large libraries), while the characterization at Stage 4 validation is much more thorough and affords a higher level of confidence (lower risk) that the compound has been correctly identified. Compounds need to be taken to the Level 4 of confidence to successfully complete compound validation.

*Level 0.* Active compounds should be prioritized for follow-up using cheminformatic filters such as PAINS (Baell, 2013), REOS (Walters & Namchuk, 2003; Walters et al., 1998), APT (Cox et al., 2012), QED (Bickerton, Paolini, Besnard, Muresan, & Hopkins, 2012), Pfizer's CNS-MPO (Wager, Hou, Verhoest, & Villalobos, 2010), Lipinski's (Lipinski, Lombardo, Dominy, & Feeney, 1997) and Veber's Rules (Veber et al., 2002), and Badapple (Yang et al., 2016). Each of these informatics methods offers different readouts as to the risk involved in progressing a compound.
*Level 1.* Repurchase the active compounds in powder form (sometimes referred to as "dry powder") and confirm the integrity (molecular ion peak in LC/ES-HRMS) and purity (greater than 90% by UV-HPLC analysis). Reassay and look for the dose dependence in bioactivity.
*Level 2.* Confirm the molecular structure by NMR analysis ($^{1}H$, $^{13}C$) and by high-resolution mass spectrometry (HRMS). The active compounds with confirmed structure and identity (HPLC purity greater than 95%, preferably in two solvent systems (American Chemical Society, 2018)) are should then be retested for concentration–response bioactivity.
*Level 3.* At this stage, we recommend performing detailed analysis to confirm molecular structure of the compounds. If the compound structure includes stereocenters, determine its absolute stereochemistry and optical purity. The compounds with complete assignment of correct chemical structure should then be evaluated in biophysical and orthogonal assays to confirm their interactions with the target protein (Mashalidis, Sledz, Lang, & Abell, 2013).
*Level 4.* Resynthesize the selected hits that passed the Level 3 criteria and subject them to stringent purification to rule out the presence of any heavy metal impurities. Quality control by LC/MS and NMR does not exclude the possibility of the presence of metal impurities. Walden and coworkers reported a detailed study on a fragment hit obtained in screening against ubiquitin-conjugating enzyme Ube2T (Hermann et al., 2013). The compound was evaluated in a range of biophysical assays and showed consistent results in a concentration-dependent manner. However, a flat SAR prompted them to investigate further. They ultimately found that the zinc contamination in the compound stock was responsible for the resulting biological activity. We recommend using commercially available functionalized resin-based filters or

silica-based metal scavengers to eliminate the metal impurities from organic compounds (Barbaras, Brozio, Johannsen, & Allmendinger, 2009). Techniques like energy-dispersive X-ray fluorescence analysis are used in pharmaceutical industry for the analytical and quantitative analysis of metals in bulk compounds (Marui et al., 2009). Assay the compounds with confirmed structure and purity (greater than 95% by HPLC and absence of metal impurities) in biophysical assays and confirm the concentration-dependent bioactivity.

## 5.4 Right Series

The compounds that are found as actives and confirmed as hits from an assay will always have to be optimized for selectivity, potency, and physicochemical properties related to permeability, stability, or solubility, or ADMET (absorption, distribution, metabolism, excretion, and toxicological) properties. Risk in this phase of drug discovery is reduced by ensuring that the right series is being pursued. First, testing close structural analogues of active compounds can help build confidence in the activity of a series of compounds. Second, demonstrated SAR and structure–physical chemical property relationships in a series raises confidence that a selected chemical series can be further optimized. To this end, selected compounds from the HTS with the reproducible $IC_{50}$ values (in the case of inhibitors) and their closely related structural analogues (either from the HTS campaign, acquired from the commercial resources, or synthesized in-house) should be reevaluated for the biological activity. Hit analogue series that show either a flat SAR (less than three- to fivefold change in activity in a series of five to seven compounds) or an abrupt loss of activity on minor modifications (single atom changes) of the parent compound should be on low priority because of limited scope for further optimization. Closely related structures should have similar activity and increasing divergence from the parent scaffold should impact the activity significantly. Such a scaffold will have better scope for further optimization to a lead candidate (see Fig. 6 for examples of series with flat SAR and responsive SAR). A set of structurally related, validated compounds (see Section 5.3) can be said to belong to a validated series if it and its analogues display greater than 10-fold spread of activity. Additional derisking can be achieved by in vitro ADMET screening and in vivo pharmacokinetic studies.

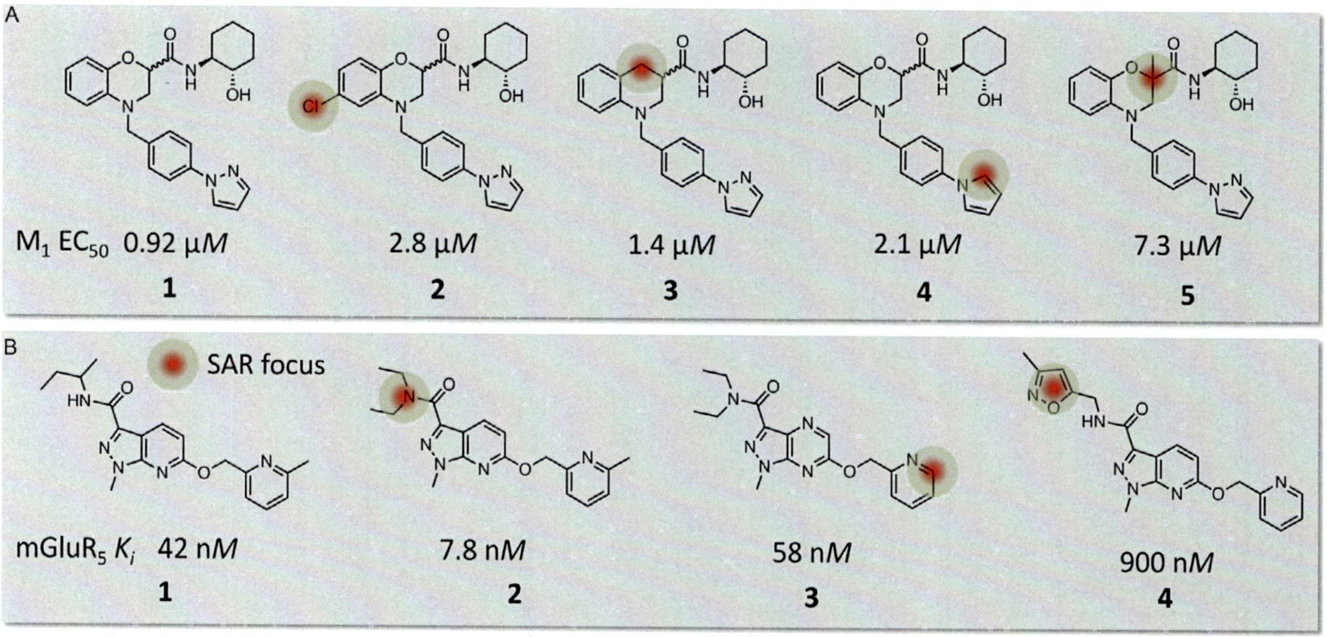


**Fig. 6** Two examples of SAR (structure–activity relationships) from the literature. (A) Flat SAR as reported for a program targeting the muscarinic acetylcholine receptor, $M_1$. It is important to note that all of the analogues of A1 have activity. Close analogues of an active compound identified in screening should have activity that is within 10-fold of the "parent" compound. However, note here that A1–A5 have essentially the same activity (flat SAR) which indicates that it may be extremely difficult to optimize this compound series (Bertron et al., 2018). (B) Analogue series based on B1. Note here that the close analogue B2 has similar activity to the "parent" compound. Additionally, minor changes to the structure vary the measured activity by up to 100-fold. This suggests that this analogue series may be amenable to optimization. This type of SAR derisks further work on this series (Zhang et al., 2014).

# 6. ADVANCED TOPICS

## 6.1 Assessing Readout Susceptibility and Specificity

Derisking for compound-mediated assay interference is becoming relatively straightforward in biochemical assays, albeit time consuming and resource intensive with equivocal cases. Identifying such interference in higher-order cellular assays presents new challenges given the complex cellular milieu, especially when testing relatively nonoptimized chemical matter at relatively high compound concentrations (10–100 μ*M*). In one example, decreases in cellular histone acetylation have been used to support on-target engagement of reported histone acetyltransferase (HAT) inhibitors (Dahlin et al., 2017). However, prototypical aggregators, redox-active, and nonspecific thiol-reactive compounds produced similar patterns of cellular proliferation and histone modification as reported HAT inhibitors. While a readout that is susceptible to compound interference is not prohibitory by any means, it should guide the design of additional cellular assays and experiments to ascertain meaningful target engagement selectivity in these more complex systems. While the above strategies should identify intractable chemical matter, they are by no means comprehensive and any derisking strategy should be customized to the specific target and project goals.

## 6.2 Risk From Irreproducible Results

Irreproducible results caused by poor reagent and equipment validation can undermine research outcomes and put many years of work and funding at risk (Freedman, Venugopalan, & Wisman, 2017; Williams, 2018). Probes that are used to interrogate drug targets must be (1) pure (typically to a level of greater than 98% to avoid confounding activity due to the presence of contaminants) and (2) structurally characterized (by techniques ranging from melting point, UV/visible light, NMR spectroscopy, and mass spectrometry, to X-ray crystallography). It is noteworthy that of 8600 compound samples in a purchased drug-screening library at the Broad Institutes' Drug Repurposing Hub, 29% were less than 85% pure (Corsello et al., 2017). Therefore, this is not an issue that is limited to screening libraries that may have been stored for years, but can affect even commercial, presumably certified libraries. Protocols should be put in place to ensure the reproducibility of experiments and to validate equipment.

**Fig. 7** Failure to manage risks upfront can lead to failure of drug discovery and development projects. Without risk management, the well-ordered architecture of an early drug discovery and development project can rapidly come crashing down, leaving nothing but rubble.

## 7. SUMMARY

Drug discovery and development is not for the risk adverse. Therefore, it is important to consider risk management in this field to minimize unnecessary losses of limited funding and time. The foundation of risk management is an embedded culture that recognizes the risks, and one in which every member of the organization feels empowered to reduce risks. Intrinsic risks, those that exist simply because getting a drug in to the body, to the target receptor, and then back out of the body, will always challenge drug discovery and development. The management of at least four sources of extrinsic risk (risks that with appropriate planning can be controlled) in early medicinal discovery can be accomplished by building on this foundation with the right library, right assays, right compounds, and right series (Fig. 7).

## REFERENCES

American Chemical Society. (2018). Guidelines for authors. ACS Publications, Retrieved from http://pubs.acs.org/paragonplus/submission/jacsat/jacsat_authguide.pdf (Revised October 2018). Accessed 10/09/2018.

Arrowsmith, C. H., Audia, J. E., Austin, C., Baell, J., Bennett, J., Blagg, J., et al. (2015). The promise and peril of chemical probes. *Nature Chemical Biology*, *11*, 536–541.

Auld, D. S., Inglese, J., & Dahlin, J. L. (2017). Assay interference by aggregation. In G. S. Sittampalam (Ed.), *Assay guidance manual*. NCBI (National Center for Biotechnology Information). Retrieved June 12, 2018, from https://www.ncbi.nlm.nih.gov. https://www.ncbi.nlm.nih.gov/books/NBK442297.

Auld, D. S., Lovell, S., Thorne, N., Lea, W. A., Maloney, D. J., Shen, M., et al. (2010). Molecular basis for the high-affinity binding and stabilization of firefly luciferase by PTC124. *Proceedings of the National Academy of Sciences of the United States of America*, *107*(11), 4878–4883.

Baell, J. B. (2013). Broad coverage of commercially available lead-like screening space with fewer than 350,000 compounds. *Journal of Chemical Information and Modeling*, *53*(1), 39–55.

Baell, J., & Nissink, J. (2018). Seven year itch: Pan-assay interference compounds (pains) in 2017—Utility and limitations. *ACS Chemical Biology*, *13*, 36–44.

Barbaras, D., Brozio, J., Johannsen, I., & Allmendinger, T. (2009). Removal of heavy metals from organic reaction mixtures: Preparation and application of functionalized resins. *Organic Process Research & Development*, *13*, 1068–1079.

Bertron, J. L., Cho, H. P., Garcia-Barrantes, P. M., Panarese, J. D., Salovich, J. M., Nance, K. D., et al. (2018). The discovery of VU0486846: Steep SAR from a series of M1 PAMs based on a novel benzomorpholine core. *Bioorganic & Medicinal Chemistry Letters*, *28*(12), 2175–2179.

Bickerton, G. R., Paolini, G. V., Besnard, J., Muresan, S., & Hopkins, A. L. (2012). Quantifying the chemical beauty of drugs. *Nature Chemistry*, *4*, 90–98.

Blagg, J., & Workman, P. (2014). Chemical biology approaches to target validation in cancer. *Current Opinion in Pharmacology*, *17*, 87–100.

Brenk, R., Schipani, A., James, D., Krasowski, A., Gilbert, I., Frearson, J., et al. (2008). Lessons learnt from assembling screening libraries for drug discovery for neglected diseases. *ChemMedChem*, *3*, 435–444.

Brenke, J. K., Salmina, E. S., Ringelstetter, L., Dornauer, S., Kuzikov, M., Rothenaigner, I., et al. (2016). Identification of small-molecule frequent hitters of glutathione S-transferase-glutathione interaction. *Journal of Biomolecular Screening*, *21*(6), 596–607.

Bunnage, M. E., Chekler, E. L., & Jones, L. H. (2013). Target validation using chemical probes. *Nature Chemical Biology*, *9*, 195–199.

Corsello, S. M., Bittker, J. A., Liu, Z., Gould, J., McCarren, P., Hirschman, J. E., et al. (2017). The Drug Repurposing Hub: A next-generation drug library and information resource. *Nature Medicine (New York, NY, United States)*, *23*(4), 405–408.

Cox, P. B., Gregg, R. J., & Vasudevan, A. (2012). Abbott physicochemical tiering (APT)—A unified approach to HTS triage. *Bioorganic & Medicinal Chemistry*, *20*, 4564–4573.

Dahlin, J. L.; Baell, J. B.; Walters, M. A. 2015. Assay interference by chemical reactivity., C. N. Sittampalam GS, Ed., Assay guidance manual, NCBI (National Center for Biotechnology Information), Retrieved June 12, 2018, from www.ncbi.nlm.nih.gov: https://www.ncbi.nlm.nih.gov/books/NBK326709/.

Dahlin, J. L., Inglese, J., & Walters, M. A. (2015). Mitigating risk in academic preclinical drug discovery. *Nature Reviews. Drug Discovery*, *14*(4), 279–294.

Dahlin, J. L., Nelson, K. M., Strasser, J. M., Barsyte-Lovejoy, D., Szewczyk, M. M., Organ, S., et al. (2017). Assay interference and off-target liabilities of reported histone acetyltransferase inhibitors. *Nature Communications*, *8*, 1–14.

Dahlin, J. L., Nissink, J. W., Strasser, J. M., Francis, S., Higgins, L., Zhou, H., et al. (2015). PAINS in the assay: Chemical mechanisms of assay interference and promiscuous enzymatic inhibition observed during a sulfhydryl-scavenging HTS. *Journal of Medicinal Chemistry*, *58*(5), 2091–2113.

Dahlin, J. L., & Walters, M. A. (2014). The essential roles of chemistry in high-throughput screening triage. *Future Medicinal Chemistry*, *6*, 1265–1290.

Dahlin, J. L., & Walters, M. A. (2016). How to triage PAINS-full research. *Assay and Drug Development Technologies*, *14*(3), 168–174.

Dosa, P. I., Georg, G. I., Gurvich, V. J., Hawkinson, J. E., Walters, M. A., & Wong, H. L. (2014). From HTS to phase I: The Institute for Therapeutics Discovery and Development at the University of Minnesota. *Combinatorial Chemistry & High Throughput Screening*, *17*, 231–240.

Ernst & Young Global Limited. (2013). *Turning risk into results:managing risk for better performance, EY-Global.* Retrieved 04. 04. 2018, from www.ey.com. http://www.ey.com/gl/en/services/advisory/turning-risk-into-results-managing-risk-for-better-performance.

Falk, H., Connor, T., Yang, H., Loft, K. J., Alcindor, J. L., Nikolakopoulos, G., et al. (2011). An efficient high-throughput screening method for myst family acetyltransferases, a new class of epigenetic drug targets. *Journal of Biomolecular Screening, 16*, 1196–1205.

Feng, B. Y., & Shoichet, B. K. (2006). A detergent-based assay for the detection of promiscuous inhibitors. *Nature Protocols, 1*(2), 550–553.

Freedman, L. P., Venugopalan, G., & Wisman, R. (2017). Reproducibility2020: Progress and priorities. version 1; referees: 2 approved *F1000Research, 6*, 604.

Frye, S., Crosby, M., Edwards, T., & Juliano, R. (2011). US academic drug discovery. *Nature Reviews. Drug Discovery, 10*, 409–410.

Galkina Cleary, E., Beierlein, J. M., Khanuja, N. S., McNamee, L. M., & Ledley, F. D. (2018). Contribution of NIH funding to new drug approvals 2010–2016. *Proceedings of the National Academy of Sciences of the United States of America, 115*, 2329–2334.

Hermann, J. C., Chen, Y., Wartchow, C., Menke, J., Gao, L., Gleason, S. K., et al. (2013). Metal impurities cause false positives in high-throughput screening campaigns. *ACS Medicinal Chemistry Letters, 4*, 197–200.

Huryn, D. M. (2013). Drug discovery in an academic setting: Playing to the strengths. *ACS Medicinal Chemistry Letters, 4*, 313–315.

Huryn, D. M., Resnick, L. O., & Wipf, P. (2013). Contributions of academic laboratories to the discovery and development of chemical biology tools. *Journal of Medicinal Chemistry, 56*(18), 7161–7176.

Huth, J. R., Mendoza, R., Olejniczak, E. T., Johnson, R. W., Cothron, D. A., Liu, Y., et al. (2005). ALARM NMR: A rapid and robust experimental method to detect reactive false positives in biochemical screens. *Journal of the American Chemical Society, 127*(1), 217–224.

Inglese, J., Auld, D., Jadhav, A., Johnson, R., Simeonov, A., Yasgar, A., et al. (2006). Quantitative high-throughput screening: A titration-based approach that efficiently identifies biological activities in large chemical libraries. *Proceedings of the National Academy of Sciences of the United States of America, 103*, 11473–11478.

Johnston, P. A. (2011). Redox cycling compounds generate $H_2O_2$ in HTS buffers containing strong reducing reagents—Real hits or promiscuous artifacts. *Current Opinion in Chemical Biology, 15*, 174–182.

Johnston, P. A., Soares, K. M., Shinde, S. N., Foster, C. A., Shun, T. Y., Takyi, H. K., et al. (2008). Development of a 384-well colorimetric assay to quantify hydrogen peroxide generated by the redox cycling of compounds in the presence of reducing agents. *Assay and Drug Development Technologies, 6*(4), 505–518.

Kenny, P. W. (2017). The ecstasy and agony of assay interference compounds. *Journal of Chemical Information and Modeling, 57*, 2640–2645.

Lipinski, C. A., Lombardo, F., Dominy, B. W., & Feeney, P. J. (1997). Experimental and computational approaches to estimate solubility and permeability in drug discovery and development settings. *Advanced Drug Delivery Reviews, 23*, 3–25.

Lombardino, J. G., & Lowe, J. A. (2004). A guide to drug discovery: The role of the medicinal chemist in drug discovery—Then and now. *Nature Reviews. Drug Discovery, 3*, 853–862.

Lor, L. A., Schneck, J., McNulty, D., Diaz, E., Brandt, M., Thrall, S. H., et al. (2007). A simple assay for detection of small-molecule redox activity. *Journal of Biomolecular Screening, 12*(6), 881–890.

Lushington, G., & Chaguturu, R. (2014). To screen or not to screen: An impassioned plea for smarter chemical libraries to improve drug lead finding. *Future Medicinal Chemistry, 6*, 497–502.

Macarron, R., Banks, M. N., Bojanic, D., Burns, D. J., Cirovic, D. A., Garyantes, T., et al. (2011). Impact of high-throughput screening in biomedical research. *Nature Reviews. Drug Discovery, 10*, 188–195.

Malo, N., Hanley, J. A., Cerquozzi, S., Pelletier, J., & Nadon, R. (2006). Statistical practice in high-throughput screening data analysis. *Nature Biotechnology*, *24*, 167–175.

Marui, E., Meel, K. V., Grieken, R. V., Buendia, A., Fontas, C., Hidalgo, M., et al. (2009). Method for the determination of Pd-catalyst residues in active pharmaceutical ingredients by means of high-energy polarized beam energy dispersive X-ray fluorescence. *Analytical Chemistry*, *81*, 1404–1410.

Mashalidis, E. H., Sledz, P., Lang, S., & Abell, C. (2013). A three-stage biophysical screening cascade for fragment-based drug discovery. *Nature Protocols*, *8*, 2309–2324.

McArthur, R., Leister, W., Veith, H., Shinn, P., Southall, N., Austin, C. P., et al. (2009). Monitoring compound integrity with cytochrome P450 assays and qHTS. *Journal of Biomolecular Screening*, *14*(5), 538–546.

McGovern, S. L., Caselli, E., Grigorieff, N., & Shoichet, B. K. (2002). A common mechanism underlying promiscuous inhibitors from virtual and high-throughput screening. *Journal of Medicinal Chemistry*, *45*(8), 1712–1722.

McGovern, S., Helfand, B., Feng, B., & Shoichet, B. K. (2003). A specific mechanism of nonspecific inhibition. *Journal of Medicinal Chemistry*, *46*, 4265–4272.

Morgan, P., Brown, D. G., Lennard, S., Anderton, M. J., Barrett, J. C., Eriksson, U., et al. (2018). Impact of a five-dimensional framework on R&D productivity at AstraZeneca. *Nature Reviews. Drug Discovery*, *17*, 167.

Morreale, F., Testa, A., Chaugule, V., Bortoluzzi, A., Ciulli, A., & Walden, H. (2017). Mind the metal: A fragment library-derived zinc impurity binds the e2 ubiquitin-conjugating enzyme ube2t and induces structural rearrangements. *Journal of Medicinal Chemistry*, *60*, 8183–8191.

Patil, S., Lis, L. G., Schumacher, R. J., Norris, B. J., Morgan, M. L., Cuellar, R. A., et al. (2015). Phosphonooxymethyl prodrug of triptolide: Synthesis, physicochemical characterization, and efficacy in human colon adenocarcinoma and ovarian cancer xenografts. *Journal of Medicinal Chemistry*, *58*, 9334–9344.

Roy, A. (2018). Early probe and drug discovery in academia: A minireview. *High-Throughput*, 7(1), 1–14.

Ryan, A. J., Gray, N. M., Lowe, P. N., & Chung, C.-W. (2003). Effect of detergent on "promiscuous" inhibitors. *Journal of Medicinal Chemistry*, *46*, 3448–3451.

Schorpp, K., Rothenaigner, I., Salmina, E., Reinshagen, J., Low, T., Brenke, J. K., et al. (2014). Identification of small-molecule frequent hitters from alphascreen high-throughput screens. *Journal of Biomolecular Screening*, *19*(5), 715–726.

Simeonov, A., & Davis, M. I. (2015). Interference with fluorescence and absorbance. In G. S. Sittampalam (Ed.), *Assay guidance manual*. NCBI (National Center for Biotechnology Information). Retrieved June 12, 2018, from Assay Guidance Manual [Internet] https://www.ncbi.nlm.nih.gov/books/NBK343429.

Simeonov, A., Jadhav, A., Thomas, C. J., Wang, Y., Huang, R., Southall, N. T., et al. (2008). Fluorescence spectroscopic profiling of compound libraries. *Journal of Medicinal Chemistry*, *51*(8), 2363–2371.

Skuta, C., Popr, M., Muller, T., Jindrich, J., Kahle, M., Sedlak, D., et al. (2017). Probes & drugs portal: An interactive, open data resource for chemical biology. *Nature Methods*, *14*, 759–760.

Slusher, B. S., Conn, P. J., Frye, S., Glicksman, M., & Arkin, M. (2013). Bringing together the academic drug discovery community. *Nature Reviews. Drug Discovery*, *12*, 811–812.

Soares, K. M., Blackmon, N., Shun, T. Y., Shinde, S. N., Takyi, H. K., Wipf, P., et al. (2010). Profiling the NIH small molecule repository for compounds that generate $H_2O_2$ by redox cycling in reducing environments. *Assay and Drug Development Technologies*, *8*(2), 152–174.

Tarnowski, M., Barozet, A., Johansson, C., Eriksson, P. O., Engkvist, O., Walsh, J., et al. (2018). Utility of resazurin, horseradish peroxidase, and NMR assays to identify redox-related false-positive behavior in high-throughput screens. *Assay and Drug Development Technologies*, *16*(3), 171–191.

Tarzia, G., Antonietti, F., Duranti, A., Tontini, A., Mor, M., Rivara, S., et al. (2007). Identification of a bioactive impurity in a commercial sample of 6-methyl-2-p-tolylaminobenzo[d][1,3]oxazin-4-one (urb754). *Annali di Chimica*, *97*, 887–894.

Thorne, N., Auld, D. S., & Inglese, J. (2010). Apparent activity in high-throughput screening: Origins of compound-dependent assay interference. *Current Opinion in Chemical Biology*, *14*, 315–324.

Turek-Etienne, T. C., Small, E. C., Soh, S. C., Xin, T. A., Gaitonde, P. V., Barrabee, E. B., et al. (2003). Evaluation of fluorescent compound interference in 4 fluorescence polarization assays: 2 kinases, 1 protease, and 1 phosphatase. *Journal of Biomolecular Screening*, *8*(2), 176–184.

Veber, D. F., Johnson, S. R., Cheng, H.-Y., Smith, B. R., Ward, K. W., & Kopple, K. D. (2002). Molecular properties that influence the oral bioavailability of drug candidates. *Journal of Medicinal Chemistry*, *45*, 2615–2623.

Vedvik, K. L., Eliason, H. C., Hoffman, R. L., Gibson, J. R., Kupcho, K. R., Somberg, R. L., et al. (2004). Overcoming compound interference in fluorescence polarization-based kinase assays using farred tracers. *Assay and Drug Development Technologies*, *2*, 193–203.

Wager, T. T., Hou, X., Verhoest, P. R., & Villalobos, A. (2010). Moving beyond rules: The development of a central nervous system multiparameter optimization (CNS MPO) approach to enable alignment of druglike properties. *ACS Chemical Neuroscience*, *1*, 435–449.

Walters, W. P., & Namchuk, M. (2003). A guide to drug discovery: Designing screens: How to make your hits a hit. *Nature Reviews Drug Discovery*, *2*, 259–266.

Walters, W. P., Stahl, M. T., & Murcko, M. A. (1998). Virtual screening—An overview. *Drug Discovery Today*, *3*(4), 160–178.

Williams, M. (2018). Reagent validation to facilitate experimental reproducibility. *Current Protocols in Pharmacology*, *81*, e40.

Xie, Y., Dahlin, J. L., Oakley, A. J., Casarotto, M. G., Board, P. G., & Baell, J. B. (2018). Reviewing hit discovery literature for difficult targets: Glutathione transferase omega-1 as an example. *Journal of Medicinal Chemistry*, *61*, 7448–7470. https://doi.org/10.1021/acs.jmedchem.8b00318.

Yang, J. J., Ursu, O., Lipinski, C. A., Sklar, L. A., Oprea, T. I., & Bologa, C. G. (2016). Badapple: Promiscuity patterns from noisy evidence. *Journal of Cheminformatics*, *8*(29), 1–14.

Zhang, L., Balan, G., Barreiro, G., Boscoe, B. P., Chenard, L. K., Cianfrogna, J., et al. (2014). Discovery and preclinical characterization of 1-methyl-3-(4-methylpyridin-3-yl)-6-(pyridin-2-ylmethoxy)-1H-pyrazolo-[3,4-b]pyrazine (PF470): A highly potent, selective, and efficacious metabotropic glutamate receptor 5 (mGluR5) negative allosteric modulator. *Journal of Medicinal Chemistry*, *57*, 861–877.

Zoghbi, H. Y. (2013). The basics of translation. *Science (Washington, DC, United States)*, *339*(6117), 250.

## FURTHER READING

Baell, J. B., & Holloway, G. A. (2010). New substructure filters for removal of pan assay interference compounds (PAINS) from screening libraries and for their exclusion in bioassays. *Journal of Medicinal Chemistry*, *53*, 2719–2740.

CHAPTER TWO

# Target Identification Using Chemical Probes

**Moses Moustakim*,†,‡, Suet Ling Felce§, Nancy Zaarour§,¶, Gillian Farnie*,§,1, Fiona E. McCann§,¶,1, Paul E. Brennan*,†,||,1**

*Nuffield Department of Medicine, Structural Genomics Consortium, University of Oxford, Oxford, United Kingdom
†Nuffield Department of Medicine, Target Discovery Institute, University of Oxford, Oxford, United Kingdom
‡Department of Chemistry, Chemistry Research Laboratory, University of Oxford, Oxford, United Kingdom
§Nuffield Department of Orthopedics, Rheumatology and Musculoskeletal Sciences, Botnar Research Centre, University of Oxford, Oxford, United Kingdom
¶Kennedy Institute of Rheumatology, University of Oxford, Oxford, United Kingdom
||Nuffield Department of Medicine, Alzheimer's Research UK Oxford Drug Discovery Institute, University of Oxford, Oxford, United Kingdom
[1]Corresponding authors: e-mail address: gillian.farnie@sgc.ox.ac.uk; fiona.mccann@kennedy.ox.ac.uk; paul.brennan@sgc.ox.ac.uk

## Contents



## Abstract

Chemical probes are small molecules with potency and selectivity for a single or small number of protein targets. A good chemical probe engages its target intracellularly and is accompanied by a chemically similar, but inactive molecule to be used as a negative

*Methods in Enzymology*, Volume 610
ISSN 0076-6879
https://doi.org/10.1016/bs.mie.2018.09.013

control in cellular phenotypic screening. The utility of these chemical probes is ultimately governed by how well they are developed and characterized. Chemical probes either as single entities, or in chemical probes sets are being increasingly used to interrogate the biological relevance of a target in a disease model. This chapter lays out the core properties of chemical probes, summarizes the seminal and emerging techniques used to demonstrate robust intracellular target engagement. Translation of target engagement assays to disease-relevant phenotypic assays using primary patient-derived cells and tissues is also reviewed. Two examples of epigenetic chemical probe discovery and utility are presented whereby target engagement pointed to novel disease associations elucidated from poorly understood protein targets. Finally, a number of examples are discussed whereby chemical probe sets, or "chemogenomic libraries" are used to illuminate new target–disease links which may represent future directions for chemical probe utility.

## ABBREVIATIONS

| | |
|---|---|
| **BCL-2** | B-cell lymphoma 2 |
| **BET** | Bromodomain and extra terminal protein |
| **BI** | Boehringer Ingelheim |
| **BRD3** | Bromodomain-containing protein 3 |
| **BRD4** | Bromodomain-containing protein 4 |
| **BRD7** | Bromodomain-containing protein 7 |
| **BRD9** | Bromodomain-containing protein 9 |
| **CLF-PED-015-T** | Rare undifferentiated pediatric patient-derived sarcoma cell line |
| **CRISPR** | Clustered regularly interspaced short palindromic repeats |
| **DMSO** | Dimethyl sulfoxide |
| **DNA** | Deoxyribose nucleic acid |
| **ELISA** | Enzyme-linked immunosorbent assay |
| **EZH2** | Enhancer of zeste homolog 2 |
| **GBM** | Glioblastoma |
| **GFP** | Green fluorescent protein |
| **GLP** | G9a-like Protein |
| **H3K9me2** | Histone-3 lysine-9 dimethylation |
| **HBL-1** | Human diffuse large B-cell lymphoma cell line |
| **HEK293** | Human embryonic kidney cells 293 |
| **HIV-1** | Human immunodeficiency virus 1 |
| **HTS** | High-throughput screen |
| **HUT-78 cells** | Human T lymphocyte 78 cells |
| **$IC_{50}$** | Half maximal inhibitory concentration |
| **IL-17A** | Interleukin 17A |
| **IV** | Intravenous |
| **$K_D$** | Equilibrium dissociation constant |
| **KRAS** | Kirsten rat Sarcoma virus gene |
| **MDA-MB-231** | Triple negative breast cancer cell line |
| **MIPE 3.0** | Mechanism Interrogation plate library 3.0 |
| **MOLM-13** | Acute myeloid leukemia cell line |

| | |
|---|---|
| **mRNA** | Messenger ribonucleic acid |
| **MTS** | (3-(4,5-dimethylthiazol-2-yl)-5-(3-carboxymethoxyphenyl)-2-(4-sulfophenyl)-2h-tetrazolium) |
| **NanoLuc** | Nano luciferase |
| **NFκB** | Nuclear factor kappa-light-chain-enhancer of activated B cells |
| **P13-AKT** | Phosphoinositide 3-kinase/Protein kinase B pathway |
| **PKMTs** | Protein lysine methyltransferases |
| **RA** | Rheumatoid arthritis |
| **REC/HTA** | Research ethics committee/human tissue authority |
| **SAM** | *S*-adenosyl methionine |
| **SAR** | Structure–activity relationships |
| **SGC Toronto** | Structural genomics consortium Toronto |
| **shRNA** | Short hairpin ribonucleic acid |
| **siRNA** | Small interference ribonucleic acid |
| **SUV4–20** | Suppressor of variegation 4–20 |
| $t_{1/2}$ | Half life |
| **TMD8** | Tokyo Medical and Dental University 8 cell line |
| **TNF** | Tumor necrosis factor |
| **TR-FRET** | Time-resolved fluorescence energy transfer |
| **U20S** | Human bone osteosarcoma epithelial cells |
| **UNC** | University of North Carolina at chapel hill |

## 1. INTRODUCTION

Late stage attrition in the drug discovery pipeline has been linked to the lack of detailed molecular understanding about how perturbation of a drug target modulates the intended disease phenotype (Hughes, Rees, Kalindjian, & Philpott, 2011; Murray & Rees, 2009; Swinney & Anthony, 2011). In order to gain a deeper molecular understanding of a drug target, there has been a surge in research focused on generating novel high quality small molecule tools or chemical probes.

In this chapter, we will endeavor to introduce the concept of using chemical probes as complementary to genetic techniques in target identification and validation. We will begin with a description of high-quality chemical probes and how they can be characterized for in-cell effects by target engagement assays. Then we will describe the application of chemical probes in phenotypic screening in patient-derived cellular models of disease. Finally, we will present case studies of where potent, selective, well-characterized chemical probes have been used to link positive effects in disease models to druggable protein targets. As opposed to a description of specific experimental methods and techniques, this chapter provides a description of how the linked disciplines of medicinal chemistry, chemical biology, in vitro biochemistry, and cell biology are being used in target discovery with chemical probes as the central vehicle. For detailed experimental methods in specific techniques, we refer readers to the included references.

A chemical probe is a reagent that is capable of discerning the biology of an intended biological target within the context of a linked disease (Bunnage, Chekler, & Jones, 2013). A number of criteria have been proposed for chemical probes and it is generally agreed that they are small molecules capable of binding their intended target with high affinity (low nanomolar $IC_{50}$ or $K_D$) in a biophysical, biochemical, or enzymatic assay. Additionally, it is expected that the probe displays selectivity for its target which is dissociable from any nonspecific interactions and evidence of being able to permeate membranes and engage with its target in a cellular context at useable concentrations (Chemical Probes Portal, 2018; SGC Chemical Probes, 2018). Additional criteria such as the availability of a structurally similar but biologically inactive negative control molecule and the use of multiple chemical probes for the same target are becoming more widely accepted. A negative control compound is often the enantiomer or other structural isomer of the chemical probe that is nearly chemical identical to the chemical probe, but is inactive on the chemical probe's primary target. An available negative control and multiple chemically distinct probes can drive confidence that any observed phenotypic effects are attributable to on-target activity and not some other unannotated off-target (Fig. 1).

A number of available chemical probes are disseminated widely through chemical vendors, institutes such as the Structural Genomics Consortium (SGC) (SGC Chemical Probes, 2018) and are curated by scientific communities such as the Chemical Probes Portal (Chemical Probes Portal, 2018). The ultimate aim is to foster the production of high-quality chemical probes and guide their optimal application (Blagg & Workman, 2017).

The use of small molecule chemical probes offers the advantage of being able to selectively modulate a target in a variety of contexts; modulation could be complete inhibition of activity of a protein or disruption of a protein–protein interaction. This can be achieved through direct binding, inhibition in an active site (orthosteric), or indeed in a remote region of the target that induces an inhibitory conformational change (allosteric). In addition, targeting can also be achieved through reversible noncovalent or irreversible covalent binding. Increasingly new modalities of small molecule probes are becoming available such as heterobivalent probes, PROTACs (proteolysis targeting chimeras) capable of directing E3 ligases to specific targets to facilitate ubiquitin-mediated protein degradation (Bondeson et al., 2015; Toure & Crews, 2016). Small molecule chemical

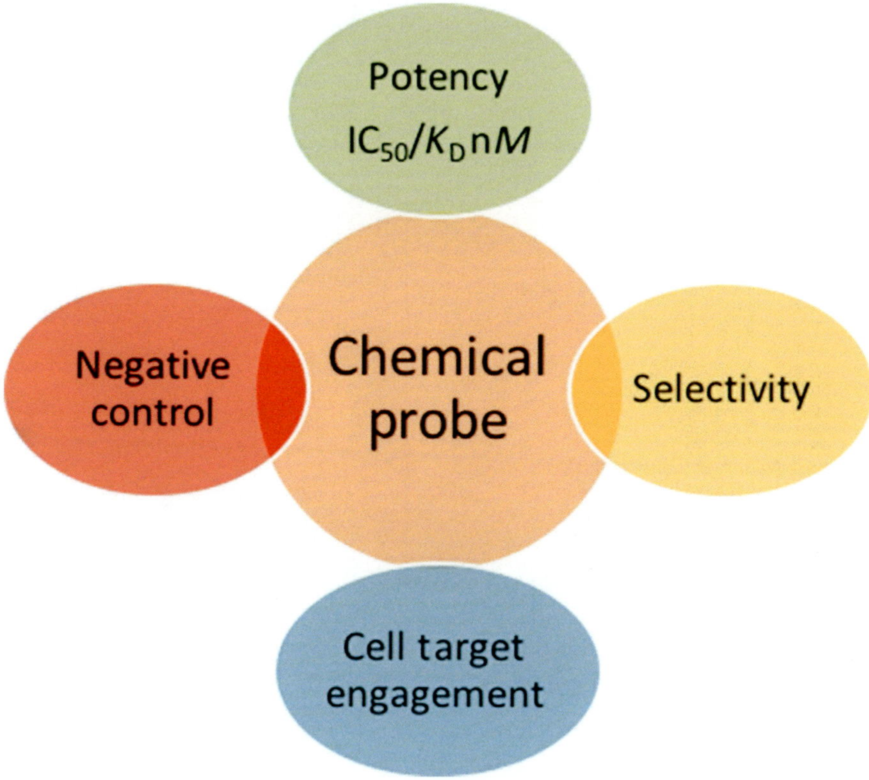


**Fig. 1** Small molecule chemical probes are potent, selective, and cell active biological modulators.

probes offer researchers the possibility of modulating a target through multiple approaches, which is in contrast to conventional siRNA/shRNA knockdown (Carthew & Sontheimer, 2009; Ghildiyal & Zamore, 2009; McManus & Sharp, 2002; Reynolds et al., 2004; Whitehead, Langer, & Anderson, 2009), genome-editing techniques such as CRISPR–Cas9 knockout (Doudna & Charpentier, 2014), antibodies (Carter & Lazar, 2017), intrabodies (Lecerf et al., 2001; Mhashilkar et al., 1995), and antibody drug conjugates (Agarwal & Bertozzi, 2015; Beck, Goetsch, Dumontet, & Corvaïa, 2017) while offering fewer pharmacokinetic challenges as compared with peptidic inhibitors (Fosgerau & Hoffmann, 2015) (Table 1). While all of these approaches may be susceptible to off-target effects, small molecule approaches benefit from the wealth of off-target selectivity panels, such as Eurofins Cerep (Eurofins, 2018) that have been developed commercially to aid drug discovery by profiling drug candidates and chemical probes. Chemical probes must also be cell permeable to

**Table 1** Target Perturbation for Target Identification

| | Small Molecules | siRNA/shRNA | Genome Editing (CRISPR–Cas9) | Antibodies | Intrabodies | Peptides | Antibody–Drug Conjugates |
|---|---|---|---|---|---|---|---|
| Advantages | • Domain specific<br>• A set precedent for translation to drug discovery<br>• Modular (orthosteric, allosteric, noncovalent, covalent inhibition)<br>• New approaches can engender whole protein chemical knockdown (PROTAC degradation) | • Highly specific and selective<br>• Highly modular<br>• Inducible/reversible modulation | • Highly specific<br>• Full target knockout or domain mutations<br>• Irreversible target modulation | • Effective for extracellular targets<br>• Highly potent<br>• Highly specific<br>• Cell type and tissue specific | • Effective for intracellular targets<br>• Highly potent<br>• Highly specific | • Efficient targeting of shallow binding sites such as protein–protein interactions<br>• Simple selection and optimization methods | • Effective for extracellular targets<br>• Highly potent<br>• Highly specific<br>• Cell type and tissue specific |
| Disadvantages | • Time and resource intensive to develop<br>• Often requires prior structural (X-ray/NMR) information about target to develop | • Limited understanding of off-target effects<br>• May imbalance posttranslational process<br>• Limited knockdown efficiency<br>• Limited by vector delivery system<br>• Lack of gene domain specificity gives whole protein knockdown | • Limited understanding of off-target effects<br>• Limited by vector delivery system | • Typically limited to extracellular modulation of whole cells<br>• Development of monoclonal specificity can be challenging<br>• Time and resource intensive in raising antibodies | • Require cell transfection | • Limited cell permeability | • Time and resource intensive to develop<br>• Can have limited cell permeability |

engage their targets intracellularly and recently numerous methods have been developed to determine this (Section 2). When a potent, selective, and cell active probe has been developed it can be used very simply in any cellular assay. Of particular, utility is the use of chemical probe profiling in assays conducted in patient-derived primary cells as highly translatable disease models (Section 3). As the cell numbers in patient-derived samples are often low, medium- or high-throughput screening of pharmacologically uncharacterized compounds is impossible, but patient-derived cells are typically ideal for screening chemical probe sets <100 compounds.

## 2. TARGET ENGAGEMENT ASSAYS

Measuring target engagement is critical for the development of chemical probes, in order to demonstrate on-target activity in cells. In the absence of such assays, the reasons behind the lack of in-cell efficacy of a probe would be difficult to determine; for example, the target could be invalid, or the probe may fail to fully engage the target in cells. On the other hand, the link between a desired phenotype and a molecular target via a chemical probe that does not show cellular target engagement should be treated as suspect. An ideal assay would also measure the extent of target engagement, which would help to determine optimal probe concentrations (Simon, Niphakis, & Cravatt, 2013). For targets with known binding partners, protein interactions can be measured using FRAP (fluorescence recovery after photobleaching) or BRET (bioluminescence resonance energy transfer). However, in cases of targets with no known interacting partner or cellular substrate, assays can be used based on the ability of an inhibitor to stabilize the target in cells during thermal denaturation in a cellular thermal shift assay (CETSA). When selecting which assay to use (Table 2), it is important to consider whether the assay in question allows for the investigation of off-target as well as on-target effects in order to assess probe selectivity. For example, if a protein target is part of a larger family, it is essential to see if the probe will bind to other family members as strongly as its target protein within a cellular context. Different assays will also require the use of negative controls for stringent testing of target engagement, to discount any false positives or off-target effects.

### 2.1 Fluorescence Recovery After Photobleaching Assays

FRAP is a powerful tool to monitor molecular dynamics in living cells, measured by confocal microscopy (Day & Kraft, 2012). Photobleaching is carried out using a high-intensity laser pulse targeting a portion of a live cell containing a fluorescently tagged target molecule. The migration (or recovery)

**Table 2** Cellular Target Engagement Assays

| Method | Type | Advantages | Disadvantages |
|---|---|---|---|
| FRAP | Proximity based | • Study binding to endogenous ligand<br>• Time course studies<br>• Can use any easy to transfect cell line | • Genetic modification of protein to add fluorescent tag<br>• Use of specialized equipment (confocal microscopy)<br>• Low throughput |
| BRET | Proximity based | • No need for specialized equipment<br>• Results easy to interpret and analyze<br>• Can use any easy to transfect cell line<br>• Medium-throughput screening | • Genetic modification of both protein and ligand<br>• Study of exogenous proteins<br>• Not as physiologically representative |
| EFC | Protein stability | • No need for specialized equipment<br>• Results easy to interpret and analyze<br>• Can use any easy to transfect cell line | • Genetic modification of protein required<br>• Assessing exogenous protein binding |
| CETSA | Thermal stability | • No protein modification required<br>• No specialized equipment necessary<br>• Measurement of endogenous protein | • Need to have specific antibodies to protein for Western blotting<br>• Need to find cell line with high endogenous protein expression levels |
| Activity-based | Protein activity | • No protein modification required<br>• No specialized equipment necessary<br>• Measurement of endogenous protein | • Need to have specific antibodies to protein with and without modifications (e.g., phosphate or methyl groups) for Western blotting<br>• Need to find cell line with high endogenous protein expression levels |
| PROTACs | Protein degradation | • No protein modification required<br>• No specialized equipment necessary<br>• Measurement of endogenous protein | • Need to have modified version of inhibitor bound to E3 ubiquitin ligase probe<br>• Need to find cell line with high endogenous protein expression levels<br>• Can require extensive chimera SAR/synthesis<br>• Large molecular weight can limit "drug-like" properties |

of labeled molecules returning into the bleached area is subsequently monitored over time. Strong protein–protein interactions between a fluorescently tagged protein and its binding partner would lead to a slow recovery time after photobleaching, due to slower dissociation rates which would take longer for unbleached fluorescent proteins to move into the bleached area. Weak or no protein–protein interactions would result in faster recovery times as bleached proteins would move away from the bleached area quicker and be replaced by unbleached proteins. This method has been used to measure cellular on-target effects of bromodomain inhibitors. Mutations in BRD3, BRD4, and tripartite motif-containing 24 affecting acetylated lysine histone binding lead to reduced FRAP recovery times compared to wild-type proteins (Philpott et al., 2014). This effect was also seen when incubating with bromodomain inhibitors such as **JQ1**. Although this method is low throughput, requires modification of target protein to add a fluorescent tag, as well as operating complex and expensive equipment (confocal microscope), it does allow assessment of inhibitor cell permeability and in-cell efficacy.

## 2.2 Bioluminescence Resonance Energy Transfer Assays

BRET can be used to study protein–protein interactions and the affinity of small molecules for target proteins. Proteins are first genetically modified to form fusion proteins with either a luciferase tag or a fluorescent tag, for example, GFP. Upon addition of the luciferase substrate, Förster resonance energy transfer occurs when the two proteins are in close proximity (1–10 nm) to one another, so that the excitation energy from the bioluminescent donor luciferase is transferred to the fluorescent receptor by nonradiative dipole–dipole coupling, resulting in the observed emission from the acceptor rather than the donor (Pfleger & Eidne, 2006). Modified versions of BRET include NanoBRET, involving the use of a much brighter and smaller luciferase (NanoLuc) paired with a long-wavelength fluorophore attached via HaloTag for increased sensitivity in detecting protein–protein interactions (Machleidt et al., 2015). BRET can be used to measure protein oligomerization through competition-based assays (Felce, Knox, & Davis, 2014) and ligand binding to G protein-coupled receptors (Stoddart et al., 2015). The testing of chemical probes in BRET assays measuring interactions of related family members and mutated versions of the target of interest can relay information on selectivity and specificity in exogenously expressed proteins. Target engagement of histone deacetylase (HDAC) inhibitors has also been investigated using BRET, in addition allowing the assessment of binding kinetics of the inhibitors (Robers et al., 2015). BRET can be optimized for medium-throughput screening in 384-well plates, requires genetic modification of both the target protein and its binding partner, and the use of a sensitive fluorescence plate reader.

## 2.3 Enzyme Fragment Complementation (EFC) Assays

EFC works in a similar method to BRET. The protein of interest is fused to a small fragment of enzyme β-galactosidase, the enhanced Prolabel. When the enzyme acceptor is added to the cells, complementation between the two fragments occurs and β-galactosidase activity can be measured. Proteins bound to a ligand will be more stable and would therefore exhibit higher levels of β-galactosidase activity than unbound protein. Different variations of this assay exist, developed by DiscoveRx such as the InCELL Hunter destabilized partner protein assay and the InCELL pulse assay. This method also requires genetic modification of the target protein and can only be used to test small molecules that are known to bind their target proteins and alter their steady-state cellular levels. EFC was used to verify **JQ1** inhibitor binding to BRD4 (Schulze et al., 2015).

## 2.4 Cellular Thermal Shift Assays

CETSA relies on changes in thermal stabilization or destabilization of a protein target that occurs upon ligand binding. Cell lysates are heated to a range of different temperatures, leading to protein unfolding and aggregation at a specific temperature. Analysis of the remaining soluble protein fraction, corresponding to nondenatured, folded protein, is carried out by Western blotting. The melting temperature ($T_m$) can be calculated by plotting the amount of soluble protein against temperature. The binding of small molecules to a protein can lead to thermal stabilization or destabilization, and the $T_m$ of protein bound to a small molecule can be compared to the $T_m$ of unbound protein; typically melting curves shift ($\Delta T_m$) to a higher temperature when ligand binding has occurred implying ligand-induced protein stabilization. CETSA has been used to interrogate inhibitors of lysine (K) and arginine (R) methyltransferases, for example, protein arginine *N*-methyltransferase 5 and its novel inhibitor **EPZ015666** (Martinez Molina & Nordlund, 2016). Although relatively simple to conduct and with no genetic modification required, small molecules can only be tested with known binding proteins and also relies on the availability of antibodies to the proteins of interest, although a proteome-wide mass-spectrometry has been developed (Huber et al., 2015).

## 2.5 Activity-Based Assays

Another method of confirming in cell target engagement is to study the activity of the protein that is being targeted. For example, the activity of kinases and methyltransferases can be measured by Western blotting, using antiphosphate- or antimethyl-specific antibodies. The chemical probe **A-196** inhibits the activity of SUV4–20 protein lysine methyltransferase; **A-196** specifically inhibits di- and trimethylation of histone 4 lysine 20, which can be shown by Western blotting (Bromberg et al., 2017). Densitometry analyses of these blots allow for calculation of $IC_{50}$ values. Although this technique does not require specialist equipment, it is only applicable to enzymes and specific antibodies to methyl or phosphate groups on the target protein. The same group also used immunofluorescence to study **A-196** inhibition on SUV4–20.

## 2.6 Proteolysis Targeting Chimeras

PROTACs are composed of heterobifunctional molecules that are able to bind E3 ubiquitin ligase and the protein of interest simultaneously, resulting in polyubiquitination of the target protein by the E3 ubiquitin ligase complex, and its subsequent degradation by the proteasome. Early studies were hampered by the lack of specific small molecule probes to E3 ubiquitin ligases; however, the discovery of thalidomide binding to the E3 ligase Cereblon has allowed the use of thalidomide analogues, which has led to increased interest and uptake of PROTACs (Neklesa, Winkler, & Crews, 2017). As previously mentioned, ligand binding to its target protein can lead to stabilization, a feature exploited by CETSA and EFC analyses. However, stabilization of the target protein can also lead to its accumulation and increasing levels of protein may lead to decreased inhibition by its small molecule inhibitor depending on protein turnover rate and compound half-life. The use of PROTAC technology overcomes this issue in part due to its catalytic nature of inhibition; incubation of BRD4 inhibitors such as **JQ1** in Burkitt's lymphoma cell lines leads to accumulation of BRD4 protein (shown by Western blotting), however, the inverse is seen with BRD4 PROTAC **ARV-825** (Lu et al., 2015). PROTAC-based probes have the advantage of removing the target of interest not just blocking the function of a domain, but current disadvantages include their large molecular weights which may limit cell permeability and carry potential side effects, e.g., thalidomide-based PROTACs. However, there is a major effort in the field to characterize and improve the pharmacokinetic properties of these PROTAC molecules, along with the identification of novel E3 ligase directing ligands.

With robust target engagement of a chemical probe established, it is suitable to be used in a phenotypic assay to decipher the biology of the target and if disease modifying, could serve as a chemical starting point in a translational drug discovery program.

## 3. PROBE SCREENING IN PATIENT-DERIVED ASSAYS

Traditionally, immortalized cell lines and animal models in rodents have been exploited to study disease-associated mechanisms for drug target identification broadly across multiple indications. However, with poor success rates in new drug discovery, in particular associated with high failure rates of clinical trials for both safety and efficacy outcome measures, there has long been a need for improved cellular assays and model systems in which to study pathobiology and interrogate effects and mode of action of candidate drugs that better mimic the typically heterogeneous and complex disease landscape, and ultimately derisk the route to the clinic. One such example is sepsis where hundreds of compounds are effective in animal models but none have proven efficacious in humans (Edwards et al., 2015; Marshall, 2014). Additionally, gaps in our understanding of disease biology and toxicity mechanisms are at the root of many drug failures and continue to exert a strong negative impact on pharmaceutical productivity (Plenge, 2016; Scannell & Bosley, 2016).

In recent years, there has been a significant shift toward using primary human cells and tissues to build robust platforms for screening compounds for their potential to regulate established disease-associated processes or phenotypes, with special efforts focused on utilizing relevant patient samples for assay development to more optimally recapitulate disease context, ex vivo (Edwards et al., 2015). Demonstrating the perceived benefits of performing target identification studies in patient cells, in 2015 the Structural Genomics Consortium (SGC) was awarded €42 million from the EU and European Federation of Pharmaceutical Industries and Associations Innovative Medicines Initiative (EU/EFPIA IMI, ULTRA-DD Grant no. 115766) to develop a drug discovery platform to prioritize new drug targets in a defined set of inflammatory diseases by screening newly generated chemical probes in assays derived only from patient cells and tissues (Fig. 2).

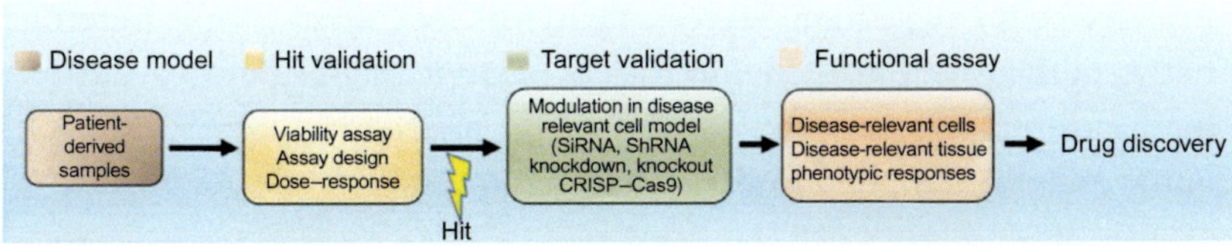


**Fig. 2** Schematic representation of target and drug discovery process using chemical probes in patient-derived cells.

While the advantages of using ex vivo patient cell-based assays for screening are obvious, there are several challenges that may not be immediately apparent.

## 3.1 Advantages of Patient-Derived Assays

- Ex vivo patient cells and tissues are disease relevant and should therefore display unique characteristics of disease phenotype, which are often difficult to recapitulate in vitro.
- Cellular complexity and diversity can be fully represented, not just a single cell subset.
- Often cells do not require exogenous stimulation to display a disease phenotype.
- Patient samples may be sufficiently annotated, and thus screening effects may be retrospectively correlated with features of the individual patient profile such as genetics and prior drug treatment.

## 3.2 Disadvantages of Patient-Derived Assays

- Screening requires at least short-term culture to allow sufficient exposure to compounds, and therefore it is necessary to ensure that cells remain viable, are metabolically active, and retain disease phenotype.
- Some compounds may require longer culture periods (>7 days) and therefore cells may dedifferentiate away from disease-associated phenotype.
- Cell numbers are often limiting, allowing screening of chemical probes sets, but prohibiting medium- to high-throughput library screening.
- Access to samples requires developing an appropriate clinical network, may require shipping of samples from distal sites adding additional time pressures.
- Ethical approval must be granted from a research ethics committee (REC), human tissue authority (HTA), and/or local biobank.

- Primary cells have a high degree of patient-dependent sample heterogeneity, perhaps due to genetics, disease activity status, and background medications.
- Large patients sample numbers may be required to see a statistically relevant response (see earlier).
- Patient medications may interfere with screening readouts.

## 3.3 Step by Step Guide for Hit Validation and Prioritization

Assuming ethical approval from REC/HTA has been granted, there are number of considerations and follow-up procedures to validate and prioritize chemical probes hits from a primary screen.

### 3.3.1 *Cellular Viability Assays*

Before studying potency and efficacy in any cell assay, a crucial first step is to determine the cytotoxicity of the compounds in the screen, in terms of antiproliferative effects and impaired cell viability. There are several commonly used approaches to address this, mostly involving the application of fluorescent dyes that bind directly to DNA (e.g., live/dead membrane impairment assays which use Sytox dyes, deep red anthraquinone 7, propidium iodide, and 7-aminoactinomycin D (7-AAD)), amine dyes (which also measure membrane impairment, binding to amine groups on intracellular proteins), cleavable vital dyes (membrane permeable that fluoresce only in metabolically active cells through measurements of adenosine triphosphate (e.g., CellTiter Glo)), or biochemical metabolites reporting on the reducing ability of living cells (e.g., PrestoBlue, resazurin). All of the above can be performed in a 384-well microtiter plate, with fluorescence or absorbance quantifiable in a fluorimeter or spectrophotometer. Cell numbers permitting, it is advisable to assay the probes at a range of concentrations to determine the highest tolerated dose. This dose may be subsequently selected for the phenotypic screen.

### 3.3.2 *Phenotypic Assay Design*

The most desirable primary assay is one that requires minimal operator intervention and manipulation to ensure a near native phenotype with the most physiologically relevant readouts. One such assay led to the initial discovery that TNF is a master regulator of proinflammatory cytokine production, and the subsequent development of TNF inhibitors for treatment of rheumatoid arthritis (RA). Whole synovial tissue excised from rheumatoid joints was enzymatically disaggregated and the resultant mixed cell culture was analyzed for the spontaneous release of cytokines in the presence of blocking antibodies (Brennan, Chantry, Jackson, Maini, & Feldmann, 1989). From this relatively simple assay, anti-TNF monoclonal antibodies were identified to regulate expression of a number of cytokines ex vivo suggestive of potential for broad reaching antiinflammatory effects, clinically.

For defining the appropriate phenotypic read out it is often helpful to have good working knowledge of prominent disease-associated mechanisms as well as the clinical landscape of any known pharmacological therapies that have demonstrated therapeutic efficacy in related indications (e.g., IL-17A in ankylosing spondylitis (AS) screen where **secukinumab**, an anti-IL-17A biologic is already licensed for AS and some autoimmune diseases).

Once the readout is defined, it is crucial that there are robust tools to support quantitation, e.g., validated antibodies for unambiguous protein detection in ELISA, or a fluorescent marker in cytometry/fluorimetry-based readouts (proliferation, metabolic activity, etc.). Cell numbers must be sufficient to accommodate a number of compounds and include appropriate negative and positive control compounds, sufficient for all samples and controls to be run in triplicate in cells from each donor, and in multiple donors. For a primary screen, a single dose previously demonstrated to engage its target in cells (Section 2) is normally applied. Once the assay has been developed and optimized, the primary screen can commence. Typically, a "hit" is regarded as a probe that inhibits the defined readout $\geq$50% relative to a control (usually DMSO at equivalent dilution) and should be reproducible in >3 donor assays.

### *3.3.3 Dose Responses and Control Compounds*

Once the primary phenotypic screen has been performed reproducibly at a single dose in triplicate and "hit" compounds have been selected, it is necessary to generate dose–response curves with 5–7 doses, DMSO control and the inactive negative control compound. This will lend greater confidence that the effects observed can be attributed to the chemical probe's annotated target and not an uncharacterized off-target. The following example highlights the potential of phenotypic screening in primary human brain cells to identify relevant drug targets. A screen comprising 24 epigenetic chemical probes from the SGC in 5 genetically distinct patient-derived brain-tumor-initiating cells lines, a population of stem-like glioma cells that are thought to contribute to GBM propagation, treatment resistance, and tumor recurrence (Bao et al., 2006; Chen et al., 2012; Singh et al., 2004), led to the identification of two *S*-adenosyl methionine (SAM) competitive EZH2 inhibitors, among which **UNC1999** exhibits low micromolar cytotoxicity in vitro and suppresses tumor growth in vivo. **UNC1999** significantly decreased viable cell numbers in a dose-dependent manner as compared to DMSO-treated cells at all time points tested (Grinshtein et al., 2016). Although inactive controls were not used in this study, the identification of two distinct EZH2 inhibitors added confidence to the role of EZH2 inhibition observed cytotoxicity.

### *3.3.4 Target Validation With Gene Modulating Techniques*

To complement any observed probe effects, a common approach is to silence or knock down the target gene. It is noteworthy that a chemical inhibitor will often only modify a functional domain of a protein and not all known functions, hence gene deletion may have a broader impact. There are currently three major forms of RNAi which are used extensively:

**(1)** siRNA (small interfering RNA) is a class of double-stranded RNA molecules for partial or near complete mRNA knockdown in cultured cells. It works by interfering with the expression of specific genes leading to degradation of complimentary RNA. Importantly, the siRNA oligonucleotide can be easily delivered by classical transfection methods.

**(2)** shRNA (short hairpin RNA) knock down, unlike siRNA, is typically accomplished by delivery through viral or bacterial vectors containing a nucleotide sequence complimentary to the gene of interest. The major advantage for using shRNA is that it provides sustained depletion of the target genes and secondly, the transfection efficiency of certain types of cells is often higher than for siRNA. The major drawback for both si- and shRNA is that a partial knockdown can fail to produce a measurable phenotype (Sachse et al., 2005) and elicit off-target effects (Marine, Bahl, Ferrer, & Buehler, 2012). Complementary to the epigenetic chemical probe screen on BITC described earlier, siRNA knockdown of EZH2 in a coculture model of human GBM cells with monocyte-derived macrophages led to its prioritization as an interesting drug target in brain cancers (Yin et al., 2017).

**(3)** CRISPR–Cas9 is a more recently developed technique whereby DNA from viruses or plasmids is cut into small fragments and incorporated into a CRISPR (clustered regularly interspaced short palindromic repeats) locus where small RNAs (guide RNA, gRNA) are generated to guide effector endonucleases (Cas9) that cleave DNA, resulting in the activation of the double-strand break (DSB) repair machinery resulting in insertions or deletions (indels) which disrupt the targeted locus. There are several advantages to using CRISPR; in addition to complete gene knockout, mutations can be inserted that may help define functional domains. Furthermore, genome edits are fully integrated into host DNA that is stable over multiple generations.

### 3.3.5 Functional Assays for Validation

Downstream mechanism-of-action studies are further required to validate cellular activities of candidate small molecules and evaluate possible off-target activities. These are usually low throughput and involve modeling a more complex cellular composition and context (tissue explants, 3D organoids, etc.) than can be sustained for medium- to high-throughput screens. Hence only lead candidates, supported by dose response and target deletion/knockdown studies are usually selected for functional assays, e.g., cell migration along chemotaxis gradients to mimic local inflammation (Hulkower & Herber, 2011; Vasaturo et al., 2012), cell migration suggestive of metastatic potential in tumor organoids (Andersson & Lendahl, 2014; Jin et al., 2013), extracellular matrix deposition on physiological substrates (Pellowe & Gonzalez, 2016), T-cell phenotype skewing and coculture with antigen presenting cells for adaptive immune responses (Pierre et al., 2017).

## 4. CASE STUDIES

The use of chemical probe screening against patient-derived primary cells is not yet a generally adopted approach due to the technical challenges in using primary cells for screening, but the following examples highlight some of the challenges and successes of this methodology in finding novel targets for unmet disease in near physiological systems.

### 4.1 G9a/GLP Histone Methyl Transferase Inhibitors

Protein lysine methyltransferases (PKMTs) have been implicated in a variety of diseases (Albert & Helin, 2010; Copeland, Solomon, & Richon, 2009). Dysregulation of the highly homologous PKMTs G9a (KMT1C/EHMT2) and GLP (G9a-like protein/KMT1D/EHMT1), which have both been identified as implementing mono- and dimethylation of histone-3 lysine-9 (H3K9) (Tachibana et al., 2002, 2005), correlates with the progression of different cancers such as hepatocellular carcinoma and lung cancer (Huang et al., 2010; Kondo et al., 2008, 2007; Watanabe et al., 2008). Furthermore, G9a/GLP has been associated with a number of neurodevelopmental and neurodegenerative disorders (Millan, 2013). Due to significant links between the homologous (80% identity in the catalytic domain) G9a/GLP proteins and disease but a paucity of potent, selective, and cell active inhibitors of these proteins to investigate their biology, research efforts have been expended over recent years to address this.

One of the first reports of small molecule inhibitors of G9a was by Jenuwein and coworkers at Boehringer Ingelheim (BI) who identified diazepin-quinazolin-amine, **BIX-01294** following a high-throughput screen using a curated BI chemical library (Kubicek et al., 2007). Compound **BIX-01294** was found to be the only G9a/GLP selective hit revealed from the HTS, showing potent activity for inhibition of G9a/GLP histone-3 lysine-9 dimethylation (H3K9me2) in a non-SAM competitive fashion. Since this report **BIX-01294** has been used in the study of cellular reprogramming (Shi, Desponts, et al., 2008; Shi, Do, et al., 2008) and HIV-1 latency (Imai, Togami, & Okamoto, 2010). In addition, using a cell-based assay, researchers at the Asan Medical center have shown that **BIX01294** induces autophagy-dependent cell death via G9a dysfunction and intracellular reactive oxygen species accumulation in breast and colon cancer cells, indicating that targeting G9a could be an effective therapeutic strategy for cancer treatment (Kim et al., 2013).

However, certain characteristics of this probe precluded extensive use, such as cellular toxicity observed at concentrations >4 μ*M*. Coworkers from the Structural Genomics Consortium Toronto (SGC Toronto) and the University of North Carolina at Chapel Hill (UNC) reported a number of efforts in order to improve this G9a/GLP inhibitor.

One of the first improved probes of G9a following **BIX-01294**, was **UNC0224**, reported in 2009 to display significant increases in G9a potency (Liu et al., 2009). In addition to this, Liu et al. also reported the first G9a crystal structure with a small molecule bound, which rationalized the peptide competitive nature of this chemical series (PDB ID 3K5K). The quinazoline **UNC0224** was later developed culminating in the discovery of the more potent and selective G9a/GLP probe molecule **UNC0321** (Liu et al., 2010). Up to this point these probe molecules had not been characterized for cellular potency and evidence of target engagement in a cellular context. Coworkers from SGC Toronto and UNC then went on to report the discovery of **UNC0638**, a cell active G9a/GLP chemical probe after discovering that **UNC0321** was in fact less potent in cells than the original reported inhibitor **BIX-01294**. In addition, a corresponding structurally similar but inactive negative control compound **UNC0737** was reported that can be used in cellular studies to attribute any functional effect as specific to G9a/GLP modulation (Liu et al., 2013). MDA-MB-231 breast cancer cells were chosen by the authors for initial testing as these cells displayed high endogenous H3K9me2 levels. Despite showing potent activity-based (Section 2.5) cellular inhibition of H3K9me2 levels using an immunofluorescence in-cell Western assay in MDA-MB-231 cells (**UNC0638** H3K9Me$_2$ cell IC$_{50}$ 81 n*M*, **UNC0737** H3K9Me$_2$ cell IC$_{50}$ > 5 μ*M*) and low cell toxicity, measured by alamar Blue assay (both inhibitors EC$_{50}$ > 5000 n*M*), **UNC0638** was found to have suboptimal pharmacokinetic properties, with high rates of clearance being exhibited upon intravenous (IV) dosing and poor bioavailability upon oral dosing. In order to further improve this, coworkers at SGC Toronto and UNC further developed the quinazoline scaffold leading to the discovery of **UNC0642** which exhibits increased metabolic stability in vitro and in vivo (Liu et al., 2013). **UNC0642** was also seen to have inhibitory effects on the clonal proliferation of PANC-1 (pancreatic epithelioid carcinoma) cells in a dose-dependent manner (50% decrease in clones at 1 μ*M* compared to untreated cells), but not on MDA-MB-231 cells, indicating that pancreatic carcinoma cells may be more sensitive to G9a/GLP inhibition than breast cancer cells, highlighting tissue-specific effects. More recently a number of other studies have reported the disclosure of a number of GLP selective probes (Xiong, Li, Babault, Dong, et al., 2017; Xiong, Li, Babault, Wu, et al., 2017).

In 2015, after the discovery of **UNC0638**, two independent research groups concomitantly reported on the use of **UNC0638** to reactivate fetal hemoglobin expression in cluster of differentiation 34 expressing cells erythroid progenitor cells from peripheral blood of healthy adult donors, providing a rationale for targeting G9a/GLP as a therapeutic intervention for blood disorders such as sickle cell anemia and β-thalassemia (Krivega et al., 2015; Renneville et al., 2015). Soon after, the epigenetic focused biopharmaceutical company Epizyme announced they were advancing a G9a/GLP inhibitor toward the clinic for treating blood disorders (Chan-Penebre et al., 2017). A related Epizyme patent discloses the structures of a number of G9a/GLP inhibitors, including **EPZ035544** that bear close resemblance to the quinazoline scaffold of the probe **UNC0638** (Campbell et al., 2017) (Fig. 3).

## 4.2 BRD9 Bromodomain Inhibitors

The bromodomain-containing protein BRD9 which forms a part of the mammalian switch/sucrose nonfermentable (SWI/SNF) complex is an acetyl reading domain shown to be required for acute myeloid leukemia (AML) (Hohmann et al., 2016). There have been a number of examples of chemical

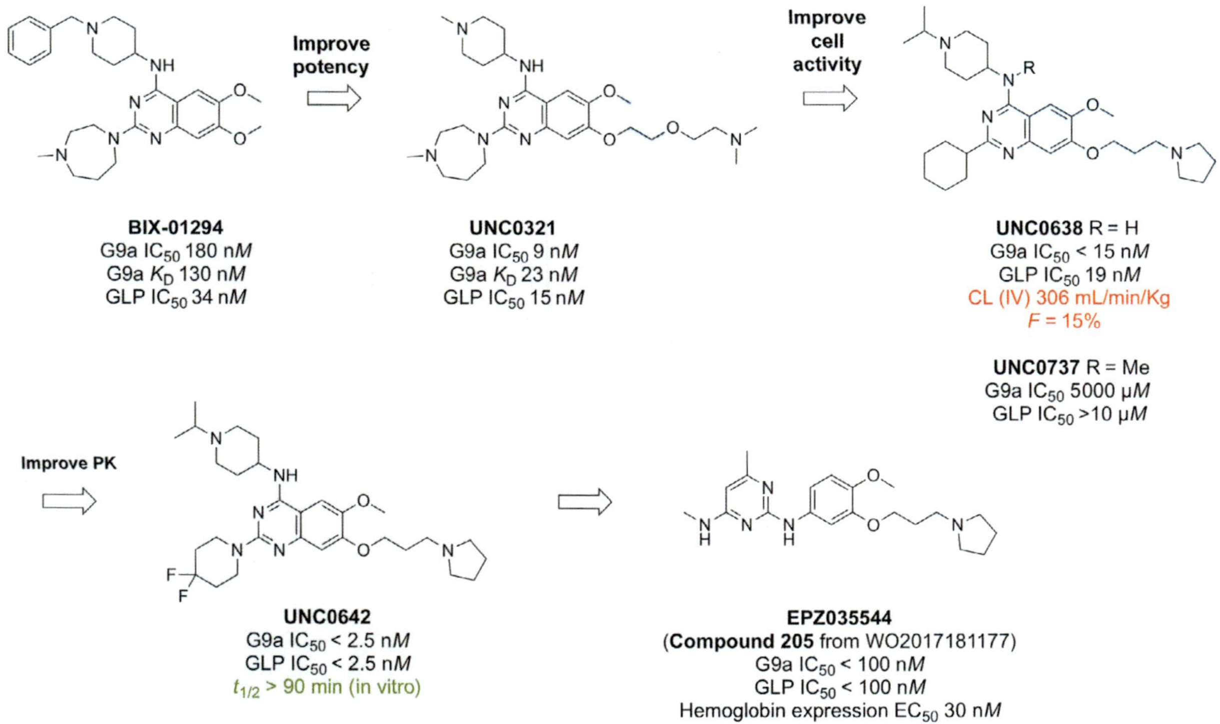


**Fig. 3** Development of G9a and GLP inhibitors.

probes developed targeting the bromodomain of BRD9 in recent years including **LP99** (Clark et al., 2015), **I-BRD9** (Theodoulou et al., 2015), **compound 28** (Hay et al., 2015), and recent probes from Boehringer Ingelheim, **BI-9564** and **BI-7273** (Martin et al., 2016).

The first reported BRD9 chemical probe was disclosed by Clark et al. (2015) in the form of **LP99** which displays potent activity against BRD9 and selectivity over a wide panel of bromodomains other than the highly homologous BRD7. Selectivity was ascertained using differential scanning fluorimetry (DSF) and target engagement was demonstrated using both a NanoBRET assay (Section 2.2) (Machleidt et al., 2015) and a FRAP assay (Section 2.1) (Deschout, Raemdonck, Demeester, De Smedt, & Braeckmans, 2014). Using FRAP, it was found that **LP99** was able to disrupt BRD9 interaction with chromatin at concentrations as low as 0.8 μ*M*. The interactions between BRD7 and BRD9 with both histones 3.3 and 4 were assessed using NanoBRET in HEK293 cells, where incubation with **LP99** led to a dose-dependent decrease in BRET signal for BRD7 and BRD9 and their target histones. The inhibitor did not exhibit toxic effects in U20S cells at 24 and 72 h at concentrations less than 33 μ*M*. In order to explore the relevance of BRD9/BRD7 inhibition to modulation of proinflammatory cytokines, lipopolysaccharide (LPS) stimulated human THP-1 monocytic cells were treated with **LP99** and treatment was found to suppress interleukin 6 (IL-6) secretion (measured by ELISA) in a dose-dependent manner, suggesting the potential of BRD9/BRD7 inhibition for treating inflammatory diseases such as RA, where the IL-6 neutralizing antibody **tocilizumab** is currently used (Scheinecker, Smolen, Yasothan, Stoll, & Kirkpatrick, 2009). Shortly afterward a number of other reports appeared in the literature including **compound 28** (Hay et al., 2015) by SGC Oxford and **I-BRD9** (Theodoulou et al., 2015) by GlaxoSmithKline and University of Strathclyde. Target engagement of **compound 28** was displayed using a FRAP assay where concentrations of 0.2–1 μ*M* decreased recovery time of BRD9-GFP, in a manner similar to the nonbinding BRD9 mutant N100F. Inhibition of BRD9 using **I-BRD9** was demonstrated using a NanoBRET assay and competitive chemoproteomic pull down studies. Interaction between NanoLuc-tagged BRD9 (bromodomain amino acids 120–240 only) and Halotagged histone 3.3 was decreased upon increasing concentrations of **I-BRD9** in HEK293 cells. Using HUT-78 cell lysate, a chemoproteomic competition binding assay was carried out to compare **I-BRD9** binding to endogenous BRD9 or BRD3; it was found that the inhibitor displayed greater than 625-fold selectivity over BRD3. While there were no functional phenotypic experiments using **compound 28** reported, Theodoulou et al. studied the effects of **I-BRD9** inhibition on global cellular gene expression. Human Kasumi-1 leukemic cells were incubated with either 10 μ*M* of **I-BRD9** or another BET family inhibitor **I-BET151** (Mirguet et al., 2012) at 1 μ*M* concentration for 6 h. Analysis of gene expression by Illumina microarray and quantitative polymerase chain reaction validation showed that while there was an expectedly high number of genes differentially regulated by both inhibitors, **I-BRD9** altered gene expression in a significant subset of genes, including cancer-relevant genes C-type lectin domain family 1, dual specificity phosphatase 6, feline sarcoma, and SAM domain, SH3 domain, and nuclear localization signals 1. This suggests that **I-BRD9** may have the potential for affecting disease-specific genes not targeted by other BET family inhibitors.

Due to high similarity between BRD9 and BRD7 in their respective bromodomains (72% sequence similarity) both **LP99** and **compound 28** display low selectivity between these proteins. However, **I-BRD9** was reported to display good selectivity for BRD9 over BRD7 in a DiscoveRx BromoSCAN assay panel. Both BRD9 and BRD7 may have distinct biology, for example, increased genomic copy number of the *BRD9* gene has been found in nonsmall cell lung cancer (NSCLC) patients (Kang, Koo, Kwon, Park, & Kim, 2008), whereas BRD7 acts as a tumor suppressor in nasopharyngeal carcinoma (Peng et al., 2007; Zhou et al., 2004). Therefore monoselective probes of BRD9 and BRD7 used alongside the dual BRD7/9 chemical probes would serve as valuable tools to discern the underlying biology of these epigenetic targets. Researchers from Boehringer Ingelheim sought to address this and following a fragment-based drug discovery effort against BRD9 they reported the discovery of **BI-9564** and **BI-7273** (Martin et al., 2016). Both probe molecules were evaluated for target engagement using a FRAP assay and despite minor structural differences between the two probes, **BI-9564** is highly BRD9 selective in vitro over BRD7 and a number of other bromodomains, whereas **BI-7273** displays excellent potency for both BRD9 and BRD7. Interestingly, both **BI-9564** and **BI-7273** were found to suppress proliferation of human eosinophilic cell line leukemic cells in a dose-dependent manner. In a parallel report domain-swapping experiments of BRD9 and BRD4 bromodomains in AML cells revealed the essentiality of BRD9 for AML stem cell maintenance, as well as demonstrating on-target effects of **BI-9564** (Hohmann et al., 2016) (Fig. 4).

In 2017, after the discovery of **I-BRD9** and **BI-9564**, a research group reported the cytotoxic and antiproliferative effects of these two specific chemical probes that target one of the bromodomains of SWI/SNF

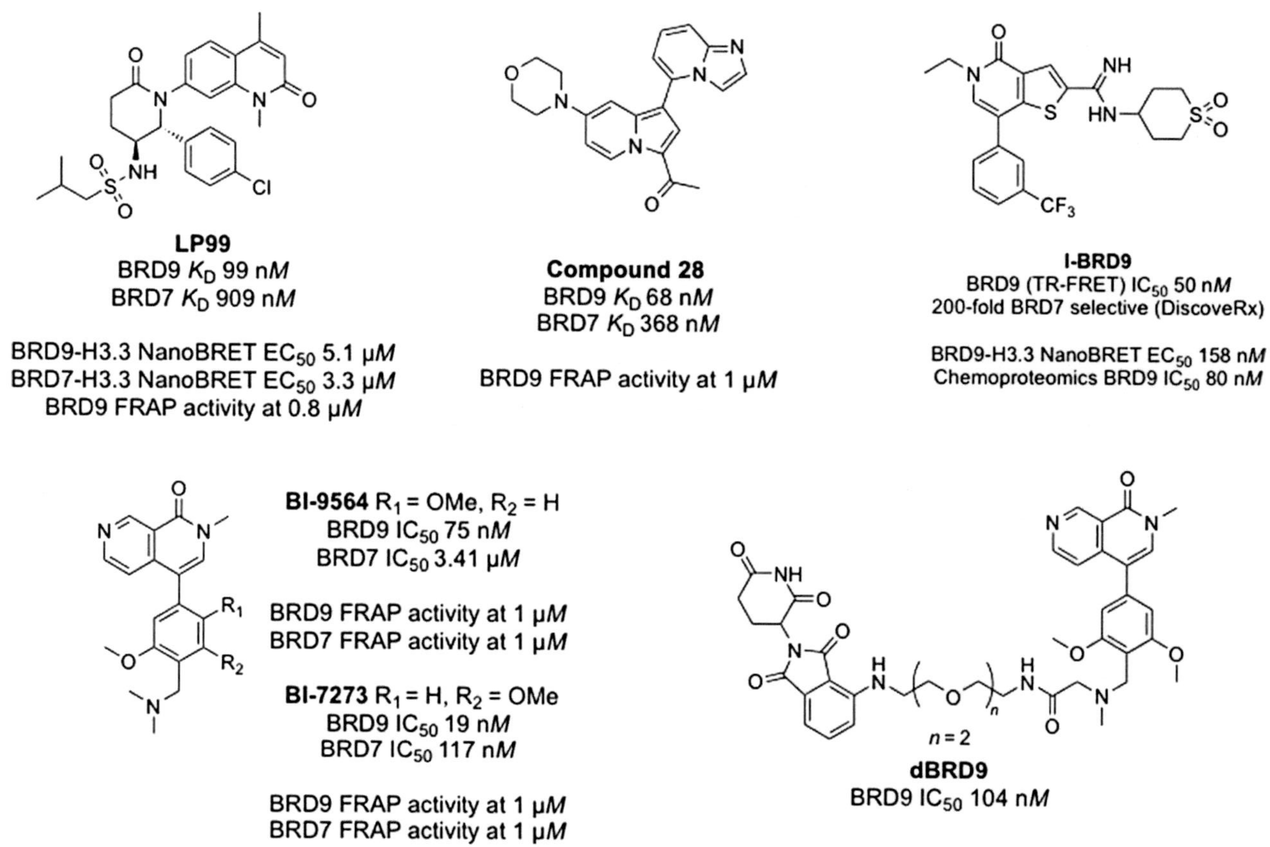


**Fig. 4** Selected BRD9 probes.

complex. As this complex has been implicated in rhabdoid tumor (RT) development, the BRD9 chemical probes were evaluated in five RT cell lines. Combinatorial effects of **I-BRD9** and cytotoxic drugs on cell proliferation were evaluated by cytotoxicity assays. Single compound treatment of RT cells with **I-BRD9** and **BI-9564** resulted in decreased cell proliferation, G1-arrest, and apoptosis (Kramer, Moreno, Fruhwald, & Kerl, 2017). These observations provide an example of benefits associated with using cytotoxicity assays in identifying target mechanism and pathways.

While the bromodomains of BRD9 and BRD7 have been the most well-studied architectural feature of these proteins, other poorly understood regions of their overall sequence may in fact have functionality relevant to BRD9/BRD7 biology. In order to explore this in the context of disease, PROTACs for BRD9 and BRD7 were created by conjugation **I-BRD9** and **BI-7273** to E3 ligase directing ligands (Remillard et al., 2017; Toure & Crews, 2016). This report culminated in the discovery of **dBRD9**, based on parent molecule **BI-7273,** which was found to be a highly potent and selective BRD9 degrader. Target engagement and degradation studies were evaluated in vitro using an AlphaScreen proximity assay and in a cellular context by immunoblot analysis of endogenous BRD9 levels in MOLM-13 leukemic cells after treatment with **dBRD9**. Through extensive characterization and additional domain swapping experiments, it was elucidated that **dBRD9-**mediated BRD9 degradation leads to a target-specific inhibitory effect on MOLM-13 cell proliferation. Interestingly despite maintaining weaker in vitro potency for recombinant BRD9 by AlphaScreen than parent probe **BI-7273**, **dBRD9** displayed significantly more potent cellular activities which may allude to the catalytic nature of proteasome-mediated degradation or highlight the necessity of other uncharacterized regions toward BRD9 functioning in pathology which underscores the difference between conventional BRD9 inhibition (**BI-7273**) and full-length BRD9 degradation (**dBRD9**).

## 4.3 Other Examples

The previous examples have explored the identification of protein targets for relevant indications using chemical probes that carry corresponding target engagement data. Many of these probes have targeted novel or emerging target proteins. Establishment of new disease links with previously studied proteins using a combination of approaches has also been explored.

### 4.3.1 CDK4 and Exportin Inhibitors

Researchers from the Broad Institute were interested in exploring a rare undifferentiated sarcoma and generated a primary cancer cell line, CLF-PED-015-T, derived from a metastatic brain lesion biopsy from a pediatric patient (Hong et al., 2016). CLF-PED-015-T was able to form tumors when subcutaneously injected into immunodeficient mice, growing at similar rates to established neuroblastoma and Ewing sarcoma cell lines. To identify therapeutic targets in this cell line, custom CRISPR–Cas9 and shRNA libraries were generated, targeting 429 genes. These 429 genes were chosen as druggable targets based on compounds used in literature on established cell lines, in development by various pharmaceutical companies and undergoing clinical trial or approval. In conjunction with these function genomic screens, a small molecule screen was also carried out in CLF-PED-015-T using a library of 440 compounds. Using these combined approaches facilitated the identification of cyclin-dependent kinase 4 (CDK4) and exportin 1 (XPO1) as important targets in this disease, as knocking down expression of these genes or targeting with small molecules lead to decreased cell viability.

### 4.3.2 KRAS Binders

While curated chemogenomic libraries offer avenues to identify novel targets and pathways in relevant indications, purposeful inclusion of promiscuous compounds can illuminate polypharmacology. Carver and colleagues from the National Institute of Health (NIH) used a chemogenomic library (MIPE 3.0) to identify novel binders of the cancer-relevant KRAS protein (Carver et al., 2014). A fusion GFP-KRAS cell line was generated for screening small molecules where compound binding would lead to KRAS mis-folding and degradation, leading to a decrease in GFP fluorescence, detected by InCELL Analyzer high content imager. Interestingly the only small molecule hits were promiscuous kinase inhibitors such as the Abelson murine leukemia kinase inhibitor **ponatinib** and the lymphocyte-specific protein tyrosine kinase inhibitor **AMG-7a**.

### 4.3.3 BTK Inhibitors

In addition to polypharmacology arising from promiscuous compound hits, cross-screening of chemogenomic libraries can also illuminate synergistic relationships in target or pathway identification. The MIPE 3.0 chemogenomic library was again used by coworkers from the NIH to find new combination therapies for the aggressive activated B cell-like subtype of diffuse large B cell lymphoma (ABC-DLBCL) (Mathews Griner et al., 2014), by measuring cell viability (Cell Titer Glo), apoptosis (Caspase-Glo 3/7), and NFκB reporter assay in ABC-DLBCL cell lines TMD8 and HBL-1. The Bruton's tyrosine kinase (BTK) inhibitor **ibrutinib**, which targets chronic B-cell receptor signaling in ABC-DLBCL, was found to act synergistically with inhibitors of the PI3K-AKT pathway, BCL-2 family inhibitors, and several common chemotherapeutic agents. Building on this work, a later study by coworkers from MD Anderson Cancer Centre revealed additional opportunities for synergy in chronic lymphocytic leukemia (CLL) through **ibrutinib-**mediated BTK inhibition in combination with **venetoclax**-mediated BCL-2 inhibition (Cervantes-Gomez et al., 2015).

## 5. SUMMARY AND OUTLOOK

The aforementioned examples of probe molecules or sets used to illuminate new biology of relevant proteins have largely relied upon the development of high-quality chemical probes (Arrowsmith et al., 2015; Bunnage et al., 2013). Probes have been used to corroborate previous disease links of targets or highlight new avenues for research with the benefit of immediate access to chemical starting points for clinical translation. Use of probe sets containing compounds with varying degrees of selectivity may also reveal new polypharmacology and synergistic relationships. Other approaches such as pooled CRISPR–Cas9 and RNAi screens for target identification present as useful orthogonal options to small molecule chemical probes. Recent reports have explored the use of probe sets in identifying targets that may modulate CRISPR–Cas9 and RNAi processes (Yu et al., 2015). It is envisaged that as multiple chemical probe chemotypes become available for novel protein families, screening of chemogenomic libraries against an ever-expanding repertoire of human disease models using primary tissue samples will offer significant inroads into target identification for unmet diseases.

## REFERENCES

Agarwal, P., & Bertozzi, C. R. (2015). Site-specific antibody-drug conjugates: The nexus of bioorthogonal chemistry, protein engineering, and drug development. *Bioconjugate Chemistry*, *26*, 176–192. https://doi.org/10.1021/bc 5004982.

Albert, M., & Helin, K. (2010). Histone methyltransferases in cancer. *Seminars in Cell and Developmental Biology*, *21*, 209–220. https://doi.org/10.1016/j.semcdb.2009.10.007.

Andersson, E. R., & Lendahl, U. (2014). Therapeutic modulation of Notch signalling—Are we there yet? *Nature Reviews. Drug Discovery*, *13*(5), 357–378. https://doi.org/10.1038/nrd4252.

Arrowsmith, C. H., Audia, J. E., Austin, C., Baell, J., Bennett, J., Blagg, J., et al. (2015). The promise and peril of chemical probes. *Nature Chemical Biology*, *11*(8), 536–541. https://doi.org/10.1038/nchembio.1867.

Bao, S., Wu, Q., McLendon, R. E., Hao, Y., Shi, Q., Hjelmeland, A. B., et al. (2006). Glioma stem cells promote radioresistance by preferential activation of the DNA damage response. *Nature*, *444*(7120), 756–760. https://doi.org/10.1038/nature05236.

Beck, A., Goetsch, L., Dumontet, C., & Corvaïa, N. (2017). Strategies and challenges for the next generation of antibody–drug conjugates. *Nature Reviews Drug Discovery*, *16*, 315. https://doi.org/10.1038/nrd.2016.268.

Blagg, J., & Workman, P. (2017). Choose and use your chemical probe wisely to explore cancer biology. *Cancer Cell*, *32*(1), 9–25. https://doi.org/10.1016/j.ccell.2017.06.005.

Bondeson, D. P., Mares, A., Smith, I. E. D., Ko, E., Campos, S., Miah, A. H., et al. (2015). Catalytic in vivo protein knockdown by small-molecule PROTACs. *Nature Chemical Biology*, *11*(8), 611–617. https://doi.org/10.1038/nchembio.1858. http://www.nature.com/nchembio/journal/v11/n8/abs/nchembio.1858.html#supplementary-information.

Brennan, F. M., Chantry, D., Jackson, A., Maini, R., & Feldmann, M. (1989). Inhibitory effect of TNF alpha antibodies on synovial cell interleukin-1 production in rheumatoid arthritis. *Lancet (London, England)*, *2*(8657), 244–247.

Bromberg, K. D., Mitchell, T. R. H., Upadhyay, A. K., Jakob, C. G., Jhala, M. A., Comess, K. M., et al. (2017). The SUV4–20 inhibitor A-196 verifies a role for epigenetics in genomic integrity. *Nature Chemical Biology*, *13*(3), 317–324. https://doi.org/10.1038/nchembio.2282.

Bunnage, M. E., Chekler, E. L. P., & Jones, L. H. (2013). Target validation using chemical probes. *Nature Chemical Biology*, *9*, 195–199. https://doi.org/10.1038/nchembio.1197.

Campbell, J. E., Duncan, K. W., Foley, M. A., Harvey, D. M., Kuntz, K. W., Mills, J. E. J., et al. (2017). *Amine-substituted aryl or heteroaryl compounds as EHMT1 and EHMT2 inhibitors and their preparation*. USA: *PCT Int. Appl.* Epizyme, Inc.

Carter, P. J., & Lazar, G. A. (2017). Next generation antibody drugs: Pursuit of the 'high-hanging fruit'. *Nature Reviews. Drug Discovery*, *17*, 197. https://doi.org/10.1038/nrd.2017.227.

Carthew, R. W., & Sontheimer, E. J. (2009). Origins and mechanisms of miRNAs and siRNAs. *Cell*, *136*, 642–655. https://doi.org/10.1016/j.cell.2009.01.035. [Copyright (C) 2018 American Chemical Society (ACS). All Rights Reserved.].

Carver, J., Dexheimer, T. S., Hsu, D., Weng, M. T., Smith, J. L., Guha, R., et al. (2014). A high-throughput assay for small molecule destabilizers of the KRAS oncoprotein. *PLoS ONE*, *9*(8), e103836. https://doi.org/10.1371/journal.pone.0103836.

Cervantes-Gomez, F., Lamothe, B., Woyach, J. A., Wierda, W. G., Keating, M. J., Balakrishnan, K., et al. (2015). Pharmacological and protein profiling suggests venetoclax (ABT-199) as optimal partner with ibrutinib in chronic lymphocytic leukemia. *Clinical Cancer Research*, *21*(16), 3705–3715. https://doi.org/10.1158/1078-0432.CCR-14-2809.

Chan-Penebre, E., Gibaja, V., Campbell, J., Olu, O., Admirand, E., Brach, D., et al. (2017). Reawakening of human fetal hemoglobin and an epigenetic path to the clinic for sickle cell disease and beta-thalassemia: Identification of an orally-available, potent, and selective euchromatic histone lysine methyltransferase 1 and 2 (EHMT1/2) inhibitor. *Blood*, *130*(Suppl 1), 537.

Chemical Probes Portal. (2018). Retrieved from http://www.chemicalprobes.org.

Chen, J., Li, Y., Yu, T. S., McKay, R. M., Burns, D. K., Kernie, S. G., et al. (2012). A restricted cell population propagates glioblastoma growth after chemotherapy. *Nature*, *488*(7412), 522–526. https://doi.org/10.1038/nature11287.

Clark, P. G. K., Vieira, L. C. C., Tallant, C., Fedorov, O., Singleton, D. C., Rogers, C. M., et al. (2015). LP99: Discovery and synthesis of the first selective BRD7/9 bromodomain inhibitor. *Angewandte Chemie, International Edition*, *54*(21), 2621–6217. https://doi.org/10.1002/anie.201501394R2.

Copeland, R. A., Solomon, M. E., & Richon, V. M. (2009). Protein methyltransferases as a target class for drug discovery. *Nature Reviews Drug Discovery*, *8*, 724–732. https://doi.org/10.1038/nrd2974.

Day, C., & Kraft, L. (2012). Analysis of protein and lipid dynamics using confocal fluorescence recovery after photobleaching (FRAP). *Current Protocols in Cytometry*, *62*, Unit 2.19, 1–29. https://doi.org/10.1002/0471142956.cy0219s62.Analysis.

Deschout, H., Raemdonck, K., Demeester, J., De Smedt, S. C., & Braeckmans, K. (2014). FRAP in pharmaceutical research: Practical guidelines and applications in drug delivery. *Pharmaceutical Research*, *31*(2), 255–270. https://doi.org/10.1007/s11095-013-1146-9.

Doudna, J. A., & Charpentier, E. (2014). The new frontier of genome engineering with CRISPR-Cas9. *Science*, *346*(6213), 1258096. https://doi.org/10.1126/science.1258096.

Edwards, A. M., Arrowsmith, C. H., Bountra, C., Bunnage, M. E., Feldmann, M., Knight, J. C., et al. (2015). Preclinical target validation using patient-derived cells. *Nature Reviews. Drug Discovery*, *14*(3), 149–150. https://doi.org/10.1038/nrd4565.

Eurofins. (2018). *Eurofins safety screen*. Retrieved from https://www.eurofinspanlabs.com/Catalog/Products/ProductDetails.aspx?prodId=WkpXTEM73wg%3D.

Felce, J. H., Knox, R. G., & Davis, S. J. (2014). Type-3 BRET, an improved competition-based bioluminescence resonance energy transfer assay. *Biophysical Journal*, *106*(12), L41–L43. https://doi.org/10.1016/j.bpj.2014.04.061.

Fosgerau, K., & Hoffmann, T. (2015). Peptide therapeutics: Current status and future directions. *Drug Discovery Today*, *20*(1), 122–128. https://doi.org/10.1016/j.drudis.2014.10.003.

Ghildiyal, M., & Zamore, P. D. (2009). Small silencing RNAs: An expanding universe. *Nature Reviews. Genetics*, *10*, 94–108. [Copyright (C) 2018 American Chemical Society (ACS). All Rights Reserved.]. https://doi.org/10.1038/nrg2504.

Grinshtein, N., Rioseco, C. C., Marcellus, R., Uehling, D., Aman, A., Lun, X., et al. (2016). Small molecule epigenetic screen identifies novel EZH2 and HDAC inhibitors that target glioblastoma brain tumor-initiating cells. *Oncotarget*, *7*(37), 59360–59376. https://doi.org/10.18632/oncotarget.10661.

Hay, D., Brennan, P., Fedorov, O., Tallant, C., Monteiro, O., Muller-Knapp, S., et al. (2015). Design and synthesis of potent and selective inhibitors of BRD9 and BRD7 bromodomains. *Medicinal Chemistry Communications*, *6*, 1381–1386. https://doi.org/10.1039/C5MD00152H.

Hohmann, A. F., Martin, L. J., Minder, J. L., Roe, J. S., Shi, J., Steurer, S., et al. (2016). Sensitivity and engineered resistance of myeloid leukemia cells to BRD9 inhibition. *Nature Chemical Biology*, *12*(9), 672–679. https://doi.org/10.1038/nchembio.2115.

Hong, A. L., Tseng, Y. Y., Cowley, G. S., Jonas, O., Cheah, J. H., Kynnap, B. D., et al. (2016). Integrated genetic and pharmacologic interrogation of rare cancers. *Nature Communications*, *7*, 11987. https://doi.org/10.1038/ncomms11987.

Huang, J., Dorsey, J., Chuikov, S., Zhang, X., Jenuwein, T., Reinberg, D., et al. (2010). G9a and Glp methylate lysine 373 in the tumor suppressor p53. *Journal of Biological Chemistry*, *285*(13), 9636–9641. https://doi.org/10.1074/jbc.M109.062588.

Huber, K. V. M., Olek, K. M., Müller, A. C., Soon Heng Tan, C., Bennett, K. L., Colinge, J., et al. (2015). Proteome-wide small molecule and metabolite interaction mapping. *Nature Methods*, *12*(11), 1055–1057. https://doi.org/10.1038/nmeth.3590.

Hughes, J. P., Rees, S., Kalindjian, S. B., & Philpott, K. L. (2011). Principles of early drug discovery. *British Journal of Pharmacology*, *162*(6), 1239–1249. https://doi.org/10.1111/j.1476-5381.2010.01127.x.

Hulkower, K. I., & Herber, R. L. (2011). Cell migration and invasion assays as tools for drug discovery. *Pharmaceutics*, *3*, 107–124. https://doi.org/10.3390/pharmaceutics3010107.

Imai, K., Togami, H., & Okamoto, T. (2010). Involvement of histone H3 lysine 9 (H3K9) methyltransferase G9a in the maintenance of HIV-1 latency and its reactivation by BIX01294. *Journal of Biological Chemistry*, *285*(22), 16538–16545. https://doi.org/10.1074/jbc.M110.103531.

Jin, R., Nakada, M., Teng, L., Furuta, T., Sabit, H., Hayashi, Y., et al. (2013). Combination therapy using Notch and Akt inhibitors is effective for suppressing invasion but not proliferation in glioma cells. *Neuroscience Letters*, *534*, 316–321. https://doi.org/10.1016/j.neulet.2012.12.008.

Kang, J. U., Koo, S. H., Kwon, K. C., Park, J. W., & Kim, J. M. (2008). Gain at chromosomal region 5p15.33, containing TERT, is the most frequent genetic event in early stages of non-small cell lung cancer. *Cancer Genetics and Cytogenetics*, *182*(1), 1–11. https://doi.org/10.1016/j.cancergencyto.2007.12.004.

Kim, Y., Kim, Y. S., Kim, D. E., Lee, J. S., Song, J. H., Kim, H. G., et al. (2013). BIX-01294 induces autophagy-associated cell death via EHMT2/G9a dysfunction and intracellular reactive oxygen species production. *Autophagy*, *9*(12), 2126–2139. https://doi.org/10.4161/auto.26308.

Kondo, Y., Shen, L., Ahmed, S., Boumber, Y., Sekido, Y., Haddad, B. R., et al. (2008). Downregulation of histone H3 lysine 9 methyltransferase G9a induces centrosome disruption and chromosome instability in cancer cells. *PLoS ONE*, *3*(4), e2037. https://doi.org/10.1371/journal.pone.0002037.

Kondo, Y., Shen, L., Suzuki, S., Kurokawa, T., Masuko, K., Tanaka, Y., et al. (2007). Alterations of DNA methylation and histone modifications contribute to gene silencing in hepatocellular carcinomas. *Hepatology Research*, *37*(11), 974–983. https://doi.org/10.1111/j.1872-034X.2007.00141.x.

Kramer, K. F., Moreno, N., Fruhwald, M. C., & Kerl, K. (2017). BRD9 inhibition, alone or in combination with cytostatic compounds as a therapeutic approach in rhabdoid tumors. *International Journal of Molecular Sciences*, *18*(7), 1537. https://doi.org/10.3390/ijms18071537.

Krivega, I., Byrnes, C., De Vasconcellos, J. F., Lee, Y. T., Kaushal, M., Dean, A., et al. (2015). Inhibition of G9a methyltransferase stimulates fetal hemoglobin production by facilitating LCR/γ-globin looping. *Blood*, *126*(5), 665–672. https://doi.org/10.1182/blood-2015-02-629972.

Kubicek, S., O'Sullivan, R. J., August, E. M., Hickey, E. R., Zhang, Q., Teodoro, M. L. L., et al. (2007). Reversal of H3K9me2 by a small-molecule inhibitor for the G9a histone methyltransferase. *Molecular Cell*, *25*(3), 473–481. https://doi.org/10.1016/j.molcel.2007.01.017.

Lecerf, J.-M., Shirley, T. L., Zhu, Q., Kazantsev, A., Amersdorfer, P., Housman, D. E., et al. (2001). Human single-chain Fv intrabodies counteract in situ huntingtin aggregation in cellular models of huntington's disease. *Proceedings of the National Academy of Sciences of the United States of America*, *98*(8), 4764–4769. https://doi.org/10.1073/pnas.071058398.

Liu, F., Barsyte-Lovejoy, D., Li, F., Xiong, Y., Korboukh, V., Huang, X.-P., et al. (2013). Discovery of an in vivo chemical probe of the lysine methyltransferases G9a and GLP. *Journal of Medicinal Chemistry*, *56*(21), 8931–8942. https://doi.org/10.1021/jm401480r.

Liu, F., Chen, X., Allali-Hassani, A., Quinn, A. M., Wasney, G. A., Dong, A., et al. (2009). Discovery of a 2,4-diamino-7-aminoalkoxyquinazoline as a potent and selective inhibitor of histone lysine methyltransferase G9a. *Journal of Medicinal Chemistry*, *52*(24), 7950–7953. https://doi.org/10.1021/jm901543m.

Liu, F., Chen, X., Allali-Hassani, A., Quinn, A. M., Wigle, T. J., Wasney, G. A., et al. (2010). Protein lysine methyltransferase g9a inhibitors: Design, synthesis, and structure activity relationships of 2,4-diamino-7-aminoalkoxy-quinazolines. *Journal of Medicinal Chemistry*, *53*(15), 5844–5857. https://doi.org/10.1021/jm100478y.

Lu, J., Qian, Y., Altieri, M., Dong, H., Wang, J., Raina, K., et al. (2015). Hijacking the E3 ubiquitin ligase cereblon to efficiently target BRD4. *Chemistry and Biology*, *22*(6), 755–763. https://doi.org/10.1016/j.chembiol.2015.05.009.

Machleidt, T., Woodroofe, C. C., Schwinn, M. K., Méndez, J., Robers, M. B., Zimmerman, K., et al. (2015). NanoBRET—a novel BRET platform for the analysis of protein-protein interactions. *ACS Chemical Biology*, *10*(8), 1797–1804. https://doi.org/10.1021/acschembio.5b00143.

Marine, S., Bahl, A., Ferrer, M., & Buehler, E. (2012). Common seed analysis to identify off-target effects in siRNA screens. *Journal of Biomolecular Screening*, *17*(3), 370–378. https://doi.org/10.1177/1087057111427348.

Marshall, J. C. (2014). Why have clinical trials in sepsis failed? *Trends in Molecular Medicine*, *20*(4), 195–203. https://doi.org/10.1016/j.molmed.2014.01.007.

Martin, L. J., Koegl, M., Bader, G., Cockcroft, X.-L., Fedorov, O., Fiegen, D., et al. (2016). Structure-based design of an in vivo active selective BRD9 inhibitor. *Journal of Medicinal Chemistry*, *15*, 4462–4475. https://doi.org/10.1021/acs.jmedchem.5b01865.

Martinez Molina, D., & Nordlund, P. (2016). The cellular thermal shift assay: A novel biophysical assay for in situ drug target engagement and mechanistic biomarker studies. *Annual Review of Pharmacology and Toxicology*, *56*(1), 141–161. https://doi.org/10.1146/annurev-pharmtox-010715-103715.

Mathews Griner, L. A., Guha, R., Shinn, P., Young, R. M., Keller, J. M., Liu, D., et al. (2014). High-throughput combinatorial screening identifies drugs that cooperate with ibrutinib to kill activated B-cell–like diffuse large B-cell lymphoma cells. *Proceedings of the National Academy of Sciences of the United States of America*, *111*(6), 2349–2354. https://doi.org/10.1073/pnas.1311846111.

McManus, M. T., & Sharp, P. A. (2002). Gene silencing in mammals by small interfering RNAs. *Nature Reviews Genetics*, *3*, 737–747. Copyright (C) 2018 American Chemical Society (ACS). All Rights Reserved. https://doi.org/10.1038/nrg908.

Mhashilkar, A. M., Bagley, J., Chen, S. Y., Sziilvay, A. M., Helland, D. G., & Marasco, W. A. (1995). Inhibition of HIV-1 Tat-mediated LTR transactivation and HIV-1 infection by anti-Tat single chain intrabodies. *EMBO Journal*, *14*(7), 1542–1551.

Millan, M. J. (2013). An epigenetic framework for neurodevelopmental disorders: From pathogenesis to potential therapy. *Neuropharmacology*, *68*, 2–82. https://doi.org/10.1016/j.neuropharm.2012.11.015.

Mirguet, O., Lamotte, Y., Donche, F., Toum, J., Gellibert, F., Bouillot, A., et al. (2012). From ApoA1 upregulation to BET family bromodomain inhibition: Discovery of I-BET151. *Bioorganic & Medicinal Chemistry Letters*, *22*(8), 2963–2967. https://doi.org/10.1016/j.bmcl.2012.01.125.

Murray, C. W., & Rees, D. C. (2009). The rise of fragment-based drug discovery. *Nature Chemistry*, *1*, 187–192. Copyright (C) 2017 American Chemical Society (ACS). All Rights Reserved. https://doi.org/10.1038/nchem.217.

Neklesa, T. K., Winkler, J. D., & Crews, C. M. (2017). Targeted protein degradation by PROTACs. *Pharmacology and Therapeutics*, *174*, 138–144. https://doi.org/10.1016/j.pharmthera.2017.02.027.

Pellowe, A. S., & Gonzalez, A. L. (2016). Extracellular matrix biomimicry for the creation of investigational and therapeutic devices. *Wiley Interdisciplinary Reviews: Nanomedicine and Nanobiotechnology*, *8*, 5–22. https://doi.org/10.1002/wnan.1349.

Peng, C., Liu, H. Y., Zhou, M., Zhang, L. M., Li, X. L., Shen, S. R., et al. (2007). BRD7 suppresses the growth of nasopharyngeal carcinoma cells (HNE1) through negatively regulating β-catenin and ERK pathways. *Molecular and Cellular Biochemistry*, *303*(1&2), 141–149. https://doi.org/10.1007/s11010-007-9466-x.

Pfleger, K. D. G., & Eidne, K. A. (2006). Illuminating insights into protein-protein interactions using bioluminescence resonance energy transfer (BRET). *Nature Methods*, *3*(3), 165–174. https://doi.org/10.1038/nmeth841.

Philpott, M., Rogers, C. M., Yapp, C., Wells, C., Lambert, J. P., Strain-Damerell, C., et al. (2014). Assessing cellular efficacy of bromodomain inhibitors using fluorescence recovery after photobleaching. *Epigenetics & Chromatin*, 7, 14. https://doi.org/10.1186/1756-8935-7-14.

Pierre, C. S., Guo, J., Shin, J. D., Engstrom, L. W., Lee, H. H., Herbert, A., et al. (2017). A human tissue-based functional assay platform to evaluate the immune function impact of small molecule inhibitors that target the immune system. *PLoS ONE*, *12*(7), e0180870. https://doi.org/10.1371/journal.pone.0180870.

Plenge, R. M. (2016). Disciplined approach to drug discovery and early development. *Science Translational Medicine*, *8*(349), 349ps15. https://doi.org/10.1126/scitranslmed.aaf2608.

Remillard, D., Buckley, D. L., Paulk, J., Brien, G. L., Sonnett, M., Seo, H. S., et al. (2017). Degradation of the BAF complex factor BRD9 by heterobifunctional ligands. *Angewandte Chemie—International Edition*, *56*(21), 5738–5743. https://doi.org/10.1002/anie.201611281.

Renneville, A., Van Galen, P., Canver, M. C., McConkey, M., Krill-Burger, J. M., Dorfman, D. M., et al. (2015). EHMT1 and EHMT2 inhibition induces fetal hemoglobin expression. *Blood*, *126*(16), 1930–1939. https://doi.org/10.1182/blood-2015-06-649087.

Reynolds, A., Leake, D., Boese, Q., Scaringe, S., Marshall, W. S., & Khvorova, A. (2004). Rational siRNA design for RNA interference. *Nature Biotechnology*, *22*, 326–330. https://doi.org/10.1038/nbt936. [Copyright (C) 2017 American Chemical Society (ACS). All Rights Reserved.].

Robers, M. B., Dart, M. L., Woodroofe, C. C., Zimprich, C. A., Kirkland, T. A., Machleidt, T., et al. (2015). Target engagement and drug residence time can be observed in living cells with BRET. *Nature Communications*, *6*, 1–10. https://doi.org/10.1038/ncomms10091.

Sachse, C., Krausz, E., Kronke, A., Hannus, M., Walsh, A., Grabner, A., et al. (2005). High-throughput RNA interference strategies for target discovery and validation by using synthetic short interfering RNAs: Functional genomics investigations of biological pathways. *Methods in Enzymology*, *392*, 242–277. https://doi.org/10.1016/S0076-6879(04)92015-0.

Scannell, J. W., & Bosley, J. (2016). When quality beats quantity: Decision theory, drug discovery, and the reproducibility crisis. *PLoS One*, *11*(2), e0147215. https://doi.org/10.1371/journal.pone.0147215.

Scheinecker, C., Smolen, J., Yasothan, U., Stoll, J., & Kirkpatrick, P. (2009). Tocilizumab. *Nature Reviews Drug Discovery*, *8*(4), 273–274. https://doi.org/10.1038/nrd2863.

Schulze, J., Moosmayer, D., Weiske, J., Fernández-Montalván, A., Herbst, C., Jung, M., et al. (2015). Cell-based protein stabilization assays for the detection of interactions between small-molecule inhibitors and BRD4. *Journal of Biomolecular Screening*, *20*, 180–189. https://doi.org/10.1177/1087057114552398.

SGC Chemical Probes. (2018). Retrieved April 1, 2018, from. http://www.thesgc.org/chemical-probes.

Shi, Y., Desponts, C., Do, J. T., Hahm, H. S., Schöler, H. R., & Ding, S. (2008). Induction of pluripotent stem cells from mouse embryonic fibroblasts by Oct4 and Klf4 with small-molecule compounds. *Cell Stem Cell*, *3*(5), 568–574. https://doi.org/10.1016/j.stem.2008.10.004.

Shi, Y., Do, J. T., Desponts, C., Hahm, H. S., Schöler, H. R., & Ding, S. (2008). A combined chemical and genetic approach for the generation of induced pluripotent stem cells. *Cell Stem Cell*, *2*, 525–528. https://doi.org/10.1016/j.stem.2008.05.011.

Simon, G. M., Niphakis, M. J., & Cravatt, B. F. (2013). Determining target engagement in living systems. *Nature Chemical Biology*, *9*, 200–205. https://doi.org/10.1038/nchembio.1211.

Singh, S. K., Hawkins, C., Clarke, I. D., Squire, J. A., Bayani, J., Hide, T., et al. (2004). Identification of human brain tumour initiating cells. *Nature*, *432*(7015), 396–401. https://doi.org/10.1038/nature03128.

Stoddart, L. A., Johnstone, E. K. M., Wheal, A. J., Goulding, J., Robers, M. B., MacHleidt, T., et al. (2015). Application of BRET to monitor ligand binding to GPCRs. *Nature Methods*, *12*(7), 661–663. https://doi.org/10.1038/nmeth.3398.

Swinney, D. C., & Anthony, J. (2011). How were new medicines discovered? *Nature Reviews Drug Discovery*, *10*, 507–519. Copyright (C) 2018 American Chemical Society (ACS). All Rights Reserved. https://doi.org/10.1038/nrd3480.

Tachibana, M., Sugimoto, K., Nozaki, M., Ueda, J., Ohta, T., Ohki, M., et al. (2002). G9a histone methyltransferase plays a dominant role in euchromatic histone H3 lysine 9 methylation and is essential for early embryogenesis. *Genes and Development*, *16*(14), 1779–1791. https://doi.org/10.1101/gad.989402.

Tachibana, M., Ueda, J., Fukuda, M., Takeda, N., Ohta, T., Iwanari, H., et al. (2005). Histone methyltransferases G9a and GLP form heteromeric complexes and are both crucial for methylation of euchromatin at H3-K9. *Genes and Development*, *19*(7), 815–826. https://doi.org/10.1101/gad.1284005.

Theodoulou, N. H., Bamborough, P., Bannister, A. J., Becher, I., Bit, R. A., Che, K. H., et al. (2015). Discovery of I-BRD9, a selective cell active chemical probe for bromodomain containing protein 9 inhibition. *Journal of Medicinal Chemistry*, *59*(4), 1425–1439. https://doi.org/10.1021/acs.jmedchem.5b00256.

Toure, M., & Crews, C. M. (2016). Small-molecule PROTACS: New approaches to protein degradation. *Angewandte Chemie International Edition*, *55*(6), 1966–1973. https://doi.org/10.1002/anie.201507978.

Vasaturo, A., Caserta, S., Russo, I., Preziosi, V., Ciacci, C., & Guido, S. (2012). A novel chemotaxis assay in 3-D collagen gels by time-lapse microscopy. *PLoS ONE*, *7*, e52251. https://doi.org/10.1371/journal.pone.0052251.

Watanabe, H., Soejima, K., Yasuda, H., Kawada, I., Nakachi, I., Yoda, S., et al. (2008). Deregulation of histone lysine methyltransferases contributes to oncogenic transformation of human bronchoepithelial cells. *Cancer Cell International*, *8*, 15. https://doi.org/10.1186/1475-2867-8-15.

Whitehead, K. A., Langer, R., & Anderson, D. G. (2009). Knocking down barriers: Advances in siRNA delivery. *Nature Reviews Drug Discovery*, *8*, 129–138. Copyright (C) 2018 American Chemical Society (ACS). All Rights Reserved. https://doi.org/10.1038/nrd2742.

Xiong, Y., Li, F., Babault, N., Dong, A., Zeng, H., Wu, H., et al. (2017). Discovery of Potent and Selective Inhibitors for G9a-Like Protein (GLP) Lysine Methyltransferase. *Journal of Medicinal Chemistry*, *60*(5), 1876–1891. https://doi.org/10.1021/acs.jmedchem.6b01645.

Xiong, Y., Li, F., Babault, N., Wu, H., Dong, A., Zeng, H., et al. (2017). Structure-activity relationship studies of G9a-like protein (GLP) inhibitors. *Bioorganic & Medicinal Chemistry*, *25*(16), 4414–4423. https://doi.org/10.1016/j.bmc.2017.06.021.

Yin, Y., Qiu, S., Li, X., Huang, B., Xu, Y., & Peng, Y. (2017). EZH2 suppression in glioblastoma shifts microglia toward M1 phenotype in tumor microenvironment. *Journal of Neuroinflammation*, *14*(1), 220. https://doi.org/10.1186/s12974-017-0993-4.

Yu, C., Liu, Y., Ma, T., Liu, K., Xu, S., Zhang, Y., et al. (2015). Small molecules enhance crispr genome editing in pluripotent stem cells. *Cell Stem Cell*, *16*(2), 142–147. https://doi.org/10.1016/j.stem.2015.01.003.

Zhou, J., Ma, J., Zhang, B. C., Li, X. L., Shen, S. R., Zhu, S. G., et al. (2004). BRD7, a novel bromodomain gene, inhibits G1-S progression by transcriptionally regulating some important molecules involved in ras/MEK/ERK and Rb/E2F pathways. *Journal of Cellular Physiology*, *200*(1), 89–98. https://doi.org/10.1002/jcp.20013.

CHAPTER THREE

# Mining the Microbiome for Drug Targets

**Qi Wang**[*,†], **Fangming Yang**[*,†], **Huijue Jia**[*,‡,§,1]
[*]BGI-Shenzhen, Shenzhen, China
[†]BGI Education Center, University of Chinese Academy of Sciences, Shenzhen, China
[‡]Shenzhen Key Laboratory of Human Commensal Microorganisms and Health Research, BGI-Shenzhen, Shenzhen, China
[§]China National Genebank-Shenzhen, BGI-Shenzhen, Shenzhen, China
[1]Corresponding author: e-mail address: jiahuijue@genomics.cn

## Contents



## Abstract

The human microbiome is our "other genome." Implicated in a growing list of complex diseases, for which genomic studies typically explain a portion of the disease susceptibility, the human gut microbiome has been at the spotlight for our understanding of human diseases. As the microbiome is intrinsically more variable than the human genome, it is important to take careful considerations at each step of a study. Here, we put forward our recommendations, which we envision would facilitate identification of true drug targets in the human microbiome in the colon as well as at other body sites.

## ABBREVIATIONS

**BMI** body mass index
**CASINO** community and systems-level interative optimization
**CCA** canonical coordinate analysis
**eggNOG** evolutionary genealogy of genes: Non-supervised Orthologous Groups
**FTA** flinders technology associates

*Methods in Enzymology*, Volume 610
ISSN 0076-6879
https://doi.org/10.1016/bs.mie.2018.09.014

| | |
|---|---|
| **GMM** | gut metabolic modules |
| **GWAS** | genome-wide association studies |
| **KEGG** | kyoto encyclopedia of genes and genomes |
| **MGC** | metagenomic clusters |
| **MGS** | metagenomic species |
| **MLG** | metagenomic linkage groups |
| **MWAS** | metagenome-wide association studies |
| **PERMANOVA** | permutational multivariate analysis of variance |

## 1. INTRODUCTION

We are never alone. The human body is home for hundreds of trillions of microorganisms, most of which are probably bacteria (Qin et al., 2010). This ensemble of stably inhabiting microorganisms is referred to as the "microbiota," emphasizing the ecosystem, and the term "microbiome" referring to the collection of microbial genomes and other related omics.

As genetics, especially genome-wide association studies (GWAS), usually only explain a portion of the risk for a complex disease and come up with an ever-growing list of variants with small effect sizes, alterations in the human microbiome have been eagerly explored in the past decade. Concomitant with developments in shotgun sequencing technologies, it becomes possible to sequence all the DNA in a metagenomic sample, regardless of which organism each DNA molecule belongs to (Qin et al., 2010; Tringe et al., 2005). As a comprehensive and unbiased approach that does not rely on culturing and isolating monoclones of microorganisms, metagenomic shotgun sequencing is regarded as perhaps the most important breakthrough in microbiology since the invention of the microscope. Functional and taxonomic annotations of metagenomic data began with direct prediction from sequencing reads (Tringe et al., 2005) and were greatly facilitated by algorithms that assemble the short sequencing reads into contigs (Qin et al., 2010) and later, genomes (Bäckhed et al., 2015; Nielsen et al., 2014; Sczyrba et al., 2017).

First established in the study of type 2 diabetes mellitus, metagenome-wide association studies (MWAS) have enabled identification of disease-associated gene clusters in the human gut and oral microbiomes (Qin et al., 2012; Wang & Jia, 2016; Zhang et al., 2015). Here, we outline the workflow and the cautions to take in MWAS, which would help pinpoint potential drug targets in the human microbiome.

## 2. SAMPLE COLLECTION AND SEQUENCING FOR MWAS

For a metagenome-wide association study to be informative, it is critical to gather a cohort that is representative of the target population, and properly collect all relevant samples (Fig. 1) and metadata (Chen et al., 2017; Falony et al., 2016; Wang & Jia, 2016; Zhernakova et al., 2016).

### 2.1 Controlling for Heterogeneity in the Cohort

When designing such a case–control study, factors such as age, geography, body mass index (BMI), medication and that are known to influence the microbiota need to be reasonably matched between the healthy and diseased groups, and extensive metadata should be obtained. Because most of the gut microbial genes are not present in all samples and vary wildly in abundance (i.e., sparse and heavy-tailed distribution), linear regression methods aimed at correcting the influencing factors would undermine the power of the case–control comparison, and are not used as a remedy. For disease samples such as colorectal cancer in feces and rheumatoid arthritis in dental plaques (Feng et al., 2015; Zhang et al., 2015), the disease signal is sufficiently strong compared to other factors.

Microbial markers with a smaller difference between groups (small effect size) might require a larger sample size to be distinguished from background variations. Existing MWAS included no more than a few hundreds of samples (Jie et al., 2017; Wang & Jia, 2016). However, larger sample sizes are expected, as low-cost sequencing platforms such as BGISEQ-500 becomes available (Fang et al., 2018) and computational capacity increases. A formal sample size estimation remains to be developed by statistical experts (Kelly et al., 2015).

Heterogeneity within groups due to different etiology, disease stage, medication, etc., or the microbiome itself, could reduce the actual sample size for a marker of interest. A routine practice would be to assess the influence from each known factor using Permutational multivariate analysis of variance (PERMANOVA) (McArdle & Anderson, 2001), canonical coordinate analysis (CCA), etc., and make sure none of the confounding factors differ between the case–control groups (e.g., in a MWAS of healthy controls, colorectal adenoma and carcinoma patients, only red meat consumption and serum ferritin differed between groups; Feng et al., 2015).

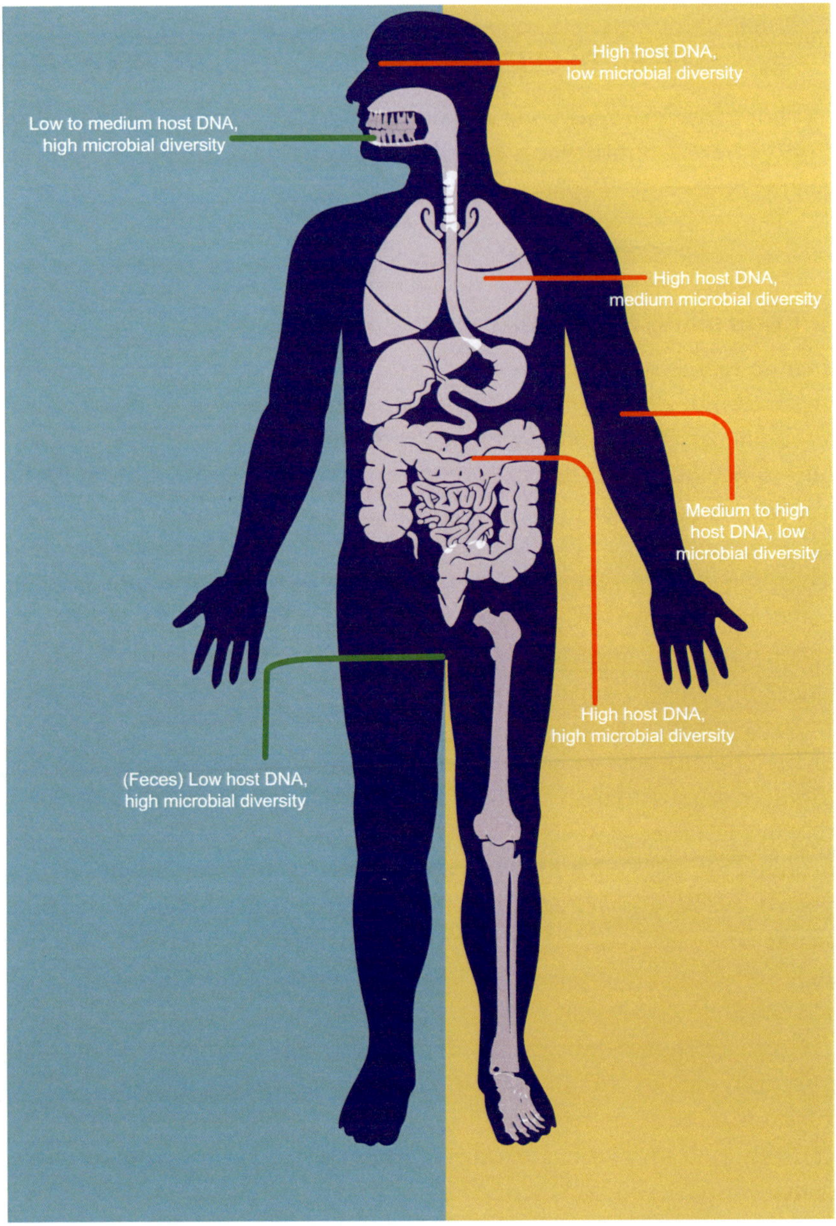


**Fig. 1** Summary of microbial biomass and diversity across the human body.

Notably, the controversial concept of enterotypes is not to be ignored (Arumugam et al., 2011; Ding & Schloss, 2014). In fecal metagenomic data, it is *Prevotella copri* vs a number of different *Bacteroides* strains (Jie et al., 2017), and stratification according to dominance by either *Prevotella* or *Bacteroides* could help identify patients who are more responsive to treatment by the diabetes drug acarbose (Gu et al., 2017).

## 2.2 Sample Preservation

The gold standard for freezing the community structure on spot is to have the metagenomic sample on ice (or better dry ice) for no more than 1 h, immediately freeze the sample at −80°C and transport the frozen sample on dry ice. This is obviously a challenge for sample collection at home, which typically allows a few days of storage at 4°C, and is also a burden for sample transportation.

A stabilizing reagent which stops microbial growth as well as prevents DNA degradation is on high demand (Han et al., 2018). Dry forms of sample preservation methods such as Flinders Technology Associates (FTA) cards could be an option as well (Song et al., 2016), although the amount of DNA preserved is likely to be lower than using liquid stabilizing reagents.

### *2.2.1 Reagents for Room-Temperature Preservation of Metagenomic Samples*

Omnigene•GUT (feces) from DNA Genotek, or MGIEasy fecal sample collection kit from MGI.

## 2.3 DNA Extraction

DNA yield, purity, and integrity are considerations for all types of DNA extraction protocols. In addition, a good DNA extraction protocol for metagenomic sequencing should preserve the community structure in the original unextracted sample, that is, the relative proportions of taxa and genes in the sample should not change after DNA extraction. In order to lyse bacterial and fungal cell walls and retrieve a sufficient amount of DNA for construction of metagenomic sequencing libraries, harsh conditions such as alkaline and mechanical disruption are typically applied to the sample. For nonfecal samples, getting a sufficient amount of DNA could be a greater challenge (Fig. 1). With hundreds of species in a metagenomic sample, differences between DNA extraction protocols have received attention from researchers (Costea et al., 2017; Kennedy et al., 2014; Wesolowska-Andersen et al., 2014), and researchers are always advised to try to correct for such differences when comparing samples from different studies (Li et al., 2014). For relatively unbiased extraction of both Gram-negative and Gram-positive bacteria, the protocol INRA, NucleoSpin™®Soil from Macherey Nagel™, and protocol MagPure Fast Stool DNA KF Kit B showed comparable performances (Table 1) (Yang et al., in press). According to evaluations using yeasts, Protocol Macherey Nagel™ NucleoSpin™®Soil is recommended when focusing on fungi or extracting DNA from fungal-enriched environments (Table 1) (Yang et al., in press).

## 2.4 Considerations in Shotgun Sequencing

With shotgun sequencing, a first thing to decide on is the sequencing amount, which should take into consideration the fraction of unusable reads on each sequencing platform. For any given sample, the number of genes detected (the apparent gene richness) increases with sequencing amount until saturation. Quantitative comparisons between samples would not be reliable in the steep regions of the curve (Fig. 2).

While sequencing reads from the human host take up no more than 1% of the total reads in fecal samples, they can dominate the sample in some oral, skin, and vaginal sites, and go over 99% in tissue samples such as the placenta (Aagaard et al., 2014; Chen et al., 2018; Li et al., 2014; Oh et al., 2014). Depending on the microbial diversity and the taxonomic resolution needed (Nayfach, Rodriguez-Mueller, Garud, & Pollard, 2016; Oh et al., 2014; Qin et al., 2010), a few gigabases of raw sequences would not be enough for such nonfecal samples (Figs. 1 and 2). We are not aware of experimental methods that could completely remove host sequences without skewing the microbiota composition (Chen et al., 2018). We currently recommend 50 million high-quality nonhuman sequencing reads for a typical adult human fecal sample, and at least 20 million high-quality nonhuman sequencing reads for dental plaques. Studies with lower or variable amounts of sequencing reads inevitably show up in the literature (Scher et al., 2013; Thaiss et al., 2016; Zhernakova et al., 2016).

Besides the fraction of high-quality sequencing reads, one should also beware of differences in biases and errors when including data from different sequencing platforms. In a recent analysis of identical fecal samples, it was reported that the Hiseq2000 platform tended to over-sequence genes with high GC contents compared to the BGISEQ-500 sequencer, while Hiseq4000 was more comparable to BGISEQ-500 (Fang et al., 2018).

**Table 1** Commonly Used DNA Extraction Protocols

| DNA Extraction Kit | Additional Processing | Key Feature for Success | Bacteria | Archaea | Fungi | Viruses | References |
|---|---|---|---|---|---|---|---|
| MOBIO DNeasy PowerSoil | — | — | Underestimates Gram-positive bacteria | Not analyzed | Underestimates yeast at least | Not analyzed | Li et al. (2013), Costea et al. (2017), and Yang et al. (in press) |
| QIAGEN's QIAampStoolMinikit | Mechanical cell lysis by running Fastprep™ instrument; Removed InhibitEX Tablet | Mechanical cell lysis with 0.1 mm zirconia beads | Good | Not analyzed | Not analyzed | Not analyzed | Costea et al. (2017) |
| QIAGEN | Repeated bead-beating | Repeated bead-beating | Not analyzed | Not analyzed | Good | Not analyzed | Huseyin et al. (2017) |
| MagPure Fast Stool DNA KF Kit B | — | Mechanical cell lysis with 0.1 mm glass beads | Good | Not analyzed | Underestimates yeast at least | Not analyzed | Yang et al. (in press) |
| Macherey Nagel™ NucleoSpin™®Soil | — | Mechanical cell lysis with 0.6–0.8 mm ceramic beads | Good | Not analyzed | Good at least for yeast | Not analyzed | Yang et al. (in press) |
| INRA(No commercial kit) | — | Mechanical cell lysis with 0.1 mm glass beads | Good | Not analyzed | Underestimates yeast at least | Not analyzed | Yang et al. (in press) |

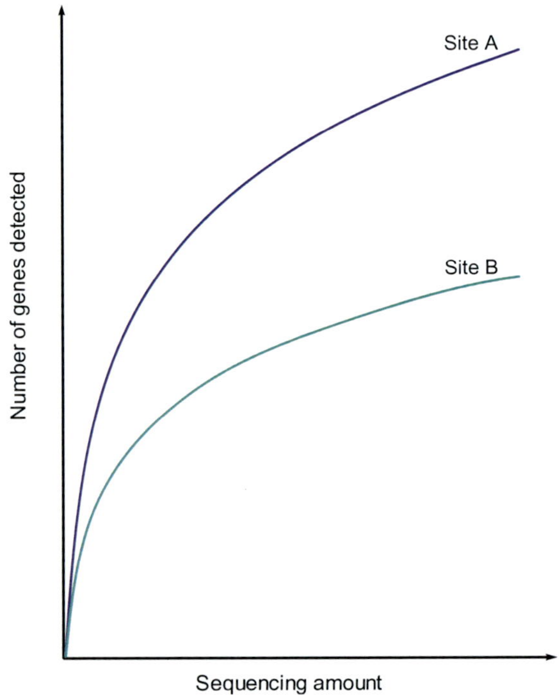


**Fig. 2** Dependence of the apparent gene richness in a sample on its sequencing amount.

#### *2.4.1 Reagents for Metagenomic Shotgun Sequencing*

TruSeq PE CLUSTER KIT v3-cBot-HS and TruSeq SBS KIT-HS v3 from Illumina (Hiseq2000) or

MGIEasy™ DNA Library Prep Kit and BGISEQ-500 High-throughput Sequencing Kit from MGI

## 3. IDENTIFYING POTENTIAL DRUG TARGETS

With the above-mentioned issues from sampling, DNA extraction to sequencing controlled for, one is ready to investigate the disease of interest using MWAS (Fig. 3).

### 3.1 Pinpointing the Disease-Associated Microbial Taxa

A major advantage of metagenomic shotgun sequencing compared to 16S rRNA gene amplicon sequencing is that the shotgun approach takes into

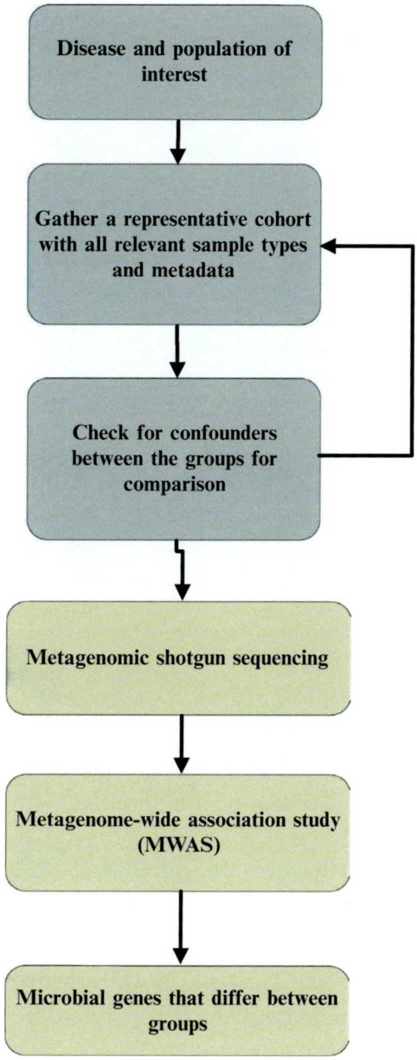


**Fig. 3** Overview of MWAS.

account both known and unknown taxa, as well as fungi, DNA viruses, protists, and worms in addition to bacteria and archaea.

To organize microbial genes into strain-level clusters, metagenomic linkage groups (MLG) (Qin et al., 2012), metagenomic clusters (MGC) (Karlsson et al., 2013), and metagenomic species (MGS) (Le Chatelier et al., 2013; Nielsen et al., 2014) have been developed (Table 2), all based on the idea that genes from the same microbial genome should display the same pattern of

**Table 2** Metagenomic Equivalents of Microbial Strains

| | Approach to Reduce Gene Number | Clustering Genes by | Correlation Coefficient | Approach to Reduce Cluster Number | References |
|---|---|---|---|---|---|
| MLG | Filter by occurrence frequency; significant difference between cohorts | Pair-wise correlation | Kendall | Filter by gene number | Qin et al. (2012) |
| MGC | Filter by occurrence frequency | Pair-wise correlation | Spearman | Filter by gene number | Karlsson et al. (2013) |
| MGS | Filter by occurrence frequency | Random seeding, canopy median | Pearson, Spearman | Merge according to Pearson's correlation; filter by gene number | Le Chatelier et al. (2013) and Nielsen et al. (2014) |

abundance variation across many samples. As this idea implies, a key to the accuracy of such coabundance binning is sample size. Incorporation of additional information such as GC content vs sequencing depth, and tetranucleotide frequency should improve the binning (Albertsen et al., 2013; Alneberg et al., 2014; Bäckhed et al., 2015; Sczyrba et al., 2017). Whereas MLGs were originally clustered from genes that were differentially enriched between cases and controls, genes are only filtered with an occurrence frequency cutoff for MGCs and MGSs, i.e., genes found in a small number of samples are not used for clustering (Table 2). With improvements in computing capacity, we expect the trade-off between accuracy and speed to favor accuracy.

Annotation of the gene clusters relies on alignment to sequenced microbial genomes or draft genomes. Instead of treating all the genes the same, annotation based on marker genes is also frequently used, e.g., mOTU (Mende et al., 2017; Sunagawa et al., 2013) and MetaPhlan2 (Truong et al., 2015). Alignment-free k-mer methods such as Kraken and CLARK may also be used (Ounit, Wanamaker, Close, & Lonardi, 2015; Sczyrba et al., 2017; Wood & Salzberg, 2014). Whether annotated to a known taxon or not, the gene cluster is equivalent to a strain-level unit, whose abundance could be calculated from the abundance of its genes. Note that as MLGs (>100 genes) and MGSs (>700 genes) typically only look at the larger clusters (Table 2), it is a good habit to check back and see how representative they are for the metagenomic data in each sample (Jie et al., 2017).

Besides the case–control comparison, all the metadata collected in the study could be analyzed for their association with microbiome. It is not uncommon for different phenotypes to associate with different bacteria, underscoring the value of a holistic understanding of the diseased microbiome (Jie et al., 2017; Zhang et al., 2015).

#### 3.1.1 Software or Algorithms

MGS (Metagenomic species) https://bitbucket.org/HeyHo/mgs-canopy-algorithm
mOTU (metagenomic operational taxonomic units) http://www.bork.embl.de/software/motu/
Metaphlan2 (metagenomic phylogenetic analysis 2) http://segatalab.cibio.unitn.it/tools/metaphlan2/
Kraken http://ccb.jhu.edu/software/kraken/
CLARK http://clark.cs.ucr.edu/

### 3.2 Mining the Gene Functions

Although there are efforts to link function to taxa using algorithms such as FishTaco (Manor & Borenstein, 2017), functional annotations for individual genes in the microbiome are more common, and extensive literature search is recommended.

In order to identify functions that are differentially enriched between healthy and diseased samples, reporter scores can be used at the KEGG (Kyoto Encyclopedia of Genes and Genomes) pathway or module level (Bäckhed et al., 2015). Requiring more than 40% of the KOs (KEGG Orthology) in a module to be present ensures confidence in the module detection.

Time series samples, especially samples before and after medication, would also be useful for confirming the relevance of the case–control differences, and for pinpointing new drug targets.

#### 3.2.1 Databases for Mining Gene Functions

KEGG (Kyoto Encyclopedia of Genes and Genomes) http://www.genome.jp/kegg/
eggNOG (evolutionary genealogy of genes: Nonsupervised Orthologous Groups) http://eggnog.embl.de/version_3.0/
GMM (Gut Metabolic Modules) http://www.raeslab.org/companion/gmms

CASINO (Community And Systems-level INterative Optimization) http://biomet-toolbox.chalmers.se/index.php?page=tools, http://www.metabolicatlas.com/downloads/micro

## 4. SUMMARY AND OUTLOOK

MWAS are still young compared to GWAS. With successful identification of biomarkers in colorectal cancer, cardiometabolic diseases and autoimmune diseases, and follow-up mechanistic studies, we are positive that microbiome-directed drugs are coming of age. Population-specific effects would likely be more pronounced in the microbiome compared to genomes. Due to their lower sequencing cost and actionable nature, however, microbiome reports may reach every patient before whole genomes do.

## REFERENCES


Aagaard, K., Ma, J., Antony, K. M., Ganu, R., Petrosino, J., & Versalovic, J. (2014). The placenta harbors a unique microbiome. *Science Translational Medicine*, *6*(237), 237ra65. https://doi.org/10.1126/scitranslmed.3008599.

Albertsen, M., Hugenholtz, P., Skarshewski, A., Nielsen, K. L., Tyson, G. W., & Nielsen, P. H. (2013). Genome sequences of rare, uncultured bacteria obtained by differential coverage binning of multiple metagenomes. *Nature Biotechnology*, *31*(6), 533–538. https://doi.org/10.1038/nbt.2579.

Alneberg, J., Bjarnason, B. S., de Bruijn, I., Schirmer, M., Quick, J., Ijaz, U. Z., et al. (2014). Binning metagenomic contigs by coverage and composition. *Nature Methods*, *11*(11), 1144–1146. https://doi.org/10.1038/nmeth.3103.

Arumugam, M., Raes, J., Pelletier, E., Le Paslier, D., Yamada, T., Mende, D. R., et al. (2011). Enterotypes of the human gut microbiome. *Nature*, *473*(7346), 174–180. https://doi.org/10.1038/nature09944.

Bäckhed, F., Roswall, J., Peng, Y., Feng, Q., Jia, H., Kovatcheva-Datchary, P., et al. (2015). Dynamics and stabilization of the human gut microbiome during the first year of life. *Cell Host & Microbe*, *17*(5), 690–703. https://doi.org/10.1016/j.chom.2015.04.004.

Chen, C., Li, F., Wei, W., Wang, Z., Dai, J., Hao, L., et al. (2018). Metagenome of the female upper reproductive tract. *GigaScience* (accepted).

Chen, C., Song, X., Wei, W., Zhong, H., Dai, J., Lan, Z., et al. (2017). The microbiota continuum along the female reproductive tract and its relation to uterine-related diseases. *Nature Communications*, *8*(1), 875. https://doi.org/10.1038/s41467-017-00901-0.

Costea, P. I., Zeller, G., Sunagawa, S., Pelletier, E., Alberti, A., Levenez, F., et al. (2017). Towards standards for human fecal sample processing in metagenomic studies. *Nature Biotechnology*. https://doi.org/10.1038/nbt.3960.

Ding, T., & Schloss, P. D. (2014). Dynamics and associations of microbial community types across the human body. *Nature*, *509*, 357–360. https://doi.org/10.1038/nature13178.

Falony, G., Joossens, M., Vieira-Silva, S., Wang, J., Darzi, Y., Faust, K., et al. (2016). Population-level analysis of gut microbiome variation. *Science*, *352*(6285), 560–564. https://doi.org/10.1126/science.aad3503.

Fang, C., Zhong, H., Lin, Y., Chen, B., Han, M., Ren, H., et al. (2018). Assessment of the cPAS-based BGISEQ-500 platform for metagenomic sequencing. *GigaScience*, 7(3). https://doi.org/10.1093/gigascience/gix133.

Feng, Q., Liang, S., Jia, H., Stadlmayr, A., Tang, L., Lan, Z., et al. (2015). Gut microbiome development along the colorectal adenoma–carcinoma sequence. *Nature Communications. 6*, 6528. https://doi.org/10.1038/ncomms7528.

Gu, Y., Wang, X., Li, J., Zhang, Y., Zhong, H., Liu, R., et al. (2017). Analyses of gut microbiota and plasma bile acids enable stratification of patients for antidiabetic treatment. *Nature Communications, 8*(1), 1785. https://doi.org/10.1038/s41467-017-01682-2.

Han, M., Hao, L., Lin, Y., Li, F., Wang, J., Yang, H., et al. (2018). A novel affordable reagent for room temperature storage and transport of fecal samples for metagenomic analyses. *Microbiome, 6*(1), 43. https://doi.org/10.1186/s40168-018-0429-0.

Huseyin, C. E., et al. (2017). Forgotten fungi—The gut mycobiome in human health and disease. FEMS Microbiology Reviews, *fuw047*. https://doi.org/10.1093/femsre/fuw047.

Jie, Z., Xia, H., Zhong, S.-L., Feng, Q., Li, S., Liang, S., et al. (2017). The gut microbiome in atherosclerotic cardiovascular disease. *Nature Communications, 8*(1), 845. https://doi.org/10.1038/s41467-017-00900-1.

Karlsson, F. H., Tremaroli, V., Nookaew, I., Bergström, G., Behre, C. J., Fagerberg, B., et al. (2013). Gut metagenome in European women with normal, impaired and diabetic glucose control. *Nature, 498*(7452), 99–103. https://doi.org/10.1038/nature12198.

Kelly, B. J., Gross, R., Bittinger, K., Sherrill-Mix, S., Lewis, J. D., Collman, R. G., et al. (2015). Power and sample-size estimation for microbiome studies using pairwise distances and PERMANOVA. *Bioinformatics, 31*(15), 2461–2468. https://doi.org/10.1093/bioinformatics/btv183.

Kennedy, N. A., Walker, A. W., Berry, S. H., Duncan, S. H., Farquarson, F. M., et al. (2014). The impact of different DNA extraction kits and laboratories upon the assessment of human gut microbiota composition by 16S rRNA gene sequencing. *PLoS ONE, 9*(2), e88982. https://doi.org/10.1371/journal.pone.0088982.

Le Chatelier, E., Nielsen, T., Qin, J., Prifti, E., Hildebrand, F., Falony, G., et al. (2013). Richness of human gut microbiome correlates with metabolic markers. *Nature, 500*(7464), 541–546. https://doi.org/10.1038/nature12506.

Li, A., et al. (2013). A pyrosequencing-based metagenomic study of methane-producing microbial community in solid-state biogas reactor. *Biotechnology for Biofuels, 6*(1), 3. https://doi.org/10.1186/1754-6834-6-3.

Li, J., Jia, H., Cai, X., Zhong, H., Feng, Q., Sunagawa, S., et al. (2014). An integrated catalog of reference genes in the human gut microbiome. *Nature Biotechnology, 32*(8), 834–841. https://doi.org/10.1038/nbt.2942.

Manor, O., & Borenstein, E. (2017). Systematic characterization and analysis of the taxonomic drivers of functional shifts in the human microbiome. *Cell Host & Microbe, 21*(2), 254–267. https://doi.org/10.1016/j.chom.2016.12.014.

McArdle, B. H., & Anderson, M. J. (2001). Fitting multivariate models to community data: A comment on distance-based redundancy analysis. *Ecology, 82*(1), 290–297. https://doi.org/10.1890/0012-9658(2001)082[0290:FMMTCD]2.0.CO;2.

Mende, D. R., Letunic, I., Huerta-Cepas, J., Li, S. S., Forslund, K., Sunagawa, S., et al. (2017). proGenomes: A resource for consistent functional and taxonomic annotations of prokaryotic genomes. *Nucleic Acids Research, 45*(D1), D529–D534. https://doi.org/10.1093/nar/gkw989.

Nayfach, S., Rodriguez-Mueller, B., Garud, N., & Pollard, K. S. (2016). An integrated metagenomics pipeline for strain profiling reveals novel patterns of bacterial transmission and biogeography. *Genome Research, 26*(11), 1612–1625. https://doi.org/10.1101/gr.201863.115.

Nielsen, H. B., Almeida, M., Juncker, A. S., Rasmussen, S., Li, J., Sunagawa, S., et al. (2014). Identification and assembly of genomes and genetic elements in complex metagenomic samples without using reference genomes. *Nature Biotechnology, 32*(8), 822–828. https://doi.org/10.1038/nbt.2939.

Oh, J., Byrd, A. L., Deming, C., Conlan, S., Barnabas, B., Blakesley, R., et al. (2014). Biogeography and individuality shape function in the human skin metagenome. *Nature*, *514*(7520), 59–64. https://doi.org/10.1038/nature13786.

Ounit, R., Wanamaker, S., Close, T. J., & Lonardi, S. (2015). CLARK: Fast and accurate classification of metagenomic and genomic sequences using discriminative k-mers. *BMC Genomics*, *16*, 236.

Qin, J., Li, Y., Cai, Z., Li, S., Zhu, J., Zhang, F., et al. (2012). A metagenome-wide association study of gut microbiota in type 2 diabetes. *Nature*, *490*(7418), 55–60. https://doi.org/10.1038/nature11450.

Qin, J., Li, R., Raes, J., Arumugam, M., Burgdorf, K. S. S., Manichanh, C., et al. (2010). A human gut microbial gene catalogue established by metagenomic sequencing. *Nature*, *464*(7285), 59–65. https://doi.org/10.1038/nature08821.

Scher, J. U., Sczesnak, A., Longman, R. S., Segata, N., Ubeda, C., Bielski, C., et al. (2013). Expansion of intestinal *Prevotella copri* correlates with enhanced susceptibility to arthritis. *eLife*. *2*, e01202. https://doi.org/10.7554/eLife.01202.

Sczyrba, A., Hofmann, P., Belmann, P., Koslicki, D., Janssen, S., Dröge, J., et al. (2017). Critical assessment of metagenome interpretation—A benchmark of metagenomics software. *Nature Methods*, *14*(11), 1063–1071. https://doi.org/10.1038/nmeth.4458.

Song, S. J., Amir, A., Metcalf, J. L., Amato, K. R., Xu, Z. Z., Humphrey, G., et al. (2016). Preservation methods differ in fecal microbiome stability, affecting suitability for field studies. mSystems. *1*(3), e00021–e00016. https://doi.org/10.1128/mSystems.00021-16.

Sunagawa, S., Mende, D. R., Zeller, G., Izquierdo-Carrasco, F., Berger, S. A., Kultima, J. R., et al. (2013). Metagenomic species profiling using universal phylogenetic marker genes. *Nature Methods*, *10*(12), 1196–1199. https://doi.org/10.1038/nmeth.2693.

Thaiss, C. A., Levy, M., Korem, T., Dohnalová, L., Shapiro, H., Jaitin, D. A., et al. (2016). Microbiota diurnal rhythmicity programs host transcriptome oscillations. *Cell*, *167*(6), 1495–1510.e12. https://doi.org/10.1016/j.cell.2016.11.003.

Tringe, S. G., von Mering, C., Kobayashi, A., Salamov, A. A., Chen, K., Chang, H. W., et al. (2005). Comparative metagenomics of microbial communities. *Science (New York, N.Y.)*, *308*(5721), 554–557. https://doi.org/10.1126/science.1107851.

Truong, D. T., et al. (2015). MetaPhlAn2 for enhanced metagenomic taxonomic profiling. *Nature Methods*, *12*(10), 902.

Wang, J., & Jia, H. (2016). Metagenome-wide association studies: Fine-mining the microbiome. *Nature Reviews Microbiology*, *14*(8), 508–522. https://doi.org/10.1038/nrmicro.2016.83.

Wesolowska-Andersen, A., Bahl, M., Carvalho, V., Kristiansen, K., Sicheritz-Pontén, T., Gupta, R., et al. (2014). Choice of bacterial DNA extraction method from fecal material influences community structure as evaluated by metagenomic analysis. *Microbiome*, *2*(1), 19. https://doi.org/10.1186/2049-2618-2-19.

Wood, D. E., & Salzberg, S. L. (2014). Kraken: Ultrafast metagenomic sequence classification using exact alignments. *Genome Biology*, *15*(3), R46. https://doi.org/10.1186/gb-2014-15-3-r46.

Yang, F., Zhou, H., Han, M., Lin, Y., Cai, K., Kristiansen, K., et al. (in press). Metagenomic extraction procotols assessed for both bacterial and fungal samples. *GigaScience*.

Zhang, X., Zhang, D., Jia, H., Feng, Q., Wang, D., Liang, D., et al. (2015). The oral and gut microbiomes are perturbed in rheumatoid arthritis and partly normalized after treatment. *Nature Medicine*, *21*(8), 895–905. https://doi.org/10.1038/nm.3914.

Zhernakova, A., Kurilshikov, A., Bonder, M. J., Tigchelaar, E. F., Schirmer, M., Vatanen, T., et al. (2016). Population-based metagenomics analysis reveals markers for gut microbiome composition and diversity. *Science*, *352*(6285), 565–569. https://doi.org/10.1126/science.aad3369.

CHAPTER FOUR

# Screening Library Design

**Stephanie Kay Ashenden[1]**
Department of Chemistry, Cambridge University, Cambridge, United Kingdom
Discovery Sciences, IMed Biotech Unit, AstraZeneca R&D, Cambridge, United Kingdom
[1]Corresponding author: e-mail addresses: ska35@cam.ac.uk; stephanie.ashenden@astrazeneca.com

## Contents



## Abstract

Thanks to technological advances and a greater understanding of the biological and chemical natures of targets and related diseases, high-throughput screening (HTS) has been allowed to be faster, cheaper, and more accessible. Yet, despite these increased technologies and understandings, the frequency of novel and drugs are being approved each year has not being increasing over the years. 2017 was considered a "bumper" year with a total of 46 approved drugs, over double that of the previous year. However, it is thought that part of the problem that HTS has not lived up to expectations is because of the contents of current chemical libraries. Therefore, new methods to design screening libraries are of great interest.

Thanks to technological advances and a greater understanding of the biological and chemical natures of targets and related diseases, high-throughput screening (HTS) has been allowed to be faster, cheaper, and more accessible. Yet, despite these increased technologies and understandings, the frequency of novel drugs are being approved each year has not being increasing over the years (Mullard, 2018). 2017 was considered a "bumper" year with a total of 46 approved drugs, over double that of the previous year (FDA, 2018; U.S. Food & Drug Administration, n.d.). However, it is thought that part of the problem that HTS has not lived up to expectations is because of the

*Methods in Enzymology*, Volume 610
ISSN 0076-6879
https://doi.org/10.1016/bs.mie.2018.09.016

contents of current chemical libraries (Irwin, 2006). Therefore, new methods to design screening libraries are of great interest.

A screening library is a set of compounds or fragments that are screened against a target that has been found to be important in a therapeutic area. By screening many compounds allows for a greater chemical space to be tested against the target of interest.

However, due to how large chemical space is estimated to be it is unlikely that it is completely represented in a screening library due to storage space and cost. Due to this, the compounds within a library need to be carefully managed. HTS is no longer just of interest to large pharmaceutical companies but also within academia (Huggins, Venkitaraman, & Spring, 2011). The overall aim to identify a novel lead compound, that will hit the target of interest, requires a suitable library of compounds in which to screen from initially. Therefore, it is important to optimize the library to reduce the selection of compounds that are unlikely to become hits while increasing the hit rate of compounds that will eventually end up as lead compounds (Huggins et al., 2011).

Through this chapter we will look at key areas and concepts in screening library design. We will consider the significance of diversity within a screening library followed by target-orientated libraries. We will then explore natural products in screening libraries as natural products have given rise to many drugs throughout history and therefore are likely a good starting point. Fragment-based screening library design is then considered as another approach to helping identify novel bioactive drugs. We will then finalize the chapter by considering the impact of cheminformatics on screening library design.

## 1. DIVERSITY IN SCREENING LIBRARY DESIGN

Diversity within a screening library is one of the key issues and considerations thought about when devising a screening library (Schnur, 2008). Yet, what do we mean by diversity when we are referring to a screening library? Diversity refers to how similar or dissimilar a set of compounds are to each other. A screening collection that has mostly very similar compounds to one or another is more likely to be associated with a smaller number of therapeutic targets. Frequently, the purpose of designing a large screening collection is intended to be able to test the compounds in a variety of assays to identify modulators of a breath of different targets (Wawer et al., 2014). Therefore, the larger the variety of compounds that cover a wider and

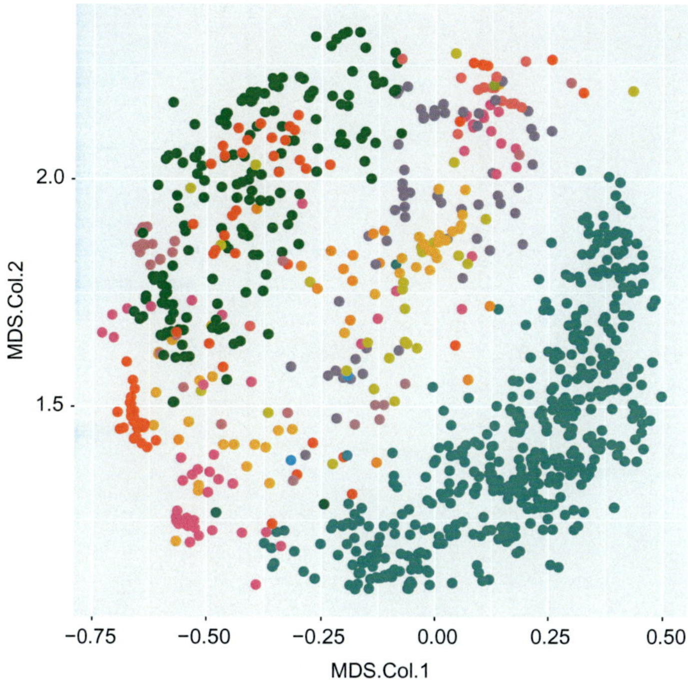


**Fig. 1** An example of compound diversity based on compound structure. Morgan fingerprints at a bit length of 2048 and radius 2 were compared via a tanimoto distance matric and then multidimensionality scaling was applied. It is *colored* by the various type of targets that the compound is associated with.

more diverse area of chemical space, the more likely will HTS methods are able to identify novel bioactive molecules (Fig. 1).

The next question is though, how does one actually measure the amounts of diversity in a screening library (several types of methods are summarized in Table 1). A number of ways were suggested in the mid-1990s including the calculation and comparison of descriptors such as physicochemical properties like lipophilicity, topological indices (where you consider the molecule shape), chemical functionality descriptors (where you assess the functional groups contained within the molecule), and receptor recognition descriptors as examples (Martin, 1995). These can be used to visualize the chemical space using techniques such as multidimensional scaling or principle component analysis. Various additional methods have been proposed to ensure that the diversity of a screening library is ensured. These methods include filtering out undesirable compounds prior to building the library. This prevents the chemical space of the screening library to be taken up with compounds that have not been deemed to have lead or drug-like properties (Willett, 2000).

**Table 1** Methods for Assessing Diversity in a Screening Library

| Method | Description | Notes |
|---|---|---|
| Maximum Average Diversity | • Multiobjective Maximum Average Diversity allows for the number of elements to be selected to be a decision variable (Vera et al., 2017) | • Optimize the average of each diversity model |
| Fingerprint Space—Nonhashed | • MACCS Keys consist of 166 bits | • Presence or absence of particular features |
| Fingerprint Space (Rogers & Hahn, 2010)—Hashed | • Extended connectivity fingerprint—ECFP 4 is most closely related to Morgan fingerprints where the number refers to the circular substructures up to the numbers maximum bonds (ECFP 6 = up to 6 bonds) (Rogers et al., 2005) compare diversity in 2D space<br>• Generated quickly<br>• Includes positive and negative structures | • Compared using similarity algorithms such as Tanimoto score |
| Scaffold Space | • Frequently analyzed by calculating Murcko molecular or topological frameworks<br>• Scaffold trees level 1 is important for characterizing the scaffold diversity (Langdon et al., 2011) | • Use a tree map to observe clusters of similar scaffolds (Langdon et al., 2011) |
| Bioactivity Space | • Bayes Activity Fingerprints (Bender et al., 2006; Koutsoukas et al., 2014)<br>• Descriptors based on atom topology correlate well whereas shape-based descriptors have a weaker correlation with other descriptors | |

The most explored method for measuring similarity is via fingerprints and descriptors. There are different types of fingerprints including hashed and nonhashed. Hashed fingerprints are represented as a bit string where the 0's and 1's represent whether a particular feature is present or not (Rogers & Hahn, 2010). The length of the bit vector can be varied and depends on the amount of information to be stored. A bit length of 512 bits can be used for both small and large databases, however, they may struggle to distinguish differences in very similar compounds and therefore a longer bit length (such as 1024 bits) is more suitable (ChemAxon, 2018a). Comparing the fingerprints via similarity algorithms can be used assessing diversity between compounds. Tanimoto similarity (also known as Jaccard coefficient) is one of the most popular algorithms used to compare the similarity between two compounds. It gives a score of between 0 and 1 with 1 representing a very similar/potentially identical compound. A score of >0.85 is often used to represent a level of high similarity between compounds (Maggiora, Vogt, Stumpfe, & Bajorath, 2014). Eq. (1) gives an example of how tanimoto is calculated. In the equation, *a* and *b* represent the features present in compounds *A* and *B*, respectively. Whereas, *c* represents the number of features that are shared between the two (Maggiora et al., 2014). To calculate the distance between two compounds, it is simply 1 − tanimoto coefficient. Laplacian-modified Bayesian approaches have been combined with extended connectivity fingerprints to analyse HTS thanks to its fast and automatic nature (Rogers, Brown, & Hahn, 2005). This approach was part of a competition and showed that using this 2D method was comparable to the 3D methodologies used to predict activity.

Tanimoto coefficient:

$$Tc(A, B) = \frac{c}{a + b - c} \tag{1}$$

MACCS (molecular access system) fingerprints of which are described as nonhashed and highlight the presence of 166 different features (rdkit.Chem.MACCSkeys, 2018). Another method is to produce a series of models and determine the number of decision variables to be selected and optimize the average of each diversity models (Prokopyev, Kong, & Martinez-Torres, 2009; Vera, Lopez-Pires, Baran, & Sandoya, 2017). However, rather than comparing the diversity of each compound, another approach is to analyze the compound molecular or topological fragments. It is possible to calculate these fragments from the compounds and is essentially the backbone and the core of the compound. Therefore, a more diverse set of compounds contained in a library that can be used to test in a variety of assays against a wide range of modulators, potentially will have a more diverse set of frameworks. A tree map can be produced to show the scaffold space and observe large clusters of similar scaffolds (Langdon, Brown, & Blagg, 2011).

Another consideration to be taken into the diversity of a screening library is that of the bioactivity. A comparison of a wide variety of fingerprints showed that those based on atom topology, instead of shape related were well correlated in rank ordering of compounds both within and between descriptor types (Koutsoukas et al., 2014). The Bayes Affinity Fingerprints had the highest coverage of 25 large human activity classes compared to 7 other molecular descriptor coverage's in an assessment to determine which descripts could select compounds with high coverage of bioactivity space (Koutsoukas et al., 2014).

Discussions have been had on what is the best way to design a screening library, with direct comparisons being made between structure- and diversity-based library design (Wyss et al., 2003).

In one example, the authors were aiming to identify novel dihydrofolate reductase inhibitors while comparing structure-based and diversity-based library design. They found that a greater number of hits as well as active compounds were found through structure-based library design as opposed to diversity-based library design (Wyss et al., 2003). Generally, compounds with similar structures may actually differ in terms of their activity profile against a given target. At the same time, a target may allow a variety of dissimilar compounds to bind to it. Therefore, it has been suggested that affinity fingerprints (which describe a compound in terms of its binding affinity to dissimilar target proteins) would be better suited to identify active compounds from a large library (Dixon & Villar, 1998).

It has been suggested that there is biases in the libraries screened and how to quantity this has been discussed (Hert, Irwin, Laggner, Keiser, & Shoichet, 2009). The reason that screening library bias has been discussed (Martin, 2001) is due to the fact that it is estimated that there is over 10 (Davis & Erlanson, 2013) drug-like molecules that can be potentially synthesized (Bohacek, McMartin, & Guida, 1996), and that because libraries that are screened are a only a fraction of the size of this, so it would seem unlikely that screening would be successful (Hert et al., 2009). So, if it is potentially considered that screening a library would be successful and although, diversity remains a key consideration, knowledge-based library design has gained a lot of attention.

Target-oriented screening is where the compounds within the library being screened are focused toward an association with a particular target. These libraries are still diverse, but within a target family. Such families may include kinases or G protein-coupled receptors (GPCRs) and often utilize the structural target information (John Harris, Hill, Sheppard, Slater, & Stouten, 2011).

This is thanks to improved technological systems, allowing for computers to screen molecules on a large scale much faster, combinatorial libraries that are smaller and more focused on a particular target class are employed more frequently nowadays (Schnur, Beno, Tebben, & Cavallaro, 2011). Library design, where the focus is on a single target rather than a family of receptors, is still important; however, it has been shown that knowledge-based screening library can make up for the lack of structural information (Evensen et al., 2003).

It is generally recognized that knowledge-based screening library design is better than just randomly searching in an "all-against-all" manner. Especially as each target class differs from the next and will have its own specific set of requirements. Examples of where the target class has been used to design a screening library have been discussed, for example, for the case of kinases (Prien, 2005). This study discusses the various and diverse approaches used by commercial vendors that are selling collections of potential kinase inhibitors. However, the study that compares these notes that a combination of the methods used to increase the probability of identifying compounds (Prien, 2005). In general, however, within a library and for each individual chemotype, a suitable scaffold is selected as a starting point of which to grow based on prior chemical knowledge (John Harris et al., 2011). It is also important to consider synthetic feasibility especially when taking into account the need for multiple parallel production and purification methods (John Harris et al., 2011). The idea being that of all the targets within a target class, one will be a good fit and that molecule will be deemed a hit (John Harris et al., 2011).

Initially, it was considered that the more molecules available for screening, the more hits would be identified. However, this has been shown to not be the case despite structural diversity. Therefore, it has been suggested that a multiapproach is key for successful library design, combining both biological and chemical knowledge to solve the issue (Nicolaou & Kannas, 2011). Such proposed combined methods include combining combinatorial chemistry methods with structure-based design where the authors used protein structural information to derive active site maps. They showed that their method outperformed both docking calculations as well as 2D similarity searching methods (Eksterowicz et al., 2002). An example can be shown from (Jimonet & Jäger, 2004) of whom discuss for the GPCR family several strategies for library design. For example, the use of ligand-based methods of which involve analyzing the compound structure and understanding its physicochemical properties. Yet as highlighted earlier, combining methods are essential to avoid gaps in the knowledge design process. Therefore, the authors also employed the use of 2D and 3D mapping of the receptor interaction sites (Jimonet & Jäger, 2004).

When that structural information is not available then chemogenomic model informing sequences and its mutagenesis data, or by using the knowledge of ligands of the target via scaffold hopping (John Harris et al., 2011). It has been shown, through experience, that removing any compounds that have been deemed to be undesirable will reduce the occurrences of false positives as well as aiding in the optimization procedures (Baell & Holloway, 2010). Undesirable compounds may include electrophiles or toxicophores (Walters, Stahl, & Murcko, 1998).

## 2. METHODS

Methods for increasing the diversity within the screening library have been discussed at length with many methods and combinations of methods being put forward (Table 2). One such method was the multiobjective genetic algorithm (MOGA) proposed by Wright and colleagues back in 2003 (Wright, Gillet, Green, & Pickett, 2003). This method can deal with multiple optimization issues such as varying the library size, its configuration as well as other properties. However, having multiple issues to optimize often means that a trade-off is required as it is not likely that all issues can be optimized to their maximum potential simultaneously without affecting another factor. In order to deal with this, MOGA, of which was developed by Wright and colleagues, is able to deliver multiple solutions, for the user to choose between, to the optimization problem while considering various trade-offs that can be applied. Another method is by means of simulated annealing (Eglese, 1990; Kirkpatrick, Gelatt, & Vecchi, 1983). The concept here is that the algorithm starts with an initial solution and then generates other solutions. Ideally, the current solution is always replaced with a better solution until at the end; only the best solution is left. However, in simulated annealing, occasionally, a solution that is perhaps not better than the current

**Table 2** Methods to Create Diverse Screening Libraries

| Method | Description | Advantages | Disadvantages |
|---|---|---|---|
| MOGA (Wright et al., 2003) (multiobjective genetic algorithm) | Optimize library configuration and library size while considering additional important factors | • Reduce the number of reactants while increasing the diversity of library<br>• The user is free to decide which compromises to make | Difficult to compare results across different runs of MOGA |
| Monte-Carlo modeling (Falcioni & Deem, 2000) | Searches variable space | • Multiple-rounds protocols are effective on difficult systems | • More data are correlated with greater success and advances |
| Simulated annealing (Kirkpatrick et al., 1983) | Simulated annealing occasionally accepting a solution that may not be better than the previous which allows a global analysis to occur. Only solutions that are slightly worse are accepted and the control parameter is greatly beneficial then these types of moves (accepting a solution that is slightly worse than the previous) will be accepted. However, if the solution is worse and the control parameter is also not optimal then it is more likely that this move would be rejected (Eglese, 1990) | Does not instantly reject a solution that may be considered worse than the previous which may then yield a more optimal solution. In addition, it can be easily extended to new problems | Needs a lot of information to work including a concise description of the system configuration, random generator of rearrangements of the configuration elements, trade-offs that have it be made as a quantitative objective function as well an annealing schedule of the temperature and time length for which the system is to be evolved |

solution is used rather than retaining the previous. The point of this is to allow a more global analysis to occur. However, in order to allow for successful trade-offs to be put in place, where again it is unlikely that all parameters can be optimized to the best they can be without having a negative effect on other properties, the event that the selected solution is not better than the previous solution, only solutions in which the control parameter is highly beneficial and those solutions that are marginally not better are selected (rather than those that are significant worse solutions).

Cheminformatic practices have played a highly important part of the birth of a screening library (Zhou, 2011). The design of compound libraries relies heavily on computational methodologies and so it has been considered that the design of a screening library is very much a cheminformatics task (Boehm, Zhang, Bodycombe, MacIejewski, & Wassermann, 2016). However, as technological methods, such as cheminformatics methods, have become more advanced, simply hoping to identify a hit from a screening collection has over time been replaced by target class focused libraries (Schnur, 2008).

In silico docking methods have been used to curate small and targeted screening libraries with the possibility of identifying new bioactive molecules at a reduced cost (due to its targeted and well-informed nature) (Cavasotto & Phatak, 2011). The docking methods require a 3D representation of the receptor in question, that allows for a library collection to be virtually screened and scored assessing the best compounds that are showing desirable steric and electrostatic properties with the binding pocket. By analyzing of ligands in the binding pocket, it can allow for a development of an enriched library (Orry, Abagyan, & Cavasotto, 2006). In addition to docking, high-throughput NMR (Villar, Yan, & Hansen, 2004) as well as crystallography (Blundell & Patel, 2004) are being used to aid screening library design due to the ability to extract useful structural biology information (Congreve, Murray, & Blundell, 2005).

## 3. CASE STUDIES

When creating a structurally diverse collection several aspects need to be considered as discussed by Huggins et al. (2011). The first area of consideration is where to source compounds. Often a screening collection will have compounds that have been derived from multiple sources. It allows for a wide variety of compound "types" in terms of their physicochemical properties. It has been shown that out of the 1.6 million unique structures from 23 suppliers, only 37% pass all of the Vernalis drug-like filters and tend to lean toward high lipophilicity (Baurin et al., 2004). This highlights the benefit of extracting compounds from multiple sources. These compounds could be from suppliers that have diverse collections to choose from, or these compounds can be sourced from natural products (discussed later on in this chapter) or compound datasets. Such a database is ZINC (Irwin & Shoichet, 2005). Additionally, consideration must be made to the types of compounds within the library. For example, removing compounds that are unlikely to be deemed hit, lead, or drug like. This can be applied by using various filtering methods such as those based on ADMET properties. It has been shown that the less complex molecules are quite often the most common starting points for drug discovery (Hann, Leach, & Harper, 2001).

Pfizer have published their fragment library design strategy (Lau et al., 2011). Their strategy involves the combination of multiple and orthogonal metrics which allow the incorporation of pharmacophores, pharmacological space diversity, and structure (Lau et al., 2011). The Pfizer strategy also considers molecular complexity. Their strategy is split into two parts with the first being to ensure pharmacophore and framework diversity and the second part focuses on framework and target diversity. Both of these strategies are split into filters (physicochemical properties), diversity (orthogonal diversity metrics), visual (such as NMR techniques), and analytical confirmation of structural integrity, solubility, etc. (Lau et al., 2011).

Many examples of knowledge-based screening libraries that have been created and used have been published. The analysis of the inhibitory action of quercetin glycosides against the target was angiotensin-converting enzyme. By using this study, they found that they could be used to treat serious conditions such as myocardial infraction and congestive heart failure (Muhammad & Fatima, 2015). The ligands were extracted from buckwheat and onions as natural metabolites, suggesting that nature is still used as a starting point for the identification of novel ligands. This is further discussed in the next paragraphs.

Another case study is that of CombiDOCK where they published a method to be able to perform docking methods on a large combinatorial library. The method orientates in the scaffold in many ways and then any potential fragments are combined with the scaffolds and the binding interactions with the receptor are scored (Sun, Ewing, Skillman, & Kuntz, 1998). They showed that their method was successful at scoring molecules and selecting them as well as fragments in comparison to control algorithms (Sun et al., 1998).

## 4. NATURAL PRODUCTS FOR SCREENING LIBRARY DESIGN

Many lead compounds give thanks to their beginnings as a natural product. A natural product is a compound that is found naturally occurring in nature (No authors, 2007). It has been suggested that over a third of new molecular entities were natural product derivatives (Patridge, Gareiss, Kinch, & Hoyer, 2016). In addition, natural products have great chemical diversity and are therefore highly desirable in a screening library due to the chance of finding a wide range of bioactive compounds (Hong, 2011).

It is important that in the generation of libraries of natural product like molecules, key lead like descriptors, namely, molecular weight and the partition coefficient are incorporated and considered (Pascolutti & Quinn, 2014). Ensuring that the scaffolds and chemistry of compounds are in line with lead-like chemical space will aid the generation of a better quality library (Pascolutti & Quinn, 2014).

Identification of scaffold structures that are not present within drugs as well as novel scaffolds, allowed for the generation of a virtual combinatorial library. It was shown that drugs and natural products have a number of pharmacophore patterns that are shared between the two of which can lead to the exploration of novel structures within a selected combinatorial library that has a base of synthetic and natural product building blocks (Lee & Schneider, 2001).

Using scaffolds instead of a full compound can allow for central chemistry that dictates much of the physicochemical properties to be maintained throughout the making of new molecules, where the molecular fragments added vary the measured results. Using fragments is another way to achieve this where you can start small and build compounds by adding molecular fragments that provided the desired effect.

In addition, it has been shown that solid-phase organic chemistry is now able to cope with complex synthesis targets thanks to the generation of natural product libraries (Breinbauer, Vetter, & Waldmann, 2002). This is an important step forward due to the complex nature of natural products. Being able to synthesis these experimentally means that we can explore further into chemical space that is more related to nature. Previously in history, pharmaceutical companies had reduced the amount of research performed on natural products due the difficulties with HTS but due to these technological advances that interested has been renewed (Koehn & Carter, 2005).

An example of generating a screening library from a natural product scaffold is that of 3-chloro-4-hydroxyphenylacetic acid. All compounds were tested for cytotoxicity as well as antiparasitic activity and showed no significant activity at 10 $\mu M$. It was also shown that those structures that were flurobenzyl analogs reduced the cellular phospholipid as well as neutral lipid levels significantly (Kumar et al., 2015). It has also been suggested that understanding the chemical space of natural products is essential to really identifying novel drugs or lead compounds. In a study of which aimed to understand the chemical space of natural products, they took over 197,000 natural product structures along with their bioactivity profiles and showed that there was a huge overlap with FDA approved drugs. Despite this, natural products do have a diverse chemical space coverage and were shown to interact with multiple cellular target proteins (Gu et al., 2013).

## 5. LIBRARY DESIGN FOR SCREENING FRAGMENTS

Fragment-based drug discovery is an important part of the drug discovery process where small fragments are grown to produce drug leads. These fragments can be either grown, merged, or linked together to produce these leads (Davis & Erlanson, 2013). In fact, there are examples of drugs that were derived from fragments that have been approved (as well as a variety of compounds that are in various stages of the clinical phases). An example of a drug that has been approved that was derived from fragments is that of Vemurafenib a B-Raf inhibitor (Jahnke & Erlanson, 2006). The Hann's model of molecular complexity says that screening with high sensitivity of small and noncomplex fragments will lead to a higher probability of binding detection to a target than screening lower sensitivity (Schuffenhauer et al., 2005). When identifying interactions between a target and a fragment, the technique needs to be tailored to the fragment collection as well as the target. Therefore, it is suggested that a key driving factor should be the targets nature to optimize fragment-based drug discovery and development of the screening library (Siegal, Ab, & Schultz, 2007). Additionally, it has been shown that you can identify specific fragments for different binding sites (Chen & Hubbard, 2009).

To aid in the success rate of identifying fragments and producing lead compounds, fragment libraries should not have compounds that are already known to be reactive, for example, (Keseru et al., 2016). In addition, physicochemical properties can be monitored and those not deemed ideal such as fragments that are >500 Da and highly lipophilic. However, the authors note that the drug discovery process is not as easy as just applying simple rules to any drugs and all drugs will abide by these simple rules. All suggestions are merely that, suggestions (Keseru et al., 2016).

Fragment-based drug discovery is not limited to discovering drugs and lead compounds, but also for identifying chemical probes for protein function. However, it has been noted that fragment-based drug discovery tends to be primarily used in pharmaceutical companies rather than academia.

As with traditional libraries, diversity is a key consideration to ensure a range of target classes can be interacted with. Therefore, identifying gaps in the chemistry of the library is highly beneficial. To identify missing chemistry, cheminformatic techniques, as well as intuition from chemists, are used (Kutchukian, So, Fischer, & Waller, 2015). This is an essential concept because it has been found that using fragments alongside other hit discovery methods and combining these with advanced computational methods can aid drug discovery (Whittaker, Law, Ichihara, Hesterkamp, & Hallett, 2010).

## 6. VIRTUAL LIBRARY DESIGN

Chemical data are not restricted to real molecules, it can also be extended to virtual molecules that may or may not be synthesizable. This chemical data can be used to build virtual libraries that allow for the exploration of chemical space and potentially find compounds that have either hit-like, lead-like, or drug-like properties. Such virtual libraries can be produced with algorithms such as the Fragment Optimized Growth algorithm (Kutchukian, Lou, & Shakhnovich, 2009). This algorithm allows for the growth of molecules in a statistically biased way, which allows for desired properties to be prioritized. This algorithm uses a Markov chain on different classes of compounds to create novel compounds that resemble the training set. Another cheminformatic method has been described by Kutchukian et al. (2015). Where they identify biologically relevant substructures that are missing from the library and these gaps are then organized hierarchically for the chemists to evaluate (Kutchukian et al., 2015).

Pharmacophore-based virtual screening is a method used in virtual library design where a group of structurally diverse ligands that bind to a receptor are used to build a model that uses the ligands information. This information is compared against a candidate ligand to determine whether it may successfully bind (Sun, 2008). In addition, searching for compounds that are similar in shape to known actives is a popular task. This is because compounds with similar shape to those known active compounds that bind to the target are likely to also bind to the target (Rush, Grant, Mosyak, & Nicholls, 2005).

This technique has been used in practice when identifying novel inhibitors of arylamine *N*-acetyltransferases (Ballester, Westwood, Laurieri, Sim, & Richards, 2010). Here the authors use a virtual screening technique called Ultrafast Shape Recognition. This method was used to screen nearly 700 million molecular conformers and of the compounds that were predicted, 40% had a confirmed hit rate (Ballester et al., 2010).

Virtual screening is not limited to 3D techniques, 2D techniques are also often used to assess a series of molecules against known active structures by looking at their chemical similarity, such as those mentioned previously (Koutsoukas et al., 2014).

## 7. CREATING A SCREENING LIBRARY

There are several key considerations that need to be implemented when designing and producing a screening library (Dandapani, Rosse, Southall, Salvino, & Thomas, 2012). These steps have been discussed previously by Dandapani et al. (2012) and are summarized here. The first is to remove functionalities that are considered problematic.

There are numerous methodologies available that can highlight and help eliminate such problematic functionalities. Such examples include PAINS (Pan Assay Interference Compounds) (Baell, 2016). PAINS are compounds that are often flag up as false positive during HTS. Notably, these compounds tend to interact with multiple targets (Baell & Walters, 2014) and often contain undesirable functional groups (Ertl, 2017). There are a number of available tools, such as a published KNIME workflow that uses SMARTS (Daylight Chemical Information System Inc, 2018) format and assesses PAINS filters (Saubern, Guha, & Baell, 2011), that allow for PAINS to be filtered out based on applying filters for particular substructures (Baell & Holloway, 2010). It is important to note, however, that these filters have had their accuracy criticized despite their frequent use (Capuzzi, Muratov, & Tropsha, 2017). Notably that there are marketed drugs that do set of PAINS alerts (approximately 5% of FDA approved drugs, both natural and synthetic). Although, it is argued that these were identified not through target-based screening (Baell & Nissink, 2018). Using SMARTs notation of known substructures that are not desired, allows for a tailored approach to designing a screening library by filtering out unwanted structures.

Following this, the authors (Dandapani et al., 2012) recommend using a number of steps to predict various properties including physicochemical and ADME to remove compounds that are potentially unsuitable as drug compounds. They note that the work of Lipinski showed that >90% of orally bioavailable compounds that had good solubility/permeability met the criteria of the well know "Lipinski's Rule of Five" (Lipinski, Lombardo, Dominy, & Feeney, 2012). In 1997 Christopher Lipinski published a paper describing Lipinksi's rule of five. The rules set out to help determine whether compound has suitable properties that aid in the ADME (Absorption, Distribution, Metabolism, and Excretion) of the drug. The rules set out that an active drug should not have more than one violation of the rules (Table 3).

However, these rules do not capture all drug-like compounds, some compounds such as Paclitaxel has a molecular weight of 853.906 Da, of which is outside Lipisnki's rule. Therefore alternative rules have been introduced including that by Ghose and colleagues which also consider properties not laid out in Lipinski's rule of five such as molar refractivity (Ghose, Viswanadhan, & Wendoloski, 1999).

Wenlock and colleagues compared the distributions of marketed oral drugs physicochemical properties to those in various phases of clinical trials. They found that the mean molecular weight decreases through the phases of clinical phases and eventually ends up in the same range as the molecular weights of marketed compounds as well as lipophilic compounds are discontinued from development (Wenlock, Austin, Barton, Davis, & Leeson, 2003).

In pharmaceutical companies, beyond these rules and filters, predictive models are used to determine ADMET filters. Notable models are made for hERG (to test cardiotoxicity), logD, permeability, and absorption. Extending on from this, however, companies can build more specific

**Table 3** Lipinski's Rule of Five

| **Lipinski's Rule of Five** |
|---|
| Less than or equal to five hydrogen bond donors |
| Less than or equal to 10 hydrogen bond acceptors |
| ClopP (Partition Coefficient) of less than or equal to five |
| A molecular weight of less than 500 Da |

predictive models dependent on the type of work they are conducting. However, the compounds in a screening library do not need to be perfect as they can be (and will be) enhanced during lead development and optimization.

Dandapani et al. (2012) then suggest a complexity analysis be performed of which is very much dependent on what the organization wants such as considering chirality in compounds. An approach to assess this, however, would be to estimate the synthetic accessibility of the compounds. This can be done by using a retrosynthetic synthesizability score such as the one proposed by Ertl and Schuffenhauer (2009). This method scores compounds between 0 (easy to make) and 10 (difficult to make) by considering features such as stereocomplexity, the molecules size, number of rings, etc. However, developing and using a more personalized score may be required as it does not consider specific organizational biases and preferences such as availability of reagents.

Finally, the next stage is to assess the diversity of the compounds. Earlier in this chapter we have discussed the need for diversity in a screening library collection, even if the library is target specific and the compounds selected are does so via knowledge-based methods. To do this, the methods explained in Table 2 can be used. As an example, MOGA can be used through the program MoSELECT (Gillet, Willett, Fleming, & Green, 2002) and its successor MoSELECT.II (Gillet, Khatib, Willett, Fleming, & Green, 2002). MoSELECT.II creates a complete picture of the compromises between diversity and library size and the user simply chooses which method they would intend to use. Alternatively, purchasing compounds from different sources will aid in library diversity and synthetic methods such as "*Libraries from Libraries*" can be applied (López-Vallejo et al., 2011; Ostresh et al., 1994). This method "*Libraries from Libraries*" is based upon solid-phase synthesis and has been shown to differ in their biological, chemical, and physical properties from where they were derived. This method can also be considered as a diversity-oriented synthetic approach (Burke & Schreiber, 2004), with "*Libraries from Libraries*" approach considering the bioactivity space, around each scaffold, by using a large number of appendages on every scaffold.

Such libraries can be created in languages such as python and examples have been shown (de la Vega de Leon, Lounkine, Vogt, & Bajorath, 2017). Fig. 2 shows a general workflow that can be used to generate a screening library. First, acquiring data and compounds either through public datasets

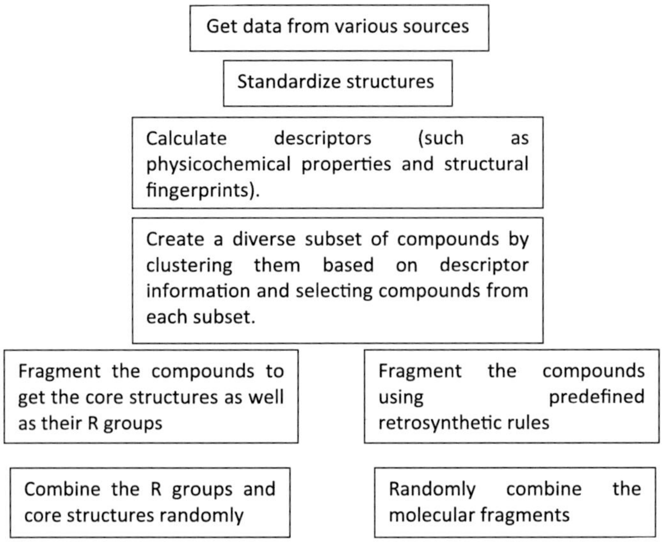


**Fig. 2** Workflow to create a screening library.

such as ChEMBL (Gaulton et al., 2012) or in-house compounds and ensure that the structures are standardized. Structures are commonly represented in SMILES (Weininger, 1988) format and the same structure can be written in multiple ways and therefore it is important to use tools such as the ChemAxon standardizer tool (ChemAxon, 2018b) or a purpose built method such as removing salts and implicit hydrogens. Following this, calculating descriptors such as the molecular weight, number of hydrogen bond donors, etc. and can be done so via tools such as RDKit (RDKit, 2017). It is also possible to calculate structural descriptors as mentioned earlier in the chapter by using methods such as extended connectivity fingerprints (Rogers & Hahn, 2010). When it comes to clustering the compounds, there are multiple ways off achieving clusters (Bacha, Brunck, & Delany, 2007). Ultimately, comparing the compounds against each other, based on the similarity (or dissimilarity) of their properties and structural nature (Table 4). Following this, the compounds are fragmented and this can be done on by methods such as the maximum common substructure approach (Cao, Jiang, & Girke, 2008) or by applying RECAP rules (Lewell, Judd, Watson, & Hann, 1998) of which fragments compounds based on chemical knowledge. Finally compounds can be created using combinatorial chemistry methods to combine fragments to create novel chemistry.

**Table 4** Key Clustering Methods

| Clustering Method | Protocol |
|---|---|
| Ward (Ward, 1963) | Hierarchical approach that is considered a "bottom up" approach and progressively merges clusters |
| k-means (Hartigan, 1975) | Partitions *n* observations into *k* clusters. Each observation belongs to the cluster with the nearest mean |
| k-modes (Carreira-Perpiñán & Wang, 2013) | Specified number of clusters and clusters are seeded with an initial structure. Other structures are added to clusters based on similarity |
| Sphere exclusion (Butina, 1999; Hudson, Hyde, Rahr, Wood, & Osman, 1996) | Picks an initial structure and then identifies all structures meeting a defined similarity threshold |
| Jarvis–Patricks (Jarvis & Patrick, 1973) | This method uses the number of nearest neighbors to consider and the similarity threshold |

## 8. FINAL THOUGHTS

Overall, much thought and effort is placed into designing a screening library with the aim of identifying a lead compound or a drug. Diversity remains a key issue in drug discovery to capture as much chemical space as possible. Despite this, efforts have moved into more structure and knowledge-based screening library design for a more focused approach. Despite this, diversity is still heavily considered.

Natural products and fragments are both used to aid in the identification of leads and drug-like compounds. Natural products for their drug like properties and fragments for the ability to grow a fragment that has been shown to have an optimal profile for a given target. However, it is important to remember that there is no hard rule to designing drugs and therefore, reinforcing methods with other techniques and knowledge is essential. Such methods that can and should be integrated for successful library design is that of cheminformatic techniques which allow for real and virtual compounds to be analyzed and compared against each other to learn from the collection in use.

## REFERENCES


Bacha, P., Brunck, T., & Delany, J. (2007). *Clustering methods and their uses.* Daylight Chemical Information Systems, Inc.

Baell, J. B. (2016). Feeling nature's PAINS: Natural products, natural product drugs, and Pan Assay Interference Compounds (PAINS). *Journal of Natural Products, 79*(3), 616–628.

Baell, J. B., & Holloway, G. A. (2010). New substructure filters for removal of Pan Assay Interference Compounds (PAINS) from screening libraries and for their exclusion in bioassays. *Journal of Medicinal Chemistry, 53*(7), 2719–2740.

Baell, J. B., & Nissink, J. W. M. (2018). Seven year itch: Pan-assay interference compounds (PAINS) in 2017-utility and limitations. *ACS Chemical Biology, 13*(1), 36–44.

Baell, J., & Walters, M. A. (2014). Chemistry: Chemical con artists foil drug discovery. *Nature, 513*(7519), 481–483.

Ballester, P. J., Westwood, I., Laurieri, N., Sim, E., & Richards, W. G. (2010). Prospective virtual screening with ultrafast shape recognition: The identification of novel inhibitors of arylamine N-acetyltransferases. *Journal of the Royal Society, Interface*, 7(43), 335–342.

Baurin, N., Baker, R., Richardson, C., Chen, I., Foloppe, N., Potter, A., et al. (2004). Drug-like annotation and duplicate analysis of a 23-supplier chemical database totalling 2.7 million compounds. *Journal of Chemical Information and Computer Sciences, 44*, 643–651.

Bender, A., Jenkins, J. L., Glick, M., Zhan, D., Nettles, J. H., & Davies, J. W. (2006). "Bayes affinity fingerprints" improve retrieval rates in virtual screening and define orthogonal bioactivity space: When are multitarget drugs a feasible concept? *Journal of Chemical Information and Modeling, 46*(6), 2445–2456.

Blundell, T. L., & Patel, S. (2004). High-throughput X-ray crystallography for drug discovery. *Current Opinion in Pharmacology, 4*(5), 490–496.

Boehm, M., Zhang, L., Bodycombe, N., MacIejewski, M., & Wassermann, A. M. (2016). The many facets of screening library design. *ACS Symposium Series, 1222*, 345–364.

Bohacek, R. S., McMartin, C., & Guida, W. C. (1996). The art and practice of structure-based drug design: A molecular modeling perspective. *Medicinal Research Reviews, 16*(1), 3–50.

Breinbauer, R., Vetter, I. R., & Waldmann, H. (2002). From protein domains to drug candidates - natural products as guiding principles in the design and synthesis of compound libraries. *Angewandte Chemie (International Ed. in English), 41*(16), 2878–2890.

Burke, M. D., & Schreiber, S. L. (2004). A planning strategy for diversity-oriented synthesis. *Angewandte Chemie (International Ed in English), 43*(1), 46–58.

Butina, D. (1999). Unsupervised data base clustering based on daylight's fingerprint and tanimoto similarity: A fast and automated way to cluster small and large data sets. *Journal of Chemical Information and Computer Science, 39*, 747–750.

Cao, Y., Jiang, T., & Girke, T. (2008). A maximum common substructure-based algorithm for searching and predicting drug-like compounds. *Bioinformatics, 24*(13), i366–i374.

Capuzzi, S. J., Muratov, E. N., & Tropsha, A. (2017). Phantom PAINS: Problems with the utility of alerts for Pan Assay Interference Compounds. *Journal of Chemical Information and Modeling, 57*(3), 417–427.

Carreira-Perpiñán, M. Á., & Wang, W. (2013). *The K-modes algorithm for clustering.* (Apr. 23; arXiv:1304.6478).

Cavasotto, C. N., & Phatak, S. S. (2011). Docking methods for structure-based library design. *Methods in Molecular Biology (Clifton, NJ), 685*, 155–174.

ChemAxon. (2018a). *Chemical Hashed Fingerprints.* http://www.chemaxon.com/jchem/doc/user/fingerprint.html. Accessed 23 May 2018.

ChemAxon, Standardizer user's guide—Documentation—ChemAxon—DOCS 2018b https://docs.chemaxon.com/display/docs/Standardizer+User%27s+Guide (accessed Sep 8, 2018).

Chen, I. J., & Hubbard, R. E. (2009). Lessons for fragment library design: Analysis of output from multiple screening campaigns. *Journal of Computer-Aided Molecular Design*, *23*(8), 603–620.

Congreve, M., Murray, C. W., & Blundell, T. L. (2005). Keynote review: Structural biology and drug discovery. *Drug Discovery Today*, *10*(13), 895–907.

Dandapani, S., Rosse, G., Southall, N., Salvino, J. M., & Thomas, C. J. S. (2012). Acquiring, and using small molecule libraries for high-throughput screening. *Current Protocols in Chemical Biology*, *4*, 177–191.

Davis, B. J., & Erlanson, D. A. (2013). Learning from our mistakes: The "unknown knowns" in fragment screening. *Bioorganic and Medicinal Chemistry Letters*, *23*(10), 2844–2852.

Daylight Chemical Information System, Inc. (2018). *Daylight theory: SMARTS—A language for describing molecular patterns*. http://www.daylight.com/dayhtml/doc/theory/theory.smarts.html. Accessed 8 September 2018.

de la Vega de Leon, A., Lounkine, E., Vogt, M., & Bajorath, J. (2017). *Part 2 library design: 5 Design of diverse and focused compound. In A. Varnek (Ed.), Tutorials in chemoinformatics*. https://www.wiley.com/en-sg/Tutorials+in+Chemoinformatics-p-9781119137962.

Dixon, S. L., & Villar, H. O. (1998). Bioactive diversity and screening library selection via affinity fingerprinting. *Journal of Chemical Information and Computer Sciences*, *38*(6), 1192–1203.

Eglese, R. W. (1990). Simulated annealing: A tool for operational research. *European Journal of Operational Research*, *46*(3), 271–281.

Eksterowicz, J. E., Evensen, E., Lemmen, C., Patrick Brady, G., Kevin Lanctot, J., Bradley, E. K., et al. (2002). Coupling structure-based design with combinatorial chemistry: Application of active site derived pharmacophores with informative library design. *Journal of Molecular Graphics & Modelling*, *20*(6), 469–477.

Ertl, P. (2017). An algorithm to identify functional groups in organic molecules. *Journal of Cheminformatics*, *9*(1), 36.

Ertl, P., & Schuffenhauer, A. (2009). Estimation of synthetic accessibility score of drug-like molecules based on molecular complexity and fragment contributions. *Journal of Cheminformatics*, *1*(1), 8.

Evensen, E., Eksterowicz, J. E., Stanton, R. V., Oshiro, C., Grootenhuis, P. D. J., & Bradley, E. K. (2003). Comparing performance of computational tools for combinatorial library design. *Journal of Medicinal Chemistry*, *46*(24), 5125–5128.

Falcioni, M., & Deem, M. W. (2000). Library design in combinatorial chemistry by Monte Carlo methods. *Physical Review. E, Statistical Physics, Plasmas, Fluids, and Related Interdisciplinary Topics*, *61*(5 Pt. B), 5948–5952.

FDA. (2018). *FDA new drug approvals more than doubled in 2017*. BusinessChemistry World. https://www.chemistryworld.com/news/fda-new-drug-approvals-more-than-doubled-in-2017/3008575.article. Accessed 23 May 2018.

Gaulton, A., Bellis, L. J., Bento, A. P., Chambers, J., Davies, M., Hersey, A., et al. (2012). ChEMBL: A large-scale bioactivity database for drug discovery. *Nucleic Acids Research*, *40*(D1), D1100–D1107.

Ghose, A. K., Viswanadhan, V. N., & Wendoloski, J. J. (1999). A knowledge-based approach in designing combinatorial or medicinal chemistry libraries for drug discovery. 1. A qualitative and quantitative characterization of known drug databases. *Journal of Combinatorial Chemistry*, *1*(1), 55–68.

Gillet, V. J., Khatib, W., Willett, P., Fleming, P. J., & Green, D. V. S. (2002). Combinatorial library design using a multiobjective genetic algorithm. *Journal of Chemical Information and Computer Sciences*, *42*(2), 375–385.

Gillet, V. J., Willett, P., Fleming, P. J., & Green, D. V. S. (2002). Designing focused libraries using MoSELECT. *Journal of Molecular Graphics & Modelling*, *20*(6), 491–498.

Gu, J., Gui, Y., Chen, L., Yuan, G., Lu, H. Z., & Xu, X. (2013). Use of natural products as chemical library for drug discovery and network pharmacology. *PLoS One*, *8*(4), e62839.

Hann, M. M., Leach, A. R., & Harper, G. (2001). Molecular complexity and its impact on the probability of finding leads for drug discovery. *Journal of Chemical Information and Computer Sciences*, *41*(3), 856–864.

Hartigan, J. A. (1975). *Clustering algorithms*. Wiley.

Hert, J., Irwin, J. J., Laggner, C., Keiser, M. J., & Shoichet, B. K. (2009). Quantifying biogenic bias in screening libraries. *Nature Chemical Biology*, *5*(7), 479–483.

Hong, J. (2011). Role of natural product diversity in chemical biology. *Current Opinion in Chemical Biology*, *15*(3), 350–354.

Hudson, B. D., Hyde, R. M., Rahr, E., Wood, J., & Osman, J. (1996). Parameter based methods for compound selection from chemical databases. *Quantitative Structure Relationships*, *15*(4), 285–289.

Huggins, D. J., Venkitaraman, A. R., & Spring, D. R. (2011). Rational methods for the selection of diverse screening compounds. *ACS Chemical Biology*, *6*(3), 208–217.

Irwin, J. J. (2006). How good is your screening library? *Current Opinion in Chemical Biology*, *10*(4), 352–356.

Irwin, J. J., & Shoichet, B. K. (2005). ZINC—A free database of commercially available compounds for virtual screening. *Journal of Chemical Information and Modeling*, *45*(1), 177–182.

Jahnke, W., & Erlanson, D. A. (Eds.), (2006). Fragment-based approaches in drug discovery. In *Methods and principles in medicinal chemistry*, Weinheim, FRG: Wiley-VCH Verlag GmbH & Co. KGaA.

Jarvis, R. A., & Patrick, E. A. (1973). Clustering using a similarity measure based on shared near neighbors. *IEEE Transactions on Computers*, *C-22*(11), 1025–1034.

Jimonet, P., & Jäger, R. (2004). Strategies for designing GPCR-focused libraries and screening sets. *Current Opinion in Drug Discovery & Development*, 7(3), 325–333.

John Harris, C., Hill, R. D., Sheppard, D. W., Slater, M. J., & Stouten, P. F. (2011). The design and application of target-focused compound libraries. *Combinatorial Chemistry & High Throughput Screening*, *14*(6), 521–531.

Keseru, G. M., Erlanson, D. A., Ferenczy, G. G., Hann, M. M., Murray, C. W., & Pickett, S. D. (2016). Design principles for fragment libraries: Maximizing the value of learnings from pharma fragment-based drug discovery (FBDD) programs for use in academia. *Journal of Medicinal Chemistry*, *59*(18), 8189–8206.

Kirkpatrick, S., Gelatt, C. D., & Vecchi, M. P. (1983). Optimization by simulated annealing. *Science*, *220*(4598), 671–680.

Koehn, F. E., & Carter, G. T. (2005). The evolving role of natural products in drug discovery. *Nature Reviews. Drug Discovery*, *4*(3), 206–220.

Koutsoukas, A., Paricharak, S., Galloway, W. R. J. D., Spring, D. R., Ijzerman, A. P., Glen, R. C., et al. (2014). How diverse are diversity assessment methods? A comparative analysis and benchmarking of molecular descriptor space. *Journal of Chemical Information and Modeling*, *54*(1), 230–242.

Kumar, R., Sadowski, M. C., Levrier, C., Nelson, C. C., Jones, A. J., Holleran, J. P., et al. (2015). Design and synthesis of a screening library using the natural product scaffold 3-chloro-4-hydroxyphenylacetic acid. *Journal of Natural Products*, *78*(4), 914–918.

Kutchukian, P. S., Lou, D., & Shakhnovich, E. I. (2009). FOG: Fragment optimized growth algorithm for the de novo generation of molecule: Occupying druglike chemical space. *Journal of Chemical Information and Modeling*, *49*(7), 1630–1642.

Kutchukian, P. S., So, S. S., Fischer, C., & Waller, C. L. (2015). Fragment library design: Using cheminformatics and expert chemists to fill gaps in existing fragment libraries. *Methods in Molecular Biology*, *1289*, 43–53.

Langdon, S. R., Brown, N., & Blagg, J. (2011). Scaffold diversity of exemplified medicinal chemistry space. *Journal of Chemical Information and Modeling*, *51*(9), 2174–2185.

Lau, W. F., Withka, J. M., Hepworth, D., Magee, T. V., Du, Y. J., Bakken, G. A., et al. (2011). Design of a multi-purpose fragment screening library using molecular complexity and orthogonal diversity metrics. *Journal of Computer-Aided Molecular Design*, *25*(7), 621–636.

Lee, M. L., & Schneider, G. (2001). Scaffold architecture and pharmacophoric properties of natural products and trade drugs: Application in the design of natural product-based combinatorial libraries. *Journal of Combinatorial Chemistry*, *3*(3), 284–289.

Lewell, X. Q., Judd, D. B., Watson, S. P., & Hann, M. M. (1998). RECAP—Retrosynthetic combinatorial analysis procedure: A powerful new technique for identifying privileged molecular fragments with useful applications in combinatorial chemistry. *Journal of Chemical Information and Computer Sciences*, *38*(3), 511–522.

Lipinski, C. A., Lombardo, F., Dominy, B. W., & Feeney, P. J. (2012). Experimental and computational approaches to estimate solubility and permeability in drug discovery and development settings. *Advanced Drug Delivery Reviews*, *64*(Suppl), 4–17.

López-Vallejo, F., Nefzi, A., Bender, A., Owen, J. R., Nabney, I. T., Houghten, R. A., et al. (2011). Increased diversity of libraries from libraries: Chemoinformatic analysis of Bis-Diazacyclic libraries. *Chemical Biology & Drug Design*, 77(5), 328–342.

Maggiora, G., Vogt, M., Stumpfe, D., & Bajorath, J. (2014). Molecular similarity in medicinal chemistry. *Journal of Medicinal Chemistry*, *57*(8), 3186–3204.

Martin, E. J. (1995). Measuring diversity: Experimental design of combinatorial libraries for drug discovery. *Journal of Medicinal Chemistry*, *38*, 1431–1436.

Martin, Y. C. (2001). Diverse viewpoints on computational aspects of molecular diversity. *Journal of Combinatorial Chemistry*, *3*(3), 231–250.

Muhammad, S., & Fatima, N. (2015). In silico analysis and molecular docking studies of potential angiotensin-converting enzyme inhibitor using quercetin glycosides. *Pharmacognosy Magazine*, *11*(42), 123.

Mullard, A. (2018). 2017 FDA drug approvals. *Nature Reviews. Drug Discovery*, *17*(2), 81–85.

Nicolaou, C. A., & Kannas, C. C. (2011). Molecular library design using multi-objective optimization methods. *Methods in Molecular Biology (Clifton, NJ)*, *685*, 53–69.

No authors. (2007). All natural. *Nature Chemical Biology*, *3*(7), 351.

Orry, A. J. W., Abagyan, R. A., & Cavasotto, C. N. (2006). Structure-based development of target-specific compound libraries. *Drug Discovery Today*, *11*(5–6), 261–266.

Ostresh, J. M., Husar, G. M., Blondelle, S. E., Dörner, B., Weber, P. A., & Houghten, R. A. (1994). "Libraries from Libraries": Chemical transformation of combinatorial libraries to extend the range and repertoire of chemical diversity. *Proceedings of the National Academy of Sciences of the United States of America*, *91*(23), 11138–11142.

Pascolutti, M., & Quinn, R. J. (2014). Natural products as lead structures: Chemical transformations to create Lead-like libraries. *Drug Discovery Today*, *19*(3), 215–221.

Patridge, E., Gareiss, P., Kinch, M. S., & Hoyer, D. (2016). An analysis of FDA-approved drugs: Natural products and their derivatives. *Drug Discovery Today*, *21*(2), 204–207.

Prien, O. (2005). Target-family-oriented focused libraries for kinases—Conceptual design aspects and commercial availability. *Chembiochem*, *6*(3), 500–505.

Prokopyev, O. A., Kong, N., & Martinez-Torres, D. L. (2009). The equitable dispersion problem. *European Journal of Operational Research*, *197*(1), 59–67.

RDKit. (2017). http://www.rdkit.org/. Accessed 4 August 2017.

rdkit.Chem.MACCSkeys. (2018). http://rdkit.org/Python_Docs/rdkit.Chem.MACCSkeys-pysrc.html. Accessed 23 May 2018.

Rogers, D., Brown, R. D., & Hahn, M. (2005). Using extended-connectivity fingerprints with Laplacian-modified Bayesian analysis in high-throughput screening follow-up. *Journal of Biomolecular Screening*, *10*(7), 682–686.

Rogers, D., & Hahn, M. (2010). Extended-connectivity fingerprints. *Journal of Chemical Information and Modeling*, *50*(5), 742–754.

Rush, T. S., Grant, J. A., Mosyak, L., & Nicholls, A. (2005). A shape-based 3-D scaffold hopping method and its application to a bacterial protein-protein interaction. *Journal of Medicinal Chemistry*, *48*(5), 1489–1495.

Saubern, S., Guha, R., & Baell, J. B. (2011). KNIME workflow to assess PAINS filters in SMARTS format. Comparison of RDKit and indigo cheminformatics libraries. *Molecular Informatics*, *30*(10), 847–850.

Schnur, D. M. (2008). Recent trends in library design: "Rational design" revisited. *Current Opinion in Drug Discovery & Development*, *11*(3), 375–380.

Schnur, D. M., Beno, B. R., Tebben, A. J., & Cavallaro, C. (2011). Methods for combinatorial and parallel library design. *Methods in Molecular Biology (Clifton, NJ)*, *672*, 387–434.

Schuffenhauer, A., Ruedisser, S., Marzinzik, A., Jahnke, W., Selzer, P., & Jacoby, E. (2005). Library design for fragment based screening. *Current Topics in Medicinal Chemistry*, *5*(8), 751–762.

Siegal, G., Ab, E., & Schultz, J. (2007). Integration of fragment screening and library design. *Drug Discovery Today*, *12*(23–24), 1032–1039.

Sun, H. (2008). Pharmacophore-based virtual screening. *Current Medicinal Chemistry*, *15*(10), 1018–1024.

Sun, Y., Ewing, T. J. A., Skillman, A. G., & Kuntz, I. D. (1998). CombiDOCK: Structure-based combinatorial docking and library design. *Journal of Computer-Aided Molecular Design*, *12*(6), 597–604.

U.S. Food & Drug Administration, Drugs@FDA: FDA approved drug products, https://www.accessdata.fda.gov/scripts/cder/daf/, n.d. Accessed 11 September 2017.

Vera, K., Lopez-Pires, F., Baran, B., & Sandoya, F. (2017). Multi-objective maximum diversity problem. In *Vol. 2017. 2017 43rd Latin American computer conference, CLEI 2017* (pp. 1–9), IEEE. Janua.

Villar, H. O., Yan, J., & Hansen, M. R. (2004). Using NMR for ligand discovery and optimization. *Current Opinion in Chemical Biology*, *8*(4), 387–391.

Walters, W. P., Stahl, M. T., & Murcko, M. A. (1998). Virtual screening—An overview. *Drug Discovery Today*, *3*(4), 160–178.

Ward, J. H. (1963). Hierarchical grouping to optimize an objective function. *Journal of the American Statistical Association*, *58*(301), 236–244.

Wawer, M. J., Li, K., Gustafsdottir, S. M., Ljosa, V., Bodycombe, N. E., Marton, M. A., et al. (2014). Toward performance-diverse small-molecule libraries for cell-based phenotypic screening using multiplexed high-dimensional profiling. *Proceedings of the National Academy of Sciences of the United States of America*, *111*(30), 10911–10916.

Weininger, D. (1988). SMILES, a chemical language and information system: 1: Introduction to methodology and encoding rules. *Journal of Chemical Information and Computer Sciences*, *28*(1), 31–36.

Wenlock, M. C., Austin, R. P., Barton, P., Davis, A. M., & Leeson, P. D. (2003). A comparison of physiochemical property profiles of development and marketed oral drugs. *Journal of Medicinal Chemistry*, *46*(7), 1250–1256.

Whittaker, M., Law, R. J., Ichihara, O., Hesterkamp, T., & Hallett, D. (2010). Fragments: Past, present and future. *Drug Discovery Today: Technologies*, 7(3), e163–e171.

Willett, P. (2000). Chemoinformatics—Similarity and diversity in chemical libraries. *Current Opinion in Biotechnology*, *11*(1), 85–88.

Wright, T., Gillet, V. J., Green, D. V. S., & Pickett, S. D. (2003). Optimizing the size and configuration of combinatorial libraries. *Journal of Chemical Information and Computer Sciences*, *43*(2), 381–390.

Wyss, P. C., Gerber, P., Hartman, P. G., Hubschwerlen, C., Locher, H., Marty, H. P., et al. (2003). Novel dihydrofolate reductase inhibitors. Structure-based versus diversity-based library design and high-throughput synthesis and screening. *Journal of Medicinal Chemistry*, *46*(12), 2304–2312.

Zhou, J. Z. (2011). Chemoinformatics and library design. *Methods in Molecular Biology*, *685*, 27–52. Humana Press.

CHAPTER FIVE

# Design of a Fragment-Screening Library

**Ashley Taylor, Bradley C. Doak, Martin J. Scanlon[1]**
Medicinal Chemistry, Monash Institute of Pharmaceutical Sciences, Monash University, Parkville, VIC, Australia
[1]Corresponding author: e-mail address: martin.scanlon@monash.edu

## Contents



## Abstract

Herein we describe a method for the design, purchase, and assembly of a fragment-screening library from a list of commercially available compounds. The computational tools used in assessment of compound properties as well as the workflow for compound selection are provided for reference as implemented in commercially available software that is free and accessible to most academic users. The workflow can be modified as necessary to generate a fit-for-purpose fragment library with the desired compound property profiles. An analytical process for assessing the quality, identity, and suitability of a purchased fragment for inclusion in a screening collection is described. Results from our in-house library are presented as an example of compound progression through this quality control process.

## 1. INTRODUCTION

A key step in any fragment-based drug design (FBDD) program is finding suitable chemical starting points for development. These are commonly

*Methods in Enzymology*, Volume 610
ISSN 0076-6879
https://doi.org/10.1016/bs.mie.2018.09.018

identified by screening a library of small compounds using biophysical assays to identify those that bind efficiently to the target of interest. Compounds typically must meet a range of different criteria in order to be considered suitable for inclusion in a fragment library. These criteria include appropriate aqueous solubility and stability at the high concentrations that are typical of fragment-screening assays, as well as synthetic tractability, which allows hit fragments to be developed into higher affinity leads. Hence, the design of a high-quality fragment library is fundamental to efficient execution of FBDD and has widespread ramifications for subsequent efforts to develop fragments into relevant chemical probes and potential drugs. Initially, chemoinformatic tools can be used in the design of a library to select fragments that are structurally diverse and suitable to be used in one or more fragment-screening assays. For example, if $^{19}F$-detected NMR spectroscopy is to be used as the primary-screening approach, fragments in the screening set should possess one or more fluorine atoms. While it can be daunting to consider the prospect of selecting a relatively small library of compounds (100–1000s) that maximally represent fragment chemical space (estimated to be in the billions; Visini, Awale, & Reymond, 2017), there are many methods to help tailor a fragment library to specific needs. Once a library of fragments has been selected and purchased, significant effort is required to obtain sufficient data to verify the identity, solubility, and purity of the compounds. The importance of generating quality control (QC) data for the fragments in the screening library is essential as it has been reported that up to ~15% of purchased compounds can be impure, poorly behaved in assays, are not consistent with the desired chemical structure, or are insoluble in aqueous buffers at concentrations required for screening. Analyzing such large datasets of QC can be time-consuming but is ultimately overcome by the amount of time that is saved from avoiding the investigation of problematic or false-positive fragment hits. Herein we provide an overview of the process of compound selection with examples and practical details from our own experience in generating fragment libraries. More extensive reviews and descriptions of different criteria that have been used in selecting compounds for fragment libraries have also been published elsewhere (Keseru et al., 2016; Na & Hu, 2011).

## 2. COMPILING A DATASET OF COMPOUNDS

Essential to the process of designing a fragment library is dealing with large datasets of compounds. With >7 million compounds in commercially available compound databases, processing such datasets requires the use of chemoinformatic software to calculate and manipulate large amounts of chemical data. Herein we outline a process for data analysis to design a fragment library, and provide a working example using datasets and tools that are available free of charge to academic groups. To access the wide variety of tools used by chemoinformations, workflow software such as KNIME (Berthold et al., 2008) provides a visual and integrated environment and gives access to many chemoinformatic packages (CDK (Willighagen et al., 2017), RDKit (Landrum, 2018), etc.) and tools required to design a library even for those with little experience (Fig. 1). The KNIME workflow and data files described in the current chapter can be downloaded from Figshare (https://doi.org/10.26180/5b6bbe778b3f9) (Taylor, Doak, & Scanlon, 2018).

Initially a list of commercially available compounds must be compiled. While there are many suppliers and compound collation services, a number of trusted suppliers are often sought for large compound purchases. In our experience Maybridge, Life, Chembridge, Enamine, InterBioScreen, VitasMlabs, Otava, Specs, and UkrOrgSyntez (UROSY) make up a large proportion of our purchased compounds and negotiation with compound purchasing and formatting services such as MolPort and eMolecules have also proved helpful. Here we exemplify the library selection process with the MolPort combined dataset of stocked compounds, which encompasses ~7.24 million commercially available compounds. Initially the compound lists from suppliers have to be standardized and formatted consistently, for example, by removing salts, grouping tautomers, etc., in order to generate a list of unique compounds. Salt forms are also discarded as upon purchase of the final set of compound an appropriate salt form can be selected. Additional filters such as price, e.g., 5 mg for $\leq$100 USD or 50 mg for $\leq$500 USD, and availability, e.g., lead time $<$ 30 days can also be applied where this information is available. Price filters are pragmatic, set by the desired number of fragments and budget for the library. If the goal is to generate a library containing ~1200 fragments in total, we would initially aim to purchase ~1400 compounds for QC analysis (assuming ~15% QC failure rate, discussed later). In our experience, the average cost of screening compounds purchased between 2015 and 2018 in amounts of 10–20 mg was ~60 USD per sample putting the total cost for fragment purchase of this scope at ~84,000 USD. A full list of unique compounds is compiled by comparing INCHI or canonical SMILES of desalted and standardized compounds.

Once the structures are standardized, initial broad filters can be applied to generate a smaller and more manageable dataset for calculations and selection. Filters based on properties of the compounds such as calculated logP (ClogP), heavy atom count (HAC), hydrogen bond donor and acceptor

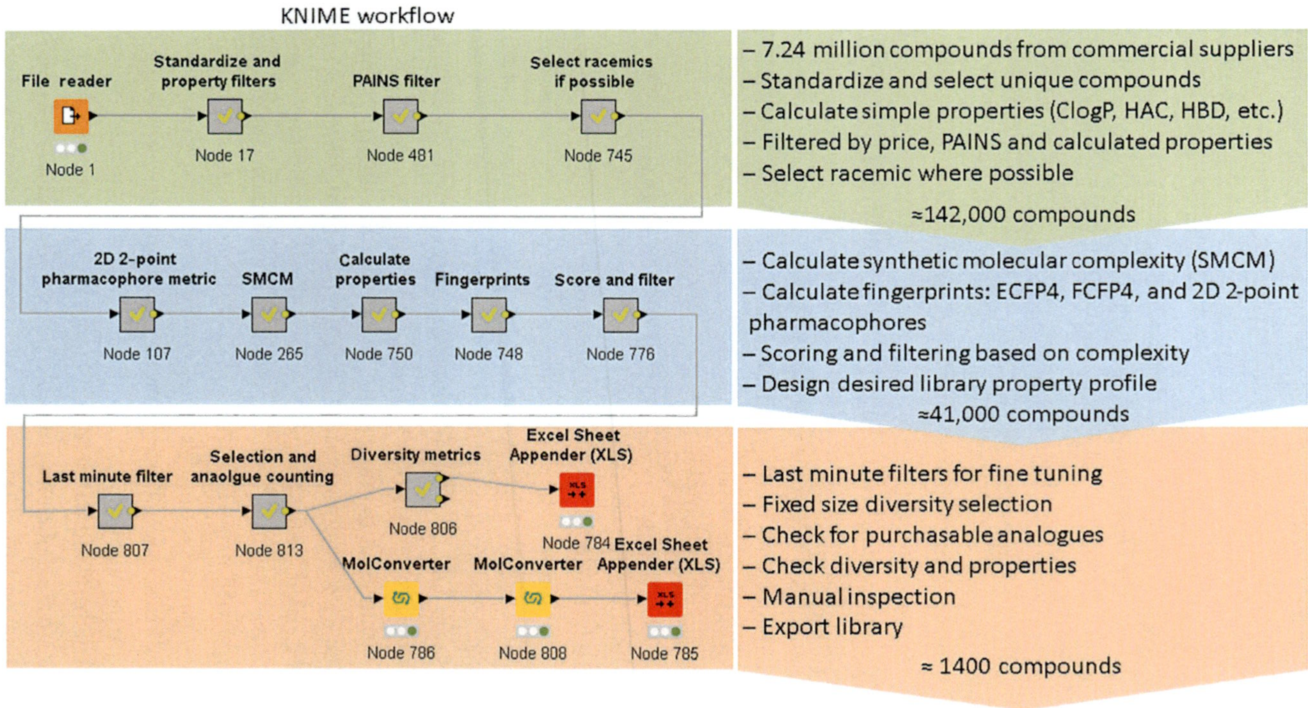


**Fig. 1** Chemoinformatic design and selection of a fragment library with a KNIME workflow and description of the major stages of the workflow.

count (HBD, HBA), topological polar surface area (TPSA), ring count, number of rotatable bonds (NRotB) can all be applied at this stage. While there are no strict guidelines on what constitutes a fragment one of the most widely adopted is the "rule of 3" which sets out approximate property guidelines for fragment chemical space of molecular weight (MW) $< 300$ Da, ClogP $\leq 3$, HBD $\leq 3$, and the HBA $\leq 3$ (Congreve, Carr, Murray, & Jhoti, 2003; Jhoti, Williams, Rees, & Murray, 2013). In our experience (Doak, Morton, Simpson, & Scanlon, 2017) and as has been detailed elsewhere (Jhoti et al., 2013; Keseru et al., 2016), the guidelines applied at this stage should reflect the intended screening strategy. For example, the majority of our fragment screens have been undertaken using ligand-detected NMR, where we assume an approximate upper limit of detection of compounds which bind with $K_D \sim 10\,\text{m}M$ and expect that the highest affinity fragments will bind with $K_D \sim 100\,\mu M$. Therefore, the lower HAC limit is set to 6 heavy atoms and upper limit to 18 heavy atoms, outside which few compound can have the expected affinity for detection based on desirable ligand efficiencies (i.e., for a fragment possessing 18 HAC that binds with $K_D = 100\,\mu M$ then LE $= 0.30\,\text{kcal}\,\text{mol}^{-1}\,\text{HAC}^{-1}$ and a fragment possessing 6 HAC binds with $K_D = 10\,\text{m}M$, then LE $= 0.45\,\text{kcal}\,\text{mol}^{-1}\,\text{HAC}^{-1}$) (Reynolds, Bembenek, & Tounge, 2007). We also select for MW $\leq 300$, ClogP $\leq 2.5$, TPSA $\leq 80$, NRotB $\leq 5$, HBA $\leq 6$, HBD $\leq 3$, rings count $\leq 3$ to find relatively small compounds with polar atoms to increase solubility in aqueous environments and decrease the likelihood of aggregation.

Compounds containing reactive and undesirable functional groups/PAINS substructures can also be removed. There are a number of excellent resources available for compiled lists of these (Benigni & Bossa, 2011; Irwin et al., 2015; Lagorce, Sperandio, Baell, Miteva, & Villoutreix, 2015; Sushko, Salmina, Potemkin, Poda, & Tetko, 2012); however, a degree of caution is required in applying these filters as they often contain generic structures, which if applied can severely limit diversity. Manual curation of the undesirable structure lists, which often contain between 100 and 2000 motifs, is time-consuming, but worthwhile in generating chemical diversity in the library. We segregate the filters into three categories, reactive functional groups (acid halides, epoxides, maleimide, Michael acceptors, etc.), PAINS motifs (dye motifs, polyenes, rhodanines, etc.), and alert structures (azides, hydroxyl amines, etc.). Compounds having $>0$ reactive functional groups, $>0$ PAINS motifs, or $>2$ alert structures are removed from the dataset. Our compiled list and classified motifs in each of these categories can be found in the associated KNIME workflow. A final grouping is performed to select racemic compounds when there is a choice of racemic or chirally pure.

Applying these initial filters to the dataset of ~7.24 million compounds reduces the number of suitable fragments down to ~142,000. However, even this condensed dataset contains a preponderance of larger, often more lipophilic compounds due to the exponential growth in the number of possible chemical structures as the number of heavy atoms in a compound increases. Therefore, secondary filters and analysis are required to obtain a diverse library with the desired physicochemical profile.

## 3. CALCULATING PROPERTIES AND SELECTING A REPRESENTATIVE SET

To further refine the dataset and establish the desired library property profile, it is necessary to calculate a number of more advanced compound properties and metrics. In the analysis presented here, these include the 2D 2-point pharmacophore fingerprint, which looks for pharmacophore topologies in the 2D representation of the structures, e.g., a HBD separated by two bonds from a HBA motif. The total number of different 2D 2-point pharmacophores in a structure can then also measure the complexity by analogy to the Hann complexity-screening model (Hann, Leach, & Harper, 2001). An additional complexity measurement in the form of the synthetic molecular complexity metric (SMCM) (Allu & Oprea, 2005) is also calculated to assess the complexity of the compounds. Molecular fingerprints that are used in the final selection of fragments are also calculated. Two fingerprints, the extended connectivity fingerprint (ECFP4) and the feature-connectivity fingerprint (FCFP4) (Morgan, 1965; Rogers & Hahn, 2010), are used for the selection due to their quick calculation and widespread application. Each compound is then given a score by summation of normalized physicochemical properties (ClogP, HAC, and number of heteroatoms normalized to the desired profile of the library) and the number of undesirable "alert" functional groups similar to previous reports (Tounge & Parker, 2011). This enables compounds that may not be outside the strict property filters mentioned earlier but having many undesirable properties to be identified, e.g., MW 280 Da, ClogP 2.4, and 2 alert structures. A set of filters is then used to eliminate these undesirable compounds by removing those with high scores, high complexity, etc., and create the desired library property profile. In the current example fragment selection was based on the following criteria: number of different 2D 2-point pharmacophores $\leq 10$,

SMCM $\leq$ 35, property score $\leq$ 2.5, and number of heteroatoms $\geq$ 1 to give a dataset of ~41,000 compounds for selection. In addition, the library was designed to have a relatively broad distribution of HAC with a peak around 10–14 HAC due to the noted enrichment in X-ray crystallography hits for molecules of this size (Hall, Mortenson, & Murray, 2014).

There are a number of ways to select representative compounds from the dataset, commonly via clustering fragments (e.g., *k*-means, hierarchical, Ward, etc.) based on some measure of similarity (commonly Tanimoto or Tversky similarities of fingerprints) before selection of representative compounds from each cluster (Tanaseichuk et al., 2015; Tounge & Parker, 2011). Some clustering methods can be slow (Tanaseichuk et al., 2015) and in our experience, the min–max selection algorithm (Mark et al., 2002) implemented in RDKit with ECFP4 and FCFP4 fingerprints provides similar results to *k*-means clustering in KNIME with significantly reduced time. The number of fragments that are selected at this stage is set higher than the desired size of the final library due to expectation that some of the selected compounds will subsequently be deemed unsuitable for inclusion as well as the expected QC failure rate. These factors are discussed in more detail later. After a preliminary selection, the library property and diversity profiles are inspected (Fig. 2). If necessary, the property scoring method and filters can be adjusted at this stage until a desired property and diversity profile is obtained. Once the desired property profile is achieved in the selected compounds, the structures are inspected manually by an experienced medicinal chemist, to assess desirable properties, substructures, synthetic accessibility, and developability (having multiple points/vectors for expansion chemistry; Cox et al., 2016; Ray et al., 2017). A "last minute filter" is used to remove any undesirable compounds that were missed by the previous PAINS, alerts, reactivity, and property filters. With very small and simple compounds such as fragments, it has been observed that large errors in prediction of some physicochemical properties can occur, resulting in apparently anomalous values being obtained (e.g., for ClogP; Mannhold, Poda, Ostermann, & Tetko, 2009). These cases can often be flagged by an experienced chemist, and the last minute filter allows rapid iteration of specific substructure and property filters, removing compounds where necessary to generate an acceptable final dataset. The first stage in the validation and development of fragments that are identified as hits in a primary screen is often to test a series of chemically similar analogues. To enable this strategy for hit validation and fragment expansion, it is desirable to have a number of larger commercially available analogues for each fragment in the screening library. Therefore, substructure searches of a commercial dataset for suitable

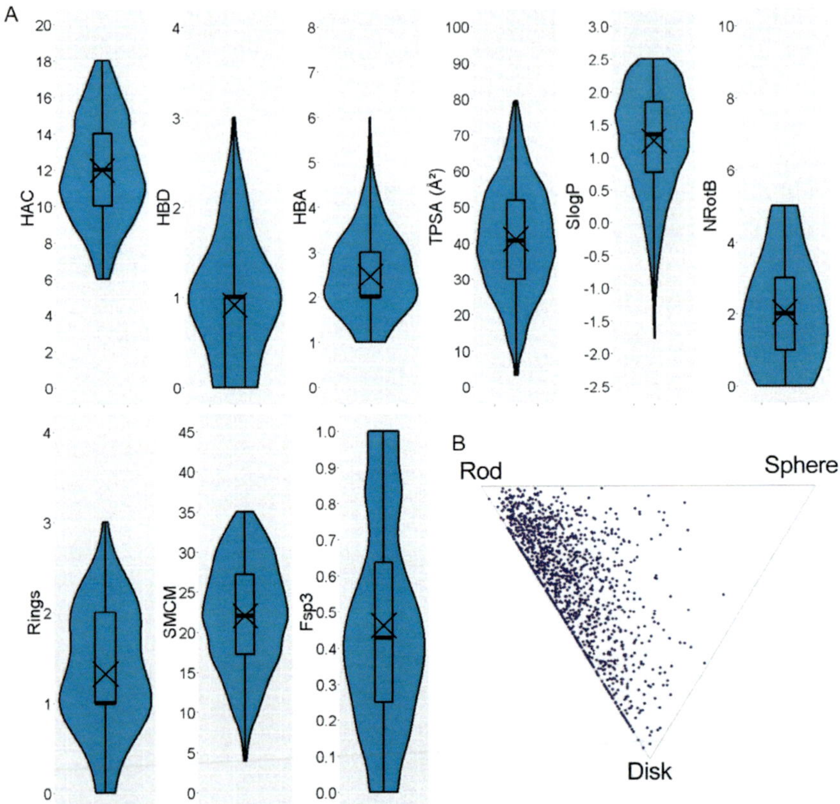


C

| Property | Number |
|---|---|
| Total compounds | 1377 |
| Unique scaffolds | 546 |

| Cell-based diversity metric | Coverage |
|---|---|
| 2D 2-point pharmacophores | 94% |
| 2D 3-point pharmacophores | 34% |

| Fingerprint diversity metric | Value |
|---|---|
| Mean Tc(ECFP4) of library compared to full library | 0.093 |
| Mean Tc(FCFP4) of library compared to full library | 0.125 |
| Mean Tc(ECFP4) of library to nearest neighbors in library | 0.330 |
| Mean Tc(FCFP4) of library to nearest neighbors in library | 0.454 |

**Fig. 2** Property and diversity profile of the selected example fragment library. (A) Violin and box plots of heavy atom count (HAC), number of hydrogen bond donors (HBD), number of hydrogen bond acceptors (HBA), topological polar surface area (TPSA), calculated logP (SlogP), number of rotatable bonds (NRotB), number of rings (Rings), synthetic molecular complexity metric (SMCM), fraction of sp3 carbons (Fsp3). Boxplots represent the min, 25th, 50th, and 75th percentiles and max, cross represents the mean. (B) Normalized principle moment of inertia (nPMI). (C) Diversity metrics.

analogues of the selected fragments (which can be filtered, e.g., for desirable physicochemical properties, removal of PAINS, etc., as desired) provide a means for elimination of compounds with <5 commercial analogues based on direct substructure, which may be deemed difficult to validate or develop.

The diversity and properties of a given library selection can be calculated for comparison to alternate selections or other libraries. Inspection of property ranges and distributions ensures that the property profiles and complexity measures are appropriate for the fragment library. Some calculated physicochemical properties have been correlated to undesirable compound properties such as toxicity, promiscuity, and low permeability (Meanwell, 2011); however, a balance has to be achieved in the selection as application of too stringent a set of filters can impact the diversity of the available fragment structures. The diversity and chemical space coverage of the selected library can be assessed by cell-based coverage metrics such as the proportion of 2D 2-point pharmacophores or 2D 3-point pharmacophores covered by the library. These metrics are based on lists of 2 or 3 pharmacophore features (HBD, HBA, aromatic, positive, negative motifs) linked by defined numbers of bonds (1–5 bonds) and hence have a finite number of possibilities and are absolute quantifications of chemical space coverage. In the example workflow a high proportion of 2D 2-point pharmacophores was obtained (94%) as well as moderate coverage of 2D 3-point pharmacophores (34%) compared to other fragment libraries (Doak et al., 2017; Lau et al., 2011). The shape representation of the library can be assessed by measures such as the normalized principal moment of inertia plot (nPMI) (Sauer & Schwarz, 2003). As is often the case with compounds selected from the commercially available chemical space, the fragments in this example library cluster along the rod-disk axis of the nPMI plot (Doak et al., 2017; Visini et al., 2017). In addition, metrics relating to the size and profile of the library can be calculated for assessment of diversity, including the number of scaffolds within the library (546 for the example library) and average fingerprint similarity (Tanimoto score, Tc) of each library compound to the whole library and to its nearest neighbor in the library (Fig. 2) (Doak et al., 2017; Schulz, Landstrom, Bright, & Hubbard, 2011).

## 4. LIBRARY QUALITY CONTROL

Once a set of fragments has been identified and purchased for inclusion in a fragment library, the identity, purity, and properties (e.g., solubility) of the fragments need to be established. Fragment libraries are often screened at high aqueous concentrations (often 200 μ*M* or greater) by using one or more biophysical or biochemical techniques (NMR, SPR, X-ray crystallography, mass spectrometry, thermal shift, high concentration biochemical assay, etc.) to determine binding/activity toward a specific target. Some requirements (e.g., aqueous solubility) are common to all of the

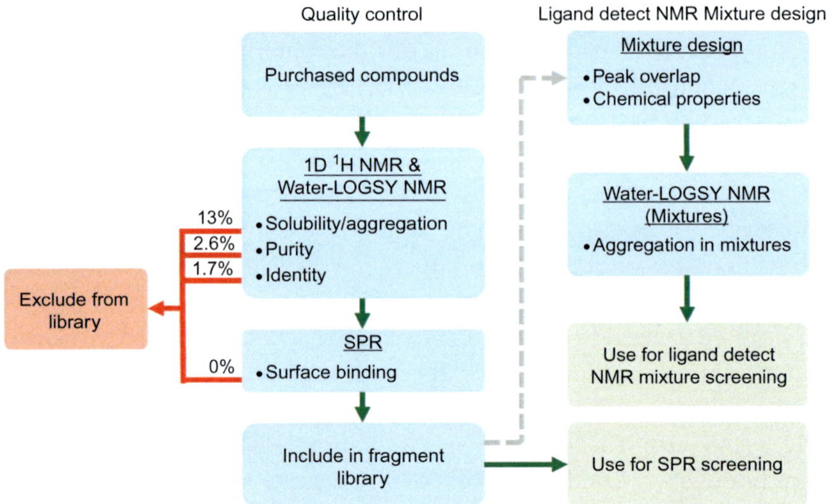


**Fig. 3** Compound quality control assessment, failure rates observed during QC of the most recent Monash fragment library, and mixture design strategy for ligand-detected NMR and SPR screening.

different approaches. In addition different screening techniques bring specific requirements, for example, ligand-detected $^{1}$H NMR screening requires at least one nonexchangeable proton to be present in the fragment, $^{19}$F NMR screening requires the presence of at least one fluorine atom, fragments that are used for SPR screening must not bind to the functionalized surfaces that are used in the SPR flow cells, and biochemical assays that use fluorescence for detection require fragments that do not interfere in the assay. An additional requirement for $^{1}$H/$^{19}$F ligand-detected NMR screening is that the primary screens are commonly undertaken on mixtures of several fragments, and this requires the design and optimization of mixtures containing fragments that do not associate with one another, and where any spectral overlap is kept to a minimum in order to allow deconvolution of the mixtures for hit identification. In addition, individual fragments and mixtures of fragments must exhibit long-term stability in the organic solvents in which they are stored. Therefore, QC and property measurements are undertaken on fragments prior to their inclusion in the library and throughout its life to ensure accurate screening data (Fig. 3).

## 5. SAMPLE STORAGE AND PREPARATION

There are many factors to be considered when deciding on storage conditions for a fragment library. Most fragments are stored as concentrated stock solutions in DMSO (100–200 m*M*). However, DMSO is a mild oxidant (Prochazkova et al., 2012) and can cause some compounds to degrade (Davis & Erlanson, 2013). Typical storage temperatures are at 4°C or −20°C to slow compound degradation (Keseru et al., 2016); however, the introduction of freeze–thaw cycles is undesirable and can be avoided if the fragments are stored as prealiquoted samples in a format that is suitable for screening. Since DMSO is a hygroscopic liquid, freeze–thaw cycles can introduce atmospheric water into stocks resulting in a higher rate of compound loss compared with libraries stored at room temperature under inert gas (Keseru et al., 2016; Kozikowski et al., 2003). The fragments in our previous libraries were stored in DMSO at 4°C for up to 5 years. Under these conditions ~11% of the library suffered measureable degradation as determined from analysis of their $^{1}H$ NMR spectra. Fragments purchased for inclusion in our current library were dissolved in $^{2}H_6$-dimethyl sulfoxide ($D_6$-DMSO) at a concentration of 100 m*M* (based on the delivered mass of the compounds) and are stored at 18°C under nitrogen. In addition to the analysis described later, fragment stocks are manually inspected and those with visible precipitate at 100 m*M* in $D_6$-DMSO either upon preparation or storage are removed from the library.

For QC purposes a semiquantitative 1D $^{1}H$ qNMR spectrum and a water-ligand observed via gradient spectroscopy (Water-LOGSY) NMR spectrum are acquired for all fragments (Fig. 4). All sample stocks are diluted to a concentration of 1 m*M* in standard NMR buffer (50 m*M* sodium phosphate buffer, 25 m*M* NaCl, pH 7.0 with 10% $D_2O$, 2% $D_6$-DMSO) with 100 μ*M* 4,4-dimethyl-4-silapentane-1-sulfonic acid (DSS) added as an internal chemical shift and concentration standard. Spectra are acquired in 5-mm NMR tubes at 298 K on a Bruker 600 MHz magnet equipped with a cryoprobe.

## 6. 1D $^{1}H$ NMR

Reference 1D $^{1}H$ qNMR spectra are acquired with 10 s relaxation delay, 16 scans, and 16k data points. The resulting spectra are manually processed and inspected. Spectra which contain resonances where the chemical shifts, multiplicity, or intensity is inconsistent with the expected structure of the fragment are considered to have failed QC. Peaks near the water frequency (e.g., deshielded benzylic protons which are expected at a chemical shift of ~4.5–5.5 ppm) are often observed to have low intensity or are missing entirely due to the water suppression scheme used for data

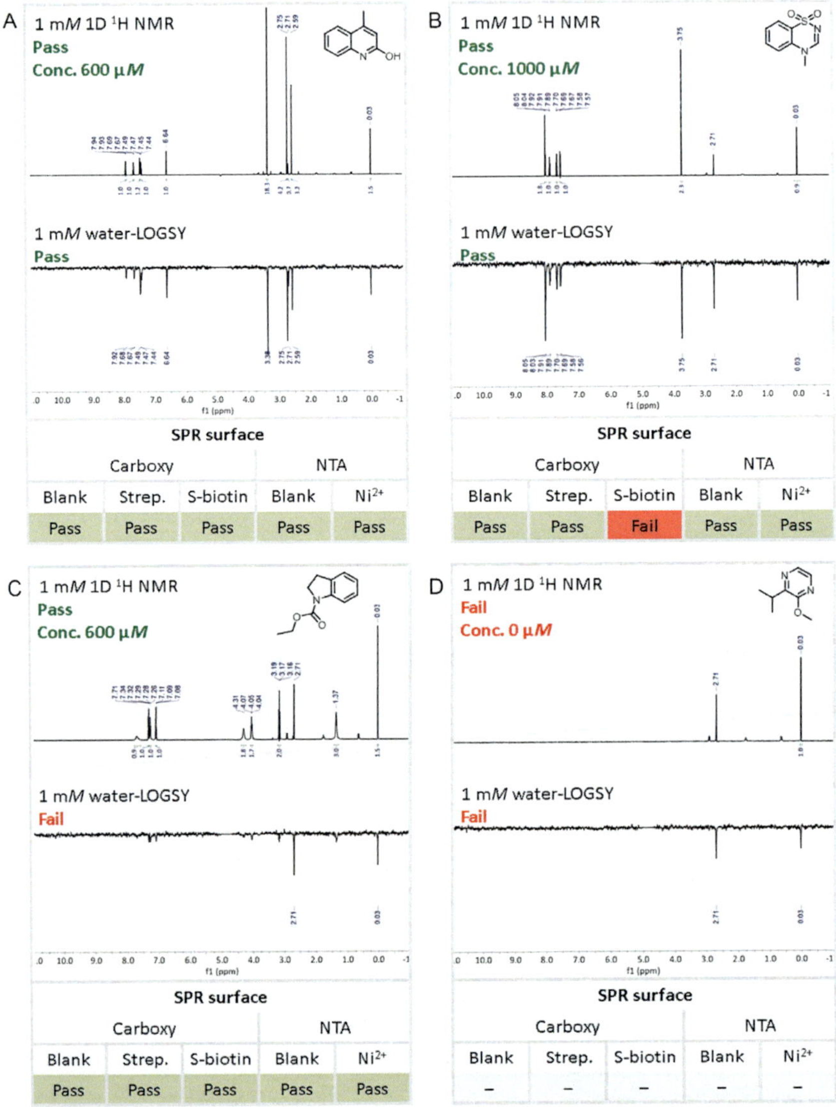


**Fig. 4** Example spectra of a compound (A) passing $^{1}$H NMR, Water-LOGSY, and SPR; (B) passing $^{1}$H NMR, Water-LOGSY, and failing on one SPR surface; (C) passing $^{1}$H NMR and SPR but failing Water-LOGSY; (D) failing $^{1}$H NMR and Water-LOGSY.

acquisition. Hence the lack of these peaks is not considered as a QC failure if this is the only inconsistency in the spectrum. Fragments are discarded on the basis of purity in cases where the expected signals are present in the NMR spectrum but there are additional signals that are not consistent with the expected structure and the relative intensities of the peaks indicate that fragment purity is <90%. The addition of a known amount of standard (100 μ*M* DSS) to the NMR sample allows for an estimation of concentration, based on the assumption that a relaxation delay of 10 s is sufficient for the majority of protons in fragments to relax fully. For each fragment an isolated resonance in the spectrum is identified and integrated to the correct relative intensity (e.g., a CH in the compound is calibrated with an intensity of 1.00 in the spectrum, a $CH_2$ would be calibrated to an intensity of 2.00, etc.), and this intensity is compared to the 9 proton singlet of DSS. The concentration is calculated using the following formula:

$$C_{\mathrm{frag}} = \frac{I_{\mathrm{frag}}}{N_{\mathrm{frag}}} \times \frac{N_{\mathrm{std}}}{I_{\mathrm{std}}} \times C_{\mathrm{std}} \quad (1)$$

where $C$ = concentration, $I$ = integral, and $N$ = number of protons for peaks of the fragment (frag) and standard (std). As the spectrum is adjusted to the correct relative fragment intensity, this simplifies to:

$$C_{\mathrm{frag}} = \frac{N_{\mathrm{std}}}{I_{\mathrm{std}}} \times C_{\mathrm{std}} \quad (2)$$

For a spectrum where the 100 μ*M* DSS signal has a relative peak intensity of 0.9, the fragment would have a concentration of 1000 μ*M*. Any fragment with a calculated concentration of less than 333 μ*M* (i.e., threefold lower than expected) fails QC on the basis of solubility or lack of sample and is removed from the library. In our most recent library generation the average calculated concentration of fragments that were included in the library was 890 μ*M* (SD = 310 μ*M*, $n$ = 1142).

## 7. $^1$H WATER-LOGSY

A second important consideration when selecting fragments is that they should have no propensity to aggregate in aqueous buffers at the concentrations used for screening, as this can lead to false positives in the screens. Fragment aggregation can be identified through the use of Water-LOGSY NMR. In a Water-LOGSY experiment magnetization is transferred from the bulk water to compounds via the nuclear Overhauser effect (NOE). The phase and signal intensity of an NOE is extremely sensitive to the rotational correlation time, hence large soluble aggregates (not visible to the naked eye), which tumble more slowly behave differently to free small molecules in solution. This is manifest in the Water-LOGSY spectrum as a change in the intensity and phase of the peaks. Compounds with a tendency to aggregate show positive signals in the Water-LOGSY spectrum, whereas well-behaved fragments show negative signals. The phase of the fragment signals can be compared with the phase of the DSS signal in the spectrum, which is expected to be negative. Water-LOGSY spectra are acquired with 3 s relaxation delay, 24 scans and 8 k data points. Fragments which show only very weak negative signals or a positive signal (with respect to the DSS peak) in the water-LOGSY spectra are deemed to be aggregators and fail QC on the basis of solubility/aggregation.

## 8. FAILURE RATE

It is common to find that a significant proportion (~15%) of compounds fail the QC process during library design (Keseru et al., 2016; Lau et al., 2011). Our experience using the library QC process described earlier resulted in similar rates of failure as those reported previously. Analysis of the QC failures provides some insights into some of the potential causes. For example, in the QC analysis of a series of fragments that were purchased to expand our current library approximately 19% of compounds failed QC because there was no evidence of compound in their $^{1}H$ NMR spectra. Analysis of the property profiles for these compounds revealed that they had low MW and low polarity, and therefore were likely more volatile and prone to evaporation. This is likely to be a common problem, and handling small quantities of volatile compounds can present a significant challenge. After eliminating these missing compounds from the library the remainder had an overall QC failure rate of 17%, which is in line with previously reported values. The reason for QC failures was lack of purity (2.6%), inconsistent structure (1.7%), and insolubility or lack of sample (13%). These values are in line with the rates expected with solubility also being noted as the main reason for failure (Davis & Erlanson, 2013; Doak et al., 2017; Keseru et al., 2016).

## 9. SPR REFERENCE SURFACE BINDING

In addition to checking compound identity, solubility, and purity a number of technique-specific checks may also be required. Where library screening is to be performed by SPR, it is necessary to check for any nonspecific binding of the fragments to unlabeled or reference labeled SPR surfaces (Chavanieu & Pugniere, 2016; Shepherd, Hopkins, & Navratilova, 2014).

**Table 1** Surface Plasmon Resonance QC Failure by Surface. 1218 Compounds Were Screened Against Each Surface

| Surface Type | Functionalization | Number of Fails |
|---|---|---|
| Carboxymethyl dextran | Blank | 2 |
| Carboxymethyl dextran | Streptavidin | 2 |
| Carboxymethyl dextran | Streptavidin and biotin | 42 |
| NTA dextran | Blank | 6 |
| NTA dextran | $Ni^{2+}$ | 16 |
| Total unique compounds | | 68 (5.6% of total) |

Fragments are therefore tested for binding to a range of different surfaces with varying functionalizations typical of standard SPR-screening methods. For example, our SPR QC involved analysis of binding to SPR sensors with two different surface chemistries, carboxy-methylated dextran and nitrilotriacetic acid (NTA). Fragments which had passed NMR QC were tested against carboxymethylated dextran blank surface, a surface to which streptavidin had been coupled using amide cross-linking and streptavidin-coupled surface with biotin bound. For the NTA sensor the fragments were tested against a blank and a $Ni^{2+}$-loaded surface (Table 1). Samples were analyzed at a concentration of 990 μ*M* on Biacore S200 or SensiQ Pioneer FE instruments using an injection of 25 μL at 100 μL/min and a gradient injection of 50% loop volume (25 μL) at 200 μL/min, respectively, with both using a dissociation time of 20 s. None of the fragments showed binding to multiple surfaces, indicating compounds passing the above NMR QC should generally also be suitable for SPR screening. However, 5% of fragments showed residual binding to one surface.

## 10. MIXTURE DESIGN

Depending on the screening method and size of fragment library it may be required to generate mixtures/cocktails of fragments to increase screening efficiency. This is particularly used in ligand-detected NMR where fragments having nonoverlapping resonances in their 1D $^1H$ NMR spectra can be screened in mixtures and deconvoluted by comparison to reference spectra of the individual fragments (Lepre, 2001). Mixtures of fragments can also be screened using X-ray crystallography where differently shaped fragments can be mixed that lead to distinctly different electron densities in cases where binding is observed. Mixtures can reduce protein consumption and data collection time but impose an additional requirement to design mixtures that are fit for purpose and introduce the potential for mixture effects. To some extent both of these potential problems can be minimized by careful design and curation of the mixtures. For ligand-detected NMR screening, there are reported procedures for designing mixtures that minimize spectral overlap in 1D $^1$H NMR (Arroyo, Goldflam, Feliz, Belda, & Giralt, 2013; Stark, Eghbalnia, Lee, Westler, & Markley, 2016), and mixture design is incorporated into at least one commercial NMR software (MestreLab, 2018). These methods involve analysis of the 1D $^1$H NMR spectra of the library compounds to design mixtures which minimize spectral overlap. This process is simplified by having a lower number of compounds in each mixture, but fewer fragments in the mixture reduces the gain in efficiency during screening (Arroyo et al., 2013; Lepre, 2011; Stark et al., 2016). In our library we treated "aliphatic only compounds"—which were designated as those with no nonexchangeable signals in the NMR spectrum downfield of the water resonance frequency as a separate subset. Aliphatic mixtures contained up to three fragments, whereas nonaliphatic mixtures contained up to five fragments due to spectral complexity in the aliphatic region commonly being higher. The designed mixtures were inspected manually to ensure little to no spectral overlap. The mixtures were then subjected to a similar QC analysis as the individual fragments. 1D $^1$H NMR and Water-LOGSY spectra were acquired for each mixture and compared to the original reference spectra to ensure that there were no significant changes in the $^1$H NMR chemical shifts of resonances for any fragments in the mixture relative to the spectra of the single compounds. Similarly Water-LOGSY spectra were acquired to ensure that there was no evidence of aggregation in the mixtures. As a result of this analysis, we selected a final fragment concentration of 300 μ*M* for each fragment in the screening mixtures. This was found to minimize any mixture effects as assessed by the conservation of chemical shifts for the fragments and evidence of negative signals in the Water-LOGSY spectra.

## 11. CONCLUSION

Building a fragment library, measuring properties and conducting QC experiments can be time-consuming and difficult; however, it is vitally important for finding good starting points for FBDD. The process outlined earlier gives practical details and highlights some of the important aspects of designing fit-for-purpose fragment libraries for ligand-detected NMR and SPR screening. Initial chemoinformatic design, diversity selection, and library property selection are undertaken with the screening techniques to be used and follow-up of hits in mind. The specific balance of diversity and properties (size, polarity, complexity, etc.) can be tailored to generate a desired profile in the fragment-screening collection that is deemed to provide optimal starting points for medicinal chemistry development. Ensuring compound quality by measurement of properties critical to screening such as solubility, aggregation, and nonspecific binding help reduce the likelihood of false positives in primary and validation screens. The ongoing evolution in fragment library design fuelled by new and more efficient screening techniques along with the desire to incorporate novelty, diversity, and "developability" criteria into fragments (Cox et al., 2016; Palmer, Peakman, Norton, & Rees, 2016) will most likely continue to shape general guidelines (Keseru et al., 2016) and methods used for fragment library design in years to come.

## REFERENCES

Allu, T. K., & Oprea, T. I. (2005). Rapid evaluation of synthetic and molecular complexity for in silico chemistry. *Journal of Chemical Information and Modeling*, *45*(5), 1237–1243.

Arroyo, X., Goldflam, M., Feliz, M., Belda, I., & Giralt, E. (2013). Computer-aided design of fragment mixtures for NMR-based screening. *PLoS One*, *8*(3), e58571.

Benigni, R., & Bossa, C. (2011). Mechanisms of chemical carcinogenicity and mutagenicity: A review with implications for predictive toxicology. *Chemical Reviews*, *111*(4), 2507–2536.

Berthold, M. R., Cebron, N., Dill, F., Gabriel, T. R., Kötter, T., Meinl, T., et al. (2008). KNIME: The Konstanz Information Miner. In: *Data Analysis, Machine Learning and Applications* (pp. 319–326). Berlin, Heidelberg: Springer Berlin Heidelberg.

Chavanieu, A., & Pugniere, M. (2016). Developments in SPR fragment screening. *Expert Opinion on Drug Discovery*, *11*(5), 489–499.

Congreve, M., Carr, R., Murray, C., & Jhoti, H. (2003). A 'rule of three' for fragment-based lead discovery? *Drug Discovery Today*, *8*(19), 876–877.

Cox, O. B., Krojer, T., Collins, P., Monteiro, O., Talon, R., Bradley, A., et al. (2016). A poised fragment library enables rapid synthetic expansion yielding the first reported inhibitors of PHIP(2), an atypical bromodomain. *Chemical Science*, 7(3), 2322–2330.

Davis, B. J., & Erlanson, D. A. (2013). Learning from our mistakes: The 'unknown knowns' in fragment screening. *Bioorganic & Medicinal Chemistry Letters*, *23*(10), 2844–2852.

Doak, B. C., Morton, C. J., Simpson, J. S., & Scanlon, M. J. (2017). Assembly of fragment screening libraries. In *Applied biophysics for drug discovery* (pp. 263–283). Chichester, UK: John Wiley & Sons, Ltd.

Hall, R. J., Mortenson, P. N., & Murray, C. W. (2014). Efficient exploration of chemical space by fragment-based screening. *Progress in Biophysics and Molecular Biology*, *116*(2–3), 82–91.

Hann, M. M., Leach, A. R., & Harper, G. (2001). Molecular complexity and its impact on the probability of finding leads for drug discovery. *Journal of Chemical Information and Computer Sciences*, *41*(3), 856–864.

Irwin, J. J., Duan, D., Torosyan, H., Doak, A. K., Ziebart, K. T., Sterling, T., et al. (2015). An aggregation advisor for ligand discovery. *Journal of Medicinal Chemistry*, *58*(17), 7076–7087.

Jhoti, H., Williams, G., Rees, D. C., & Murray, C. W. (2013). The 'rule of three' for fragment-based drug discovery: Where are we now? *Nature Reviews. Drug Discovery*, *12*(8), 644–645.

Keseru, G. M., Erlanson, D. A., Ferenczy, G. G., Hann, M. M., Murray, C. W., & Pickett, S. D. (2016). Design principles for fragment libraries: Maximizing the value of learnings from pharma fragment-based drug discovery (FBDD) programs for use in academia. *Journal of Medicinal Chemistry*, *59*(18), 8189–8206.

Kozikowski, B. A., Burt, T. M., Tirey, D. A., Williams, L. E., Kuzmak, B. R., Stanton, D. T., et al. (2003). The effect of freeze/thaw cycles on the stability of compounds in DMSO. *Journal of Biomolecular Screening*, *8*(2), 210–215.

Lagorce, D., Sperandio, O., Baell, J. B., Miteva, M. A., & Villoutreix, B. O. (2015). FAF-Drugs3: A web server for compound property calculation and chemical library design. *Nucleic Acids Research*, *43*(W1), W200–W207.

Landrum, G. (2018). *RDKit: Open-Source Cheminformatics*.

Lau, W. F., Withka, J. M., Hepworth, D., Magee, T. V., Du, Y. J., Bakken, G. A., et al. (2011). Design of a multi-purpose fragment screening library using molecular complexity and orthogonal diversity metrics. *Journal of Computer-Aided Molecular Design*, *25*(7), 621–636.

Lepre, C. A. (2001). Library design for NMR-based screening. *Drug Discovery Today*, *6*(3), 133–140.

Lepre, C. A. (2011). Practical aspects of NMR-based fragment screening. *Methods in Enzymology*, *493*, 219–239.

Mannhold, R., Poda, G. I., Ostermann, C., & Tetko, I. V. (2009). Calculation of molecular lipophilicity: State-of-the-art and comparison of log P methods on more than 96,000 compounds. *Journal of Pharmaceutical Sciences*, *98*(3), 861–893.

Mark, A., John, B., Florence, C., Michael, C., Geoffrey, D., Dominique, G., et al. (2002). Identification of diverse database subsets using property-based and fragment-based molecular descriptions. *Quantitative Structure-Activity Relationships*, *21*(6), 598–604.

Meanwell, N. A. (2011). Improving drug candidates by design: A focus on physicochemical properties as a means of improving compound disposition and safety. *Chemical Research in Toxicology*, *24*(9), 1420–1456.

MestreLab. (2018). *MNova, 12.0.1*.

Morgan, H. L. (1965). The generation of a unique machine description for chemical structures—A technique developed at chemical abstracts service. *Journal of Chemical Documentation*, *5*(2), 107–113.

Na, J., & Hu, Q. (2011). Design of screening collections for successful fragment-based lead discovery. In J. Z. Zhou (Ed.), *Chemical library design* (pp. 219–240). Totowa, NJ: Humana Press.

Palmer, N., Peakman, T. M., Norton, D., & Rees, D. C. (2016). Design and synthesis of dihydroisoquinolones for fragment-based drug discovery (FBDD). *Organic & Biomolecular Chemistry*, *14*(5), 1599–1610.

Prochazkova, E., Jansa, P., Brezinova, A., Cechova, L., Mertlikova-Kaiserova, H., Holy, A., et al. (2012). Compound instability in dimethyl sulphoxide, case studies with 5-aminopyrimidines and the implications for compound storage and screening. *Bioorganic & Medicinal Chemistry Letters*, *22*(20), 6405–6409.

Ray, P. C., Kiczun, M., Huggett, M., Lim, A., Prati, F., Gilbert, I. H., et al. (2017). Fragment library design, synthesis and expansion: Nurturing a synthesis and training platform. *Drug Discovery Today*, *22*(1), 43–56.

Reynolds, C. H., Bembenek, S. D., & Tounge, B. A. (2007). The role of molecular size in ligand efficiency. *Bioorganic & Medicinal Chemistry Letters*, *17*(15), 4258–4261.
Rogers, D., & Hahn, M. (2010). Extended-connectivity fingerprints. *Journal of Chemical Information and Modeling*, *50*(5), 742–754.
Sauer, W. H., & Schwarz, M. K. (2003). Molecular shape diversity of combinatorial libraries: A prerequisite for broad bioactivity. *Journal of Chemical Information and Computer Sciences*, *43*(3), 987–1003.
Schulz, M. N., Landstrom, J., Bright, K., & Hubbard, R. E. (2011). Design of a fragment library that maximally represents available chemical space. *Journal of Computer-Aided Molecular Design*, *25*(7), 611–620.
Shepherd, C. A., Hopkins, A. L., & Navratilova, I. (2014). Fragment screening by SPR and advanced application to GPCRs. *Progress in Biophysics and Molecular Biology*, *116*(2–3), 113–123.
Stark, J. L., Eghbalnia, H. R., Lee, W., Westler, W. M., & Markley, J. L. (2016). NMRmix: A tool for the optimization of compound mixtures in 1D (1)H NMR ligand affinity screens. *Journal of Proteome Research*, *15*(4), 1360–1368.
Sushko, I., Salmina, E., Potemkin, V. A., Poda, G., & Tetko, I. V. (2012). ToxAlerts: A web server of structural alerts for toxic chemicals and compounds with potential adverse reactions. *Journal of Chemical Information and Modeling*, *52*(8), 2310–2316.
Tanaseichuk, O., Khodabakshi, A. H., Petrov, D., Che, J., Jiang, T., Zhou, B., et al. (2015). An efficient hierarchical clustering algorithm for large datasets. *Austin Journal of Proteomics, Bioinformatics & Genomics*, *2*(1), 1008.
Taylor, A., Doak, B. C., & Scanlon, M. J. (2018). *Fragment library design KNIME workflow*. Figshare. https://figshare.com/articles/Fragment_library_design_KNIME_workflow/6948437. https://doi.org/10.26180/5b6bbe778b3f9.
Tounge, B. A., & Parker, M. H. (2011). Designing a diverse high-quality library for crystallography-based FBDD screening. *Methods in Enzymology*, *493*, 3–20.
Visini, R., Awale, M., & Reymond, J. L. (2017). Fragment database FDB-17. *Journal of Chemical Information and Modeling*, *57*(4), 700–709.
Willighagen, E. L., Mayfield, J. W., Alvarsson, J., Berg, A., Carlsson, L., Jeliazkova, N., et al. (2017). The Chemistry Development Kit (CDK) v2.0: Atom typing, depiction, molecular formulas, and substructure searching. *Journal of Cheminformatics*, *9*(1), 33.

CHAPTER SIX

# Genetically Encoded Cyclic Peptide Libraries: From Hit to Lead and Beyond

**Jacob Valentine, Ali Tavassoli[1]**
School of Chemistry, University of Southampton, Southampton, United Kingdom
[1]Corresponding author: e-mail address: a.tavassoli@soton.ac.uk

## Contents



## Abstract

With the increasing utilization of high-throughput screening for lead identification in drug discovery, the need for easily constructed and diverse libraries which cover significant chemical space is greater than ever. Cyclic peptides address this need; they combine the advantageous properties of peptides (ease of production, high diversity, high potential specificity) with increased resistance to proteolysis and often increased biological activity (due to conformational locking). There are a number of techniques for the generation and screening of cyclic peptide libraries. As drug discovery moves toward tackling challenging targets, such as protein–protein interactions, cyclic peptide libraries are expected to continue producing hits where small molecule libraries may be stymied. However, it is important to design robust systems for the generation and screening of these large libraries, and to be able to make sense of structure–activity relationships in these highly variable scaffolds. There are a plethora of possible modifications that can be made to cyclic peptides, which is both a weakness and a strength of these scaffolds; high variability will allow more precise tuning of leads to targets, but exploring the whole range of modifications may become an overwhelming challenge.

*Methods in Enzymology*, Volume 610
ISSN 0076-6879
https://doi.org/10.1016/bs.mie.2018.09.020

## 1. INTRODUCTION

There has been a significant resurgence in the development of peptide drugs (Lau & Dunn, 2018) with 28 new peptide drugs making it through the clinical approval process since 2000 (Henninot, Collins, & Nuss, 2017). Despite this, the long standing objections to peptide drugs such as poor cell permeability and rapid clearance and proteolysis still persists (Gupta et al., 2013). The utilization of cyclic peptides can help to overcome these challenges; they show superior resistance to proteolysis (Baeriswyl & Heinis, 2012) and the conformational locking of their structures decreases entropic changes in binding (Joo, 2012), which is thought to lead to improved activity and increased specificity for the target. It should be noted that compounds of this class that have progressed furthest toward the clinic are derived from natural products, most commonly from fungi (e.g., cyclosporine from *Tolypocladium inflatum*), plants (e.g., SFTI-1 from *Helianthus annuus*) or bacteria (Gramicidin S from *Brevibacillus brevis*).

There is therefore significant promise in high-throughput screening (HTS) techniques that allow the rapid preparation, screening, and deconvolution of cyclic peptide libraries. The issue of deconvolution is of particular importance; while libraries of linear peptides can be readily characterized by Edman degradation sequencing, the cyclic peptide has no free N-terminus for Edman reagent to hydrolyze. While it is possible, with suitable tools, to determine cyclic peptide sequences from tandem mass spectrometry, this is far from trivial. Due to the possibility of multiple fragmentation patterns and sequence scrambling recyclizations within the instrument, it would be highly challenging to use MS/MS to deconvolute and assign structures to the volume of hits that a HTS system may provide (Liu et al., 2009). Thus, the most common methods for generating and screening large cyclic peptide libraries include in-built codes (e.g., DNA) to enable simple deconvolution.

Libraries may be interfaced with a variety of screens, depending on the technology used to generate them (either intracellular or in vitro) and the nature of the target. Cyclic peptide libraries generated in cells have been combined with reverse two-hybrid systems (RTHS) for the identification of a variety of protein–protein interaction inhibitors (Lennard & Tavassoli, 2014; Tavassoli, 2017) Intracellular libraries have also been combined with relevant assays to identify inhibitors of protein–DNA interactions (Joung, Ramm, & Pabo, 2000) or protein–RNA interactions (Hook, 2005). One strength of such libraries is that they may be used in any cell-based assay;

for instance, a screen for HIV protease inhibitors, whereby an HIV protease recognition site was used to link cell survival to inhibition of this enzyme by a cyclic peptide inhibitor (Young et al., 2011). In contrast, cyclic peptide libraries generated in vitro (synthetic libraries, and those generate by mRNA display or phage display) are screened using affinity based assays that identify members of the library that bind most tightly to the target protein (Goto, Katoh, & Suga, 2011). Given the above differences between these techniques for generating cyclic peptide libraries, one may reasonably assume that each may be optimally deployed against a subset of target proteins (e.g., intracellular proteins, extracellular protein, protein–protein interactions). Given the potential breadth of targets and application, we have opted to provide a set of case studies as a guide, in place of a recipe for a single method. This review will discuss these techniques, focusing on two genetically encoded methods for the generation of cyclic peptide libraries. We go on to discuss potential paths for taking cyclic peptide hits to leads and beyond.

## 2. SICLOPPS

Inteins (internal proteins) are naturally occurring protein domains, capable of spontaneously excising themselves from the surrounding extein (external protein) sequence after translation. These protein domains have been found across the three major domains of life (eukaryotes, bacteria, and archaea) (Shah & Muir, 2014) and even in viral proteins (Pietrokovski, 1998). Inteins can be described as either *cis*-splicing or *trans*-splicing; this depends on whether the intein is a single contiguous protein domain (*cis*-splicing inteins) or two fragments which come together to excise (*trans*-splicing or split-inteins). While *cis*-splicing inteins are the more common naturally occurring inteins, it is *trans*-splicing split-inteins which are utilized by a technique named split-intein circular ligation of peptides and proteins (SICLOPPS) for the generation of cyclic peptide libraries (Scott, Abel-Santos, Wall, Wahnon, & Benkovic, 1999).

In SICLOPPS, DNA encoding the peptide sequence to be cyclized is flanked with the N-terminal ($I_N$) and C-terminal ($I_C$) inteins. The inteins used are most frequently from *Synechocystis* species (Tavassoli & Benkovic, 2007), *Nostoc punctiforme* (Townend & Tavassoli, 2016), or derived from naturally occurring inteins by protein engineering (Stevens et al., 2017).

Transforming or transfecting a SICLOPPS plasmid in a cell results in the production of split-inteins that process to give the cyclic peptide/protein encoded by the extein region (Fig. 1). A library may be incorporated into

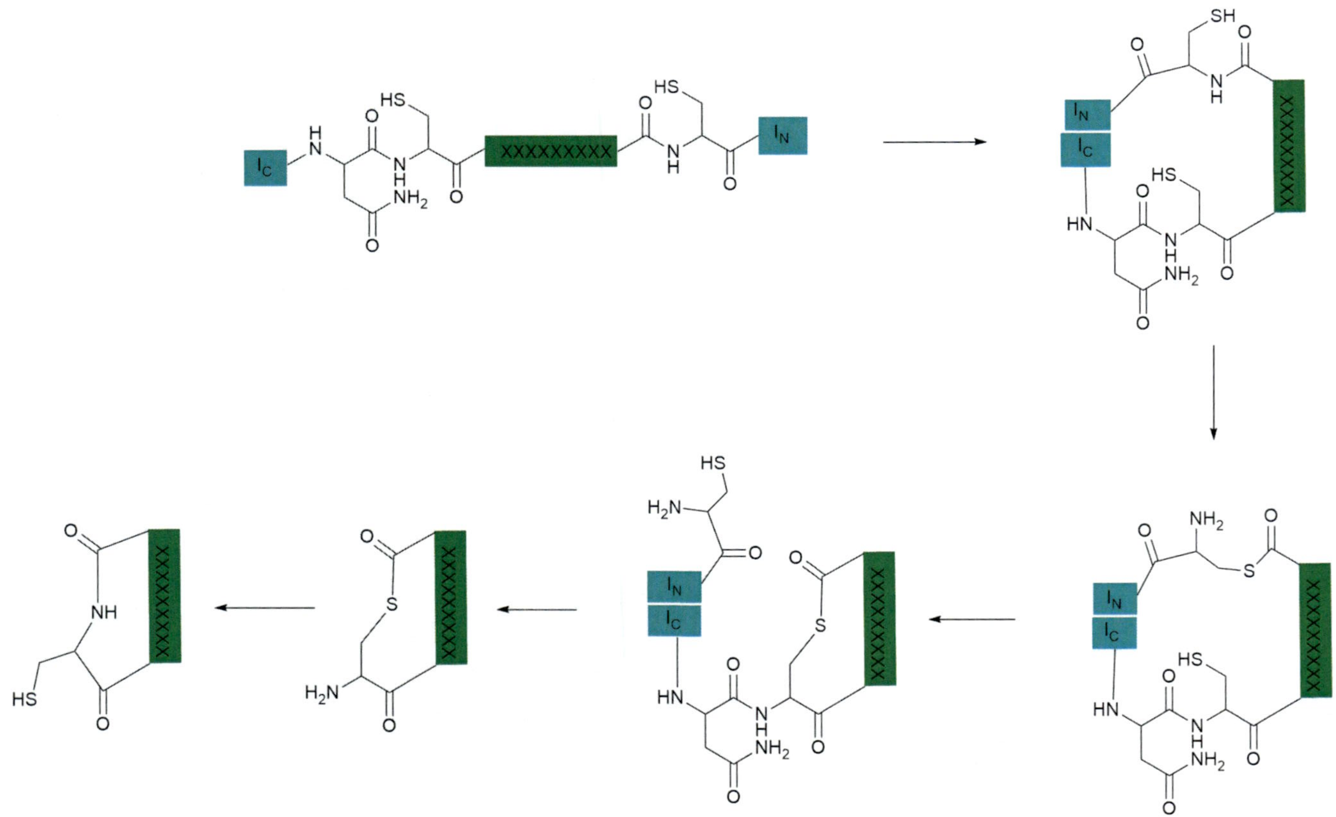

**Fig. 1** SICLOPPS mechanism.

the extein region using standard molecular biology techniques (Tavassoli & Benkovic, 2007), making SICLOPPS a readily accessible method for library generation. Given that these libraries are generated in cells, SICLOPPS may be interfaced with any cell-based assay or screen, for the identification of an active molecule against the given target. The deconvolution of hits from a SICLOPPS library is straightforward due to the colocalization of the genetic material coding the random peptide in the same cell. The plasmid encoding active molecules can be recovered and sequenced relatively easily, eliminating the need for the more complicated deconvolution techniques required to directly identify cyclic peptides once expressed.

While the size of the cyclic peptides produced by SICLOPPS is not inherently limited by the split-intein ligation technique (molecules as large as cyclic-GFP have been produced by SICLOPPS) (Townend & Tavassoli, 2016), library size cannot be indefinitely expanded. The predominant limiting factor in determining maximum library size is the transformation efficiency of the target cell type; in *Escherichia coli* this is typically up to $\sim 10^9$ members (Dower, Miller, & Ragsdale, 1988) which corresponds to a cyclic peptide with 7 fully randomized positions ($20^7 = 1.28 \times 10^9$ members). As a result, many SICLOPPS screens utilize libraries of cyclic peptides with 5 or 6 random positions ($3.2 \times 10^6$ or $6.4 \times 10^7$ members, respectively) to allow coverage of all library members rather than only partially covering larger libraries. It is possible, using limited codon sets, to prepare libraries of larger peptides which, rather than being fully randomized across all 20 proteogenic amino acids, cover only a representative subset of all amino acids. However, there is mounting evidence that smaller cyclic peptides are more readily converted to small molecules than their larger counterparts (Spurr et al., 2012). The only requirement for SICLOPPS is a cysteine, serine, or threonine as the first amino acid of the extein, which is required for cyclization to proceed. This is due to the need for a nucleophilic side chain which can attack a thioester formed in the primary step of cyclization.

The splicing efficiency of all inteins utilized thus far shows a level of bias toward certain residues at the second cyclic position; specifically, this bias leans toward phenylalanine and to a lesser extent other bulky hydrophobic residues (Cheriyan, Pedamallu, Tori, & Perler, 2013). This bias can be somewhat overcome by utilizing more promiscuous inteins. The first example of this was the use of Npu inteins from *N. punctiforme* in SICLOPPS instead of Ssp inteins from *Synechocystis* species (Townend & Tavassoli, 2016). While the primary drive of this work was the increased rate of *trans*-splicing, Npu inteins also showed an improved tolerance to extein sequence. This tolerance has been further improved by mutating the wild-type ERD sequence at positions 122–124 to a GEP motif. While this mutation does come at a price to the splicing rate of Phe and other normally favored residues at +2, significant increases (fivefold or better) in splicing activity were seen for Ala, Gly, Arg, Asp, Thr, Cys, Val, Glu, and Pro (Stevens et al., 2017). Further improvement in intein splicing rate and decreased intein bias have recently been achieved by consensus sequence generation from a list of currently utilized inteins; the resulting intein, named Cfa, showed rapid *trans*-splicing and maintained a high level of promiscuity across a number of extein sequences. Further change to the GEP122–124 loop mutation in Cfa showed similar improvements in splicing promiscuity, and the result is the current state of the art in intein engineering when it comes to *trans*-splicing inteins (Stevens et al., 2016).

However, these improvements in splicing rates and extein composition tolerance do not come without drawbacks. In particular, Npu inteins have shown significant levels of toxicity to *E. coli* which can severely impact the viability of colonies. This could eliminate otherwise viable hits from the screen. This toxicity appears to be directly mediated by the resulting spliced intein; as a result, there has been success in mitigating this effect by the addition of protein degradation tags (SsrA) to the C-terminal of the C-terminal intein. This targets the spliced intein to the ClpXA degradation mechanism, decreasing exposure to the cell and reducing toxicity as a result (Townend & Tavassoli, 2016). While SICLOPPS is limited to using the 20 canonical amino acids, it has been combined with genetic code expansion to allow libraries containing an additional, unnatural amino acid (Young et al., 2011).

SICLOPPS libraries are frequently coupled to a reverse two-hybrid system (RTHS) to detect disruptors of protein–protein interactions (Birts et al., 2013; Leitch et al., 2018; Male et al., 2017; Miranda et al., 2013; Osher et al., 2018; Yang et al., 2018). Using this approach requires the generation of a bespoke RTHS for the targeted PPI to link survival of the host cell to disruption of the target PPI (Fig. 2—RTHS), which requires the integration of two separate genetic cassettes onto the chromosome of an *E. coli* host (Horswill, Savinov, & Benkovic, 2004; Tavassoli & Benkovic, 2005; Tavassoli et al., 2008). The first encodes two proteins, each of which consists of one of the two protein domains of interest fused to one of two heterodimeric partners which form the inhibitor; the second encodes reporter genes under the control of a promoter which is repressed by the heterodimeric repressor. Upon the association of the two protein fragments of interest, the two heterodimeric partners form a functional repressor and shut off the transcription of a series of downstream genes (Di Lallo, Paolozzi, Castagnoli, & Ghelardini, 2001). The reporter genes are

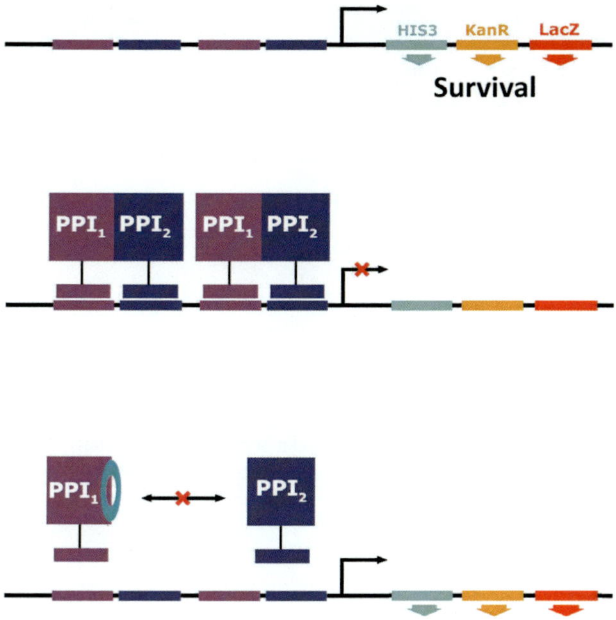


**Fig. 2** Reverse two-hybrid system.

selected so that they are required for survival to enable life/death screening; these are HIS3, which allows for the direct synthesis of histidine on otherwise histidine deficient media; KanR conferring kanamycin resistance, and LacZ, as the third gene to allow assessment of the PPI via ONPG assays. More recently fluorescent proteins (mCherry, tdTomato) have been incorporated as the third reporter gene to allow dose–response reporting of inhibition in initial screening and in target validation (Nordgren & Tavassoli, 2014).

After the transformation of SICLOPPS plasmids into the target RTHS, transformed cells are spread onto large plates of selective media containing arabinose, IPTG, 3-AT, and suitable antibiotics. After incubation for 2–3 days, surviving colonies are picked and assessed for activity by drop-spotting 10-fold serial dilutions of an overnight stock onto selective plates. The most potent peptides allow the furthest growth in this assay, thus allowing the activity of each hit to be easily validated and ranked against other hits (Fig. 3). This functional screening approach has been used to identify a number of first in class PPI inhibitors (Birts et al., 2013; Leitch et al., 2018; Male et al., 2017; Miranda et al., 2013; Osher et al., 2018) and is essentially a plug-and-play method with a proven track record of identifying inhibitors of PPIs considered "undruggable." It should be noted that SICLOPPS

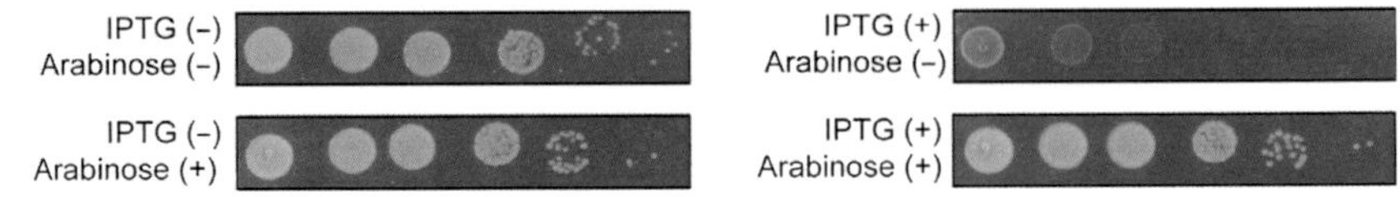


**Fig. 3** Drop spotting.

libraries have been successfully combined with screens in a variety of organisms including yeast (Kritzer et al., 2009) and mammalian cells (Kinsella et al., 2002; Mistry & Tavassoli, 2017).

## 3. mRNA DISPLAY

The two main genetically encoded approaches for affinity-based screening of cyclic peptide libraries are phage display and mRNA display. Recent advances in phage display have driven this technology toward production and screening of bicyclic peptides (Baeriswyl & Heinis, 2012; Rentero Rebollo & Heinis, 2013); we will focus on mRNA display, which is capable of producing large libraries of cyclic peptides in vitro. This links a peptide (synthesized by in vitro translation) to its parent mRNA via a puromycin molecule attached to the 3′ end of this mRNA. The peptide is translated from the mRNA, but the puromycin (imitating tyrosine tRNA) is taken into the active site of the ribosome and covalently links the peptide to the mRNA (Roberts & Szostak, 1997). A number of techniques have been applied for cyclizing the peptide, to furnish a library of cyclic peptides covalently attached to their respective mRNA (Bashiruddin & Suga, 2015). This flexibility of cyclization technique allows for a greater chemical space to be covered, with each technique presenting its own advantages and difficulties. For instance, the in vitro translation allows for the incorporation of alkynyl- and azido amino acids into the linear peptide. These residues are suited to click ligation, resulting in a linking 1,2,3-triazole between the two side chains. While this linkage is a good bioisostere for the amide linkage with restricted conformational freedom, it lacks the ability to function as a hydrogen bond donor; this may improve cell permeability, but also may be a missed opportunity for increased binding interaction. Additionally, triazoles are known to chelate metals, which may limit the utility of these molecules in cells.

Cysteine-mediated cyclizations have been widely deployed for this chemical cyclization step, with disulfide linkage formation, Michael addition to dehydroalanine (Gieselman, Zhu, Zhou, Galonic, & van der Donk, 2002), chloroacetyl thioether formation (Iwasaki, Goto, Katoh, & Suga, 2012), and dibromoxylene thioalkylation (Guillen Schlippe, Hartman, Josephson, & Szostak, 2012). However, sulfur-based linkage does not tend to be redox stable; disulfide linkages are readily reduced in the cytoplasm, limiting their targeting to extracellular targets. An internal lysine can also be coupled to the N-terminal amine via a bis-NHS glutarate; the result is a linker which relies only on amide bonds and thus shares similar properties and stability as the rest of the peptide backbone (Millward, Takahashi, & Roberts, 2005). However, this linker extends the backbone by ~4 amino acid equivalents, causing increased backbone flexibility, potentially negating one of the key advantages of cyclic peptides, the resulting loss of rigidity may compromise both activity and specificity in the resulting peptides (Josephson, Ricardo, & Szostak, 2014). It should be noted that head–tail cyclization with an amide bond (achieved by SICLOPPS) is not compatible with mRNA display due to the presence of the puromycin molecule. However, head-to-tail analogues of cyclic peptides identified by mRNA display may be chemically synthesized postselection.

A main advantage of mRNA display is that it allows for the production of cyclic peptide libraries containing multiple noncanonical amino acids. This relies on the precharging of tRNA with noncanonical amino acids and can be performed chemoenzymatically, by coupling a tRNA lacking its final two nucleotides to a dinucleotide conjugated amino acid of choice (Forster, Weissbach, & Blacklow, 2001), or fully enzymatically utilizing flexizymes; tRNA synthetases with no capacity for side chain recognition (Goto et al., 2011). Additionally, several techniques for the chemical modification of presynthesized tRNA–amino acid conjugates exist, though these are mostly employed to prepare *N*-methyl amino acids. Other display technologies use other means of anchoring a nucleic acid transcript to proteins; phage and phagemid display utilizes proteins which fuse the viral coat protein to the randomized protein sequence (Qi, Lu, Qiu, Petrenko, & Liu, 2012); ribosomal display utilizes isolated ribosomes to translate a library of mRNA transcripts which have a spacer at the 3′-terminal which lacks a stop codon, leading to a stalled ribosome which links the peptide and mRNA transcript (Zahnd, Amstutz, & Plückthun, 2007).

After preparation of an mRNA-peptide library by in vitro translation and cyclization via a suitable method, the next step toward screening is producing a cDNA–mRNA dimer from the attached mRNA. This helps to minimize mRNA interference in the screening by increasing the rigidity of the nucleic acid portion of the conjugate and allows deconvolution of hits by PCR amplification (Wang & Liu, 2011). The mRNA display cyclic peptide library is screened by incubating with the immobilized target proteins (typically biotinylated and immobilized on streptavidin beads) and washed to remove all nonbinding constructs present. The cDNA can then be dissociated from the mRNA, amplified by PCR, and sequenced to identify binding compounds. Several rounds of selection are typically conducted to narrow the library down to only the most promising targets (Valencia, Cotten, Dong, & Liu, 2008). This entails reverse transcribing the cDNA from the PCR back to an mRNA, so that puromycination, translation and rescreening can occur. This technique has been very successful at identifying potent binders of a wide number of proteins. (Bashiruddin & Suga, 2015).

## 4. THE PATH FROM PEPTIDE HIT TO THE CLINIC

There are currently a number of well-established methods for identifying cyclic peptide hits against the most challenging targets, as discussed above. The major challenge currently facing the field is the translation of these hits to leads and into the clinic. The path to the clinic for cyclic peptides identified from SICLOPPS libraries is likely to differ from those identified using display technologies. We tend to limit the ring size of our SICLOPPS cyclic peptides to six amino acids, whereas display technologies tend to use 15 amino acids or larger cyclic peptides. Therefore while we see a continuous motif of 2 or 3 amino acids being responsible for target engagement (unpublished data and Spurr et al., 2012), hits from mRNA display tend to fully engage their target. This results in significantly higher potency of hits from mRNA display, however, conversely makes their conversion to a small molecule extremely challenging. In contrast, small molecule inhibitors may be readily derived from SICLOPPS hits, with these compounds showing activity in vivo (Asby et al., 2015). Thus, any therapeutics derived from hits identified from mRNA display or phage display libraries are likely to retain their cyclic peptide backbone, this the challenge with progressing these molecules is using synthetic chemistry to generate analogues with appropriate pharmacokinetic and pharmacodynamic properties.

Generally speaking, a set series of optimization techniques may be utilized to improve a cyclic peptide hit's affinity and identify the residues of primary interest. Foremost is alanine scanning, the serial substitution of sequential residues with alanine. The sterically minimal and mildly hydrophobic residue is a good stand in for the bare peptide backbone. It may initially seem as though a superior choice would be glycine, but the lack of any steric bulk or chirality to lock the correct conformation may lead to changes in affinity not directly related to the loss of a particular residue (Cunningham & Wells, 1989). These changes may instead be related to increased flexibility, necessitating increased changes in entropy upon binding due to an increased need for ligand organization. By performing an alanine scan, residues can be ranked in terms of importance to binding affinity, and motifs of two or more important residues can then be isolated and optimized themselves as a move toward a more traditional small molecule inhibitor. An example is Compound 14, an inhibitor of aminoimidazole carboxamide ribonucleotide transformylase (AICAR Tfase) homodimerization, developed from cycloCRYFNV, identified by SICLOPPS screening (Spurr et al., 2012). Alanine scanning identified two adjacent residues (arginine and tyrosine) as vitally important for activity. This motif was excised from the cyclic peptide scaffold and capped. The resulting dipeptide was active against the target, and the activity was further improved by incorporation of noncanonical amino acids (Fig. 4). The ability to cut down the molecular weight of the lead to less than half the starting peptide was a significant step toward finding a small molecule inhibitor and demonstrates our earlier point about converting SICLOPPS-derived cyclic peptides to small molecules.

There is further scope for improving the activity of this peptide by altering the N- and C-terminal caps. Terminal capping may lead to a peptide which more closely mimics eukaryotic proteins; approximately 85% of human proteins are N-terminal acetylated (Arnesen et al., 2009), and over 50% of peptide hormones are C-terminal amidated (Kim & Seong, 2001).

In addition to the elimination of charges, and the subsequent improvements in biophysical properties, terminal modifications may also confer additional resistance to exopeptidases which would otherwise recognize the free terminal. This means a longer biological half-life and possible greater efficacy in vivo even in cases where capping might reduce affinity for the target in vitro (Brinckerhoff et al., 1999). In addition, backbone N-alkylation can help to reduce free rotation of the backbone and return some rigidity to the molecule. However, this does not guarantee the locked conformation will be the active conformation, and so in some cases this modification will decrease activity at the target. N-alkylation has a secondary purpose, in decreasing the number of hydrogen bond donors; this decreases polarity and improves membrane permeability, and many naturally occurring cyclic peptides contain an N-alkylated peptide backbone. For instance, cyclosporin forms intramolecular hydrogen bonds to present a more hydrophobic, *N*-methylated backbone toward the solvent and constrain nonmethylated amide nitrogens to an internal cavity. This solvent dependent conformation

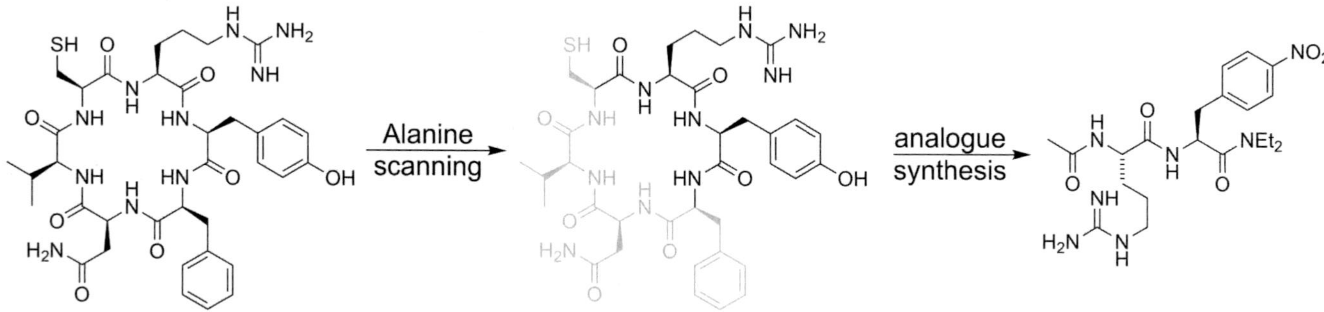


**Fig. 4** Deriving compound 14 from its parent cyclic peptide.

and hydrophilicity change allows cyclosporin to more easily cross cell membranes without active transport (El Tayar et al., 1993).

Replacement with nonproteinogenic amino acids is another common approach for improving affinity and biostability. These may be of benefit in conformational investigation and locking of side chains (replacing phenylalanine with *E*/*Z*-dehydrophenylalanine), homologation of side chains (Phe for phenylglycine or homo-Phe), replacement of unstable moieties with more stable but functionally similar ones or the incorporation of residues which can mimic standard amino acids but shield polarity through intramolecular hydrogen bonding; e.g., 2-pyridylalanine is structurally similar to phenylalanine but forms a N–H–N hydrogen bond to a backbone amide (Rafi, Hearn, Vedantham, Jacobson, & Renslo, 2012).

L-Amino acids may also be replaced with their D-amino acid counterparts to improve activity and/or confer increased stability to enzymatic degradation (Mathur et al., 2016). However, the change in chirality upon incorporation of D-amino acids into cyclic peptides can alter backbone conformation and may reduce binding affinity.

Finally, one might consider fundamental changes to the backbone itself; for instance, the replacement of one or more amide bonds for esters to form depsipeptides, removing an H-bond donor; movement of the side chain residues from the α-carbon to the amide N to form a peptoid, again reducing the number of H-bond donors (Simon et al., 1992); replacement of the amide motif completely by heterocycles like oxazole or thiazoles (Abell, 2001); swapping out the amide carbonyl for a suitable bioisostere like sulfinamide, sulfonamide (Moree, van der Marel, & Liskamp, 1995), or phosphinic amide (Collinsova & Jiracek, 2000); an α-carbon may be substituted by a nitrogen to give an azapeptide (Proulx et al., 2011); side chains may be incorporated into cyclic structures in order to rigidify and lock conformation, with either α,α-spirocycles, azepinone, or piperidine like structures (Van der Poorten, Knuhtsen, Sejer Pedersen, Ballet, & Tourwé, 2016).

## 5. CONCLUSIONS

Cyclic peptides have come into the fore in recent years as being particularly promising when deployed against some of the most challenging targets. Given the differences in the techniques described above it would be interesting to screen the same target using both approaches and compare the binding sites targeted by peptides from each approach. Furthermore, it is likely that each technique is best suited to a given class of targets. For example,

SICLOPPS screening is not compatible with extracellular protein targets (as the screen occurs inside a cell), whereas display technologies seem ideally suited for this target set. Conversely, SICLOPPS seems particularly well suited to intracellular targets, especially protein–protein interactions. In terms of developing hits, the two techniques also diverge as discussed above. As the understanding of structural elements that are responsible for the bioavailability of some naturally occurring cyclic peptides improves, the potential for unraveling a set of rules for translating cyclic peptide hits into drugs improves. However, at this point in time, this remains a highly challenging endeavor. Though there may be a temptation to focus on increasing affinity for the target using in vitro screening techniques once a lead peptide has been identified, affinity improvements often come at the expense of worsening solubility or decreasing cell permeability (both of which are already often poor). Instead, it may be better to keep a balance between in vitro and cell-based assays in order to avoid optimizing affinity at the cost of reduced efficacy. Importantly, a good balance between affinity, solubility, and permeability needs to be maintained in order to maximize the chances of taking any lead through to work in clinical models. These considerations must be foremost when deciding which paths to take when exploring structure–activity relationships and optimizing the key residues. In summary, while the first major hurdle toward the widespread use of cyclic peptide libraries has been overcome by the development of several techniques for the rapid generation of very large libraries, and good progress is being made toward overcoming the second hurdle of converting leads to candidates in the clinic.

## REFERENCES

Abell, A. D. (2001). Heterocyclic-based peptidomimetics. *Letters in Peptide Science*, *8*(3–5), 267–272. https://doi.org/10.1007/bf02446527.

Arnesen, T., Van Damme, P., Polevoda, B., Helsens, K., Evjenth, R., Colaert, N., et al. (2009). Proteomics analyses reveal the evolutionary conservation and divergence of N-terminal acetyltransferases from yeast and humans. *Proceedings of the National Academy of Sciences of the United States of America*, *106*(20), 8157–8162. https://doi.org/10.1073/pnas.0901931106.

Asby, D. J., Cuda, F., Beyaert, M., Houghton, F. D., Cagampang, F. R., & Tavassoli, A. (2015). AMPK activation via modulation of de novo purine biosynthesis with an inhibitor of ATIC homodimerization. *Chemistry & Biology*, *22*(7), 838–848. https://doi.org/10.1016/j.chembiol.2015.06.008.

Baeriswyl, V., & Heinis, C. (2012). Phage selection of cyclic peptide antagonists with increased stability toward intestinal proteases. *Protein Engineering Design and Selection*, *26*(1), 81–89. https://doi.org/10.1093/protein/gzs085.

Bashiruddin, N. K., & Suga, H. (2015). Construction and screening of vast libraries of natural product-like macrocyclic peptides using in vitro display technologies. *Current Opinion in Chemical Biology*, *24*, 131–138. https://doi.org/10.1016/j.cbpa.2014.11.011.

Birts, C. N., Nijjar, S. K., Mardle, C. A., Hoakwie, F., Duriez, P. J., Blaydes, J. P., et al. (2013). A cyclic peptide inhibitor of C-terminal binding protein dimerization links metabolism with mitotic fidelity in breast cancer cells. *Chemical Science*, *4*(8), 3046–3057.

Brinckerhoff, L. H., Kalashnikov, V. V., Thompson, L. W., Yamshchikov, G. V., Pierce, R. A., Galavotti, H. S., et al. (1999). Terminal modifications inhibit proteolytic degradation of an immunogenic mart-127-35 peptide: Implications for peptide vaccines. *International Journal of Cancer*, *83*(3), 326–334. https://doi.org/10.1002/(sici)1097-0215(19991029)83:3<326::aid-ijc7>3.0.co;2-x.

Cheriyan, M., Pedamallu, C. S., Tori, K., & Perler, F. (2013). Faster protein splicing with theNostoc punctiforme DnaE intein using non-native extein residues. *The Journal of Biological Chemistry*, *288*(9), 6202–6211. https://doi.org/10.1074/jbc.m112.433094.

Collinsova, M., & Jiracek, J. (2000). Phosphinic acid compounds in biochemistry, biology and medicine. *Current Medicinal Chemistry*, 7(6), 629–647. https://doi.org/10.2174/0929867003374831.

Cunningham, B., & Wells, J. (1989). High-resolution epitope mapping of hGH-receptor interactions by alanine-scanning mutagenesis. *Science*, *244*(4908), 1081–1085. https://doi.org/10.1126/science.2471267.

Di Lallo, G., Paolozzi, L., Castagnoli, L., & Ghelardini, P. (2001). A two-hybrid system based on chimeric operator recognition for studying protein homo/heterodimerization in Escherichia coli. *Microbiology*, *147*(6), 1651–1656. https://doi.org/10.1099/00221287-147-6-1651.

Dower, W. J., Miller, J. F., & Ragsdale, C. W. (1988). High efficiency transformation of E. coli by high voltage electroporation. *Nucleic Acids Research*, *16*(13), 6127–6145. https://doi.org/10.1093/nar/16.13.6127.

El Tayar, N., Mark, A. E., Vallat, P., Brunne, R. M., Testa, B., & van Gunsteren, W. F. (1993). Solvent-dependent conformation and hydrogen-bonding capacity of cyclosporin A: Evidence from partition coefficients and molecular dynamics simulations. *Journal of Medicinal Chemistry*, *36*(24), 3757–3764. https://doi.org/10.1021/jm00076a002.

Forster, A. C., Weissbach, H., & Blacklow, S. C. (2001). A simplified reconstitution of mRNA-directed peptide synthesis: Activity of the epsilon enhancer and an unnatural amino acid. *Analytical Biochemistry*, *297*(1), 60–70. https://doi.org/10.1006/abio.2001.5329.

Gieselman, M. D., Zhu, Y., Zhou, H., Galonic, D., & van der Donk, W. A. (2002). Selenocysteine derivatives for chemoselective ligations. *ChembioChem*, *3*(8), 709. https://doi.org/10.1002/1439-7633(20020802)3:8<709::aid-cbic709>3.0.co;2-8.

Goto, Y., Katoh, T., & Suga, H. (2011). Flexizymes for genetic code reprogramming. *Nature Protocols*, *6*(6), 779–790. https://doi.org/10.1038/nprot.2011.331.

Guillen Schlippe, Y. V., Hartman, M. C. T., Josephson, K., & Szostak, J. W. (2012). In vitro selection of highly modified cyclic peptides that act as tight binding inhibitors. *Journal of the American Chemical Society*, *134*(25), 10469–10477. https://doi.org/10.1021/ja301017y.

Gupta, S., Jain, A., Chakraborty, M., Sahni, J. K., Ali, J., & Dang, S. (2013). Oral delivery of therapeutic proteins and peptides: A review on recent developments. *Drug Delivery*, *20*(6), 237–246. https://doi.org/10.3109/10717544.2013.819611.

Henninot, A., Collins, J. C., & Nuss, J. M. (2017). The current state of peptide drug discovery: Back to the future? *Journal of Medicinal Chemistry*, *61*(4), 1382–1414. https://doi.org/10.1021/acs.jmedchem.7b00318.

Hook, B. (2005). RNA-protein interactions in the yeast three-hybrid system: Affinity, sensitivity, and enhanced library screening. *RNA*, *11*(2), 227–233. https://doi.org/10.1261/rna.7202705.

Horswill, A. R., Savinov, S. N., & Benkovic, S. J. (2004). A systematic method for identifying small-molecule modulators of protein-protein interactions. *Proceedings of the National Academy of Sciences of the United States of America*, *101*(44), 15591–15596. https://doi.org/10.1073/pnas.0406999101.

Iwasaki, K., Goto, Y., Katoh, T., & Suga, H. (2012). Selective thioether macrocyclization of peptides having the N-terminal 2-chloroacetyl group and competing two or three cysteine residues in translation. *Organic & Biomolecular Chemistry*, *10*(30), 5783. https://doi.org/10.1039/c2ob25306b.

Joo, S.-H. (2012). Cyclic peptides as therapeutic agents and biochemical tools. *Biomolecules & Therapeutics*, *20*(1), 19–26. https://doi.org/10.4062/biomolther.2012.20.1.019.

Josephson, K., Ricardo, A., & Szostak, J. W. (2014). mRNA display: From basic principles to macrocycle drug discovery. *Drug Discovery Today*, *19*(4), 388–399. https://doi.org/10.1016/j.drudis.2013.10.011.

Joung, J. K., Ramm, E. I., & Pabo, C. O. (2000). A bacterial two-hybrid selection system for studying protein-DNA and protein-protein interactions. *Proceedings of the National Academy of Sciences of the United States of America*, *97*(13), 7382–7387. https://doi.org/10.1073/pnas.110149297.

Kim, K.-H., & Seong, B. L. (2001). Peptide amidation: Production of peptide hormonesin vivo and in vitro. *Biotechnology and Bioprocess Engineering*, *6*(4), 244–251. https://doi.org/10.1007/bf02931985.

Kinsella, T. M., Ohashi, C. T., Harder, A. G., Yam, G. C., Li, W., Peelle, B., et al. (2002). Retrovirally delivered random cyclic peptide libraries yield inhibitors of interleukin-4 signaling in human B cells. *The Journal of Biological Chemistry*, *277*(40), 37512–37518. https://doi.org/10.1074/jbc.M206162200.

Kritzer, J. A., Hamamichi, S., McCaffery, J. M., Santagata, S., Naumann, T. A., Caldwell, K. A., et al. (2009). Rapid selection of cyclic peptides that reduce alpha-synuclein toxicity in yeast and animal models. *Nature Chemical Biology*, *5*(9), 655–663. https://doi.org/10.1038/nchembio.193.

Lau, J. L., & Dunn, M. K. (2018). Therapeutic peptides: Historical perspectives, current development trends, and future directions. *Bioorganic & Medicinal Chemistry*, *26*(10), 2700–2707. https://doi.org/10.1016/j.bmc.2017.06.052.

Leitch, E. K., Elumalai, N., Friden-Saxin, M., Dahl, G., Wan, P., Clarkson, P., et al. (2018). Inhibition of low-density lipoprotein receptor degradation with a cyclic peptide that disrupts the homodimerization of IDOL E3 ubiquitin ligase. *Chemical Science*, *9*(27), 5957–5966.

Lennard, K. R., & Tavassoli, A. (2014). Peptides come round: Using SICLOPPS libraries for early stage drug discovery. *Chemistry (Weinheim an der Bergstrasse, Germany)*, *20*(34), 10608–10614. https://doi.org/10.1002/chem.201403117.

Liu, W.-T., Ng, J., Meluzzi, D., Bandeira, N., Gutierrez, M., Simmons, T. L., et al. (2009). Interpretation of tandem mass spectra obtained from cyclic nonribosomal peptides. *Analytical Chemistry*, *81*(11), 4200–4209. https://doi.org/10.1021/ac900114t.

Male, A. L., Forafonov, F., Cuda, F., Zhang, G., Zheng, S., Oyston, P. C. F., et al. (2017). Targeting Bacillus anthracis toxicity with a genetically selected inhibitor of the PA/CMG2 protein-protein interaction. *Scientific Reports*, 7(1), 3104. https://doi.org/10.1038/s41598-017-03253-3.

Mathur, D., Prakash, S., Anand, P., Kaur, H., Agrawal, P., Mehta, A., et al. (2016). PEPlife: A repository of the half-life of peptides. *Scientific Reports*, *6*(1), 36617. https://doi.org/10.1038/srep36617.

Millward, S. W., Takahashi, T. T., & Roberts, R. W. (2005). A general route for post-translational cyclization of mRNA display libraries. *Journal of the American Chemical Society*, *127*(41), 14142–14143. https://doi.org/10.1021/ja054373h.

Miranda, E., Nordgren, I. K., Male, A. L., Lawrence, C. E., Hoakwie, F., Cuda, F., et al. (2013). A cyclic peptide inhibitor of HIF-1 heterodimerization that inhibits hypoxia signaling in cancer cells. *Journal of the American Chemical Society*, *135*(28), 10418–10425. https://doi.org/10.1021/ja402993u.

Mistry, I. N., & Tavassoli, A. (2017). Reprogramming the transcriptional response to hypoxia with a chromosomally encoded cyclic peptide HIF-1 inhibitor. *ACS Synthetic Biology*, *6*(3), 518–527. https://doi.org/10.1021/acssynbio.6b00219.

Moree, W. J., van der Marel, G. A., & Liskamp, R. J. (1995). Synthesis of peptidosulfinamides and peptidosulfonamides: Peptidomimetics containing the sulfinamide or sulfonamide transition-state isostere. *The Journal of Organic Chemistry*, *60*(16), 5157–5169. https://doi.org/10.1021/jo00121a038.

Nordgren, I. K., & Tavassoli, A. (2014). A bidirectional fluorescent two-hybrid system for monitoring protein-protein interactions. *Molecular BioSystems*, *10*(3), 485–490. https://doi.org/10.1039/C3MB70438F.

Osher, E. L., Castillo, F., Elumalai, N., Waring, M. J., Pairaudeau, G., & Tavassoli, A. (2018). A genetically selected cyclic peptide inhibitor of BCL6 homodimerization. *Bioorganic & Medicinal Chemistry*, *26*(11), 3034–3038. https://doi.org/10.1016/j.bmc.2018.03.012.

Pietrokovski, S. (1998). Identification of a virus intein and a possible variation in the protein-splicing reaction. *Current Biology*, *8*(18), R634–R638. https://doi.org/10.1016/s0960-9822(07)00409-5.

Proulx, C., Sabatino, D., Hopewell, R., Spiegel, J., García Ramos, Y., & Lubell, W. D. (2011). Azapeptides and their therapeutic potential. *Future Medicinal Chemistry*, *3*(9), 1139–1164. https://doi.org/10.4155/fmc.11.74.

Qi, H., Lu, H., Qiu, H.-J., Petrenko, V., & Liu, A. (2012). Phagemid vectors for phage display: Properties, characteristics and construction. *Journal of Molecular Biology*, *417*(3), 129–143. https://doi.org/10.1016/j.jmb.2012.01.038.

Rafi, S. B., Hearn, B. R., Vedantham, P., Jacobson, M. P., & Renslo, A. R. (2012). Predicting and improving the membrane permeability of peptidic small molecules. *Journal of Medicinal Chemistry*, *55*(7), 3163–3169. https://doi.org/10.1021/jm201634q.

Rentero Rebollo, I., & Heinis, C. (2013). Phage selection of bicyclic peptides. *Methods*, *60*, 46–54. https://doi.org/10.1016/j.ymeth.2012.12.008.

Roberts, R. W., & Szostak, J. W. (1997). RNA-peptide fusions for the in vitro selection of peptides and proteins. *Proceedings of the National Academy of Sciences of the United States of America*, *94*(23), 12297–12302. https://doi.org/10.1073/pnas.94.23.12297.

Scott, C. P., Abel-Santos, E., Wall, M., Wahnon, D. C., & Benkovic, S. J. (1999). Production of cyclic peptides and proteins in vivo. *Proceedings of the National Academy of Sciences of the United States of America*, *96*, 13638–13643. https://doi.org/10.1073/pnas.96.24.13638.

Shah, N. H., & Muir, T. W. (2014). Inteins: Nature's gift to protein chemists. *Chemical Science*, *5*(2), 446–461. https://doi.org/10.1039/c3sc52951g.

Simon, R. J., Kania, R. S., Zuckermann, R. N., Huebner, V. D., Jewell, D. A., Banville, S., et al. (1992). Peptoids: A modular approach to drug discovery. *Proceedings of the National Academy of Sciences of the United States of America*, *89*(20), 9367–9371. https://doi.org/10.1073/pnas.89.20.9367.

Spurr, I. B., Birts, C. N., Cuda, F., Benkovic, S. J., Blaydes, J. P., & Tavassoli, A. (2012). Small molecule inhibitor of AICAR transformylase homodimerization. *Chembiochem: A European Journal of Chemical Biology*, *13*(11), 1628–1634. https://doi.org/10.1002/cbic.201200279.

Stevens, A. J., Brown, Z. Z., Shah, N. H., Sekar, G., Cowburn, D., & Muir, T. W. (2016). Design of a split intein with exceptional protein splicing activity. *Journal of the American Chemical Society*, *138*(7), 2162–2165. https://doi.org/10.1021/jacs.5b13528.

Stevens, A. J., Sekar, G., Shah, N. H., Mostafavi, A. Z., Cowburn, D., & Muir, T. W. (2017). A promiscuous split intein with expanded protein engineering applications. *Proceedings of the National Academy of Sciences of the United States of America*, *114*(32), 8538–8543. https://doi.org/10.1073/pnas.1701083114.

Tavassoli, A. (2017). SICLOPPS cyclic peptide libraries in drug discovery. *Current Opinion in Chemical Biology*, *38*, 30–35. https://doi.org/10.1016/j.cbpa.2017.02.016.
Tavassoli, A., & Benkovic, S. J. (2005). Genetically selected cyclic-peptide inhibitors of AICAR transformylase homodimerization. *Angewandte Chemie (International Ed in English)*, *44*(18), 2760–2763. https://doi.org/10.1002/anie.200500417.
Tavassoli, A., & Benkovic, S. J. (2007). Split-intein mediated circular ligation used in the synthesis of cyclic peptide libraries in E. coli. *Nature Protocols*, *2*, 1126–1133. https://doi.org/10.1038/nprot.2007.152.
Tavassoli, A., Lu, Q., Gam, J., Pan, H., Benkovic, S. J., & Cohen, S. N. (2008). Inhibition of HIV budding by a genetically selected cyclic peptide targeting the Gag-TSG101 interaction. *ACS Chemical Biology*, *3*(12), 757–764. https://doi.org/10.1021/cb800193n.
Townend, J. E., & Tavassoli, A. (2016). Traceless production of cyclic peptide libraries in E. coli. *ACS Chemical Biology*, *11*(6), 1624–1630. https://doi.org/10.1021/acschembio.6b00095.
Valencia, C. A., Cotten, S. W., Dong, B., & Liu, R. (2008). mRNA-display-based selections for proteins with desired functions: A protease-substrate case study. *Biotechnology Progress*, *24*(3), 561–569. https://doi.org/10.1021/bp070473a.
Van der Poorten, O., Knuhtsen, A., Sejer Pedersen, D., Ballet, S., & Tourwé, D. (2016). Side chain cyclized aromatic amino acids: Great tools as local constraints in peptide and peptidomimetic design. *Journal of Medicinal Chemistry*, *59*(24), 10865–10890. https://doi.org/10.1021/acs.jmedchem.6b01029.
Wang, H., & Liu, R. (2011). Advantages of mRNA display selections over other selection techniques for investigation of protein–protein interactions. *Expert Review of Proteomics*, *8*(3), 335–346. https://doi.org/10.1586/epr.11.15.
Yang, X., Lennard, K. R., He, C., Walker, M. C., Ball, A. T., Doigneaux, C., et al. (2018). A lanthipeptide library used to identify a protein–protein interaction inhibitor. *Nature Chemical Biology*, *14*(4), 375–380. https://doi.org/10.1038/s41589-018-0008-5.
Young, T. S., Young, D. D., Ahmad, I., Louis, J. M., Benkovic, S. J., & Schultz, P. G. (2011). Evolution of cyclic peptide protease inhibitors. *Proceedings of the National Academy of Sciences of the United States of America*, *108*(27), 11052–11056. https://doi.org/10.1073/pnas.1108045108.
Zahnd, C., Amstutz, P., & Plückthun, A. (2007). Ribosome display: Selecting and evolving proteins in vitro that specifically bind to a target. *Nature Methods*, *4*(3), 269–279. https://doi.org/10.1038/nmeth1003.

CHAPTER SEVEN

# A Guide to Run Affinity Screens Using Differential Scanning Fluorimetry and Surface Plasmon Resonance Assays

**Christian Bergsdorf**[*,1], **S. Kirk Wright**[†]
[*]Novartis Institutes for BioMedical Research, Basel, Switzerland
[†]Novartis Institutes for BioMedical Research, Cambridge, MA, United States
[1]Corresponding author: e-mail address: christian.bergsdorf@novartis.com

## Contents



## Abstract

Over the past 30 years, drug discovery has evolved from a pure phenotypic approach to an integrated target-based strategy. The implementation of high-throughput biochemical and cellular assays has enabled the screening of large compound libraries which has become an important and often times the main source of new chemical matter that serve as starting point for medicinal chemistry efforts. In addition, biophysical methods measuring the physical interaction (affinity) between a low molecular weight ligand and a target protein became an integral part of hit validation/optimization to rule out false positives due to assay artifacts. Recent advances in throughput, robustness, and sensitivity of biophysical affinity screening methods have broadened their application in hit identification and validation such that they can now complement classical functional readouts. As a result, new target classes can be accessed that have not been amenable to functional assays. In this chapter, two affinity screening methods, differential scanning fluorimetry and surface plasmon resonance, which are broadly utilized in both academia and pharmaceutical industry are discussed in respect to their use in hit identification

*Methods in Enzymology*, Volume 610
ISSN 0076-6879
https://doi.org/10.1016/bs.mie.2018.09.015

and validation. These methods exemplify how assays which differ in complexity, throughput, and information content can support the hit identification/validation process. This chapter focuses on the practical aspects and caveats of these techniques in order to enable the reader to establish their own affinity-based screens in both formats.

## 1. INTRODUCTION

The early days of drug discovery started with testing a limited number of low molecular weight (LMW) ligands by phenotypic approaches using animal models, ex vivo tissue-based or phenotypic cellular assays mimicking a human disease situation. This strategy enabled identification and optimization of LMW compounds improving disease conditions and the development of commercial drugs. In the late 1980s and early 1990s, the introduction of genomics, molecular biology, and recombinant proteins offered the opportunity to identify and directly tackle disease-relevant targets with a broad variety of functional biochemical and cellular assays in drug discovery. The combination of these new techniques with advances in automation resulted in high-throughput screening (HTS) methods which became an integral part of early hit finding (Janzen, 2014; Mayr & Bojanic, 2009). HTS was routinely used to identify new chemical entities by analyzing large compound collections, typically 1 million or more compounds, which directly modulated the target of interest. In the beginning, HTS was mainly driven by (i) increasing sample throughput, (ii) assay robustness, (iii) automated workflows, (iv) multiparallel sample handling, (v) miniaturization, (vi) homogeneous assay formats, and (vii) development of data evaluation tools. It was soon recognized that high-throughput screens were prone to assay artifacts resulting in compounds that have unspecific or undesired biological effects; thus, it was critical to validate HTS hit compounds by orthogonal assay techniques (Baell & Holloway, 2010). Fortunately, advances in both sensitivity and throughput of biophysical and structural techniques have facilitated the analysis of molecular interactions between LMW ligands and target of interest in regard to covalent/noncovalent nature, stoichiometry, binding affinity/selectivity, kinetic/thermodynamic parameters, and mode of action. To answer these questions a broad selection of different biophysical methods, e.g., nuclear magnetic resonance spectroscopy (NMR), surface plasmon resonance (SPR), protein denaturation assay formats, X-ray crystallography, calorimetry, thermophoresis, and mass spectrometry (MS) can be applied (reviewed in Renaud et al., 2016). The combination of these biophysical or affinity screening methods with other technologies, e.g., biochemical, cellular readouts, makes them a powerful tool in identifying and selecting the most promising compounds for chemistry follow-up. Affinity screening methods offer the opportunities (i) to directly tackle protein/protein interactions without the need of detection reagents, e.g., antibodies, fluorescent dyes, etc., (ii) to address targets without a functional read-out, and (iii) to investigate very low affinity interactions.

A major challenge is knowing when to apply which biophysical technologies during lead discovery process. The decision is driven by the following questions: which (i) sample throughput, (ii) information content, and (iii) data quality (robustness, sensitivity) are needed? The answers to these questions are closely linked to the techniques and to different project phases: (i) hit identification, (ii) hit confirmation/validation, (iii) hit expansion/optimization, and (iv) hit to lead which require distinct throughput, information content, and data quality (Fig. 1). In addition to throughput, information content, and data quality, the choice of technique in industrial

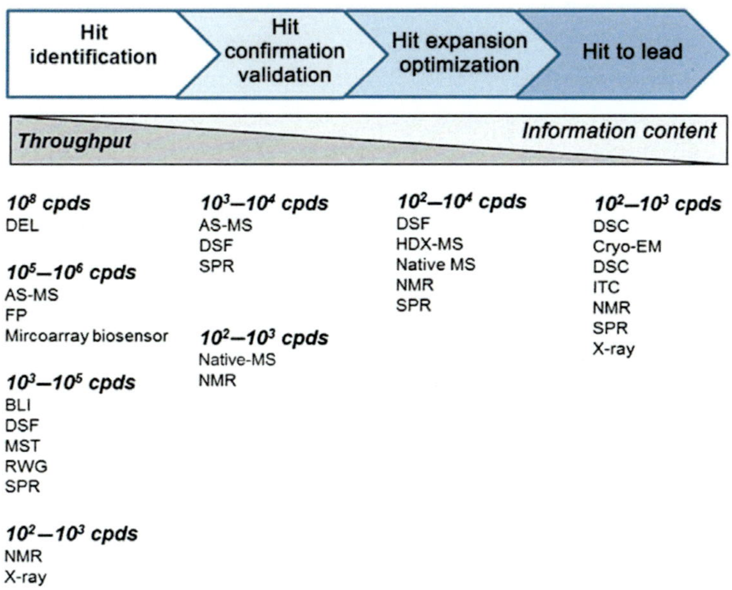


**Fig. 1** Affinity techniques in lead finding flowchart. In the *top panel* the four phases of lead finding are displayed in different colors. Progressing in the leading flowchart from hit finding to hit-to-lead phase sample throughput is dramatically decreased and the information content of used techniques is increased. Below the different affinity techniques are displayed which can be applied in the different phases including the sample throughput.

drug discovery strongly influences the costs per data point, protein consumption, needs of resources/expertise, and data handling/analysis (reviewed in Bergsdorf & Ottl, 2010; Ciulli, 2013; Folmer, 2016; Genick & Wright, 2017; Renaud et al., 2016; Silvestre, Blundell, Abell, & Ciulli, 2013).

In this chapter we focus on two affinity screening methods, a thermal shift assay format, differential scanning fluorimetry (DSF) and surface plasmon resonance (SPR), and their application in hit identification and validation. The practical aspects and caveats of these methods will be discussed in detail to give the reader a solid guidance to establish affinity-based screens in both formats. Both techniques were chosen due to (i) their broad utility in academia and pharmaceutical industry and (ii) their differences in regard to complexity, throughput, and information content. DSF represents a fast, simple, and cost-efficient technology which is applicable to a broad range of target classes and can enable screens of large compound sets (1–100 K) (Pantoliano et al., 2001; Simeonov, 2013). In contrast to, SPR binding assays offer increased information content answering questions like yes/no binding, target selectivity, affinity ranking, and kinetic/thermodynamic binding parameters at the cost of higher complexity (Olaru, Bala, Jaffrezic-Renault, & Aboul-Enein, 2015).

We are aware that the chosen techniques only represent a small portion of all affinity-based technologies which can be used to identify and validate hits. For a comprehensive overview and deeper insight in affinity-based technologies in drug discovery please refer Bergsdorf and Ottl (2010), Ciulli (2013), Folmer (2016), Genick and Wright (2017), and Renaud et al. (2016).

## 2. PROTEIN DENATURATION ASSAYS IN HIT IDENTIFICATION

In the 1950s, Christian Anfinsen's work showed that conformational stability is an intrinsic characteristic of proteins which is determined by its primary protein sequence and the environmental conditions. The stability of a protein is the net balance between opposing forces stabilizing and destabilizing a folded protein. The major forces stabilizing the native protein structures are noncovalent interactions as hydrophobic, ionic hydrogen bonds, and van der Waals, and covalent ones as disulfide bridges, whereas unfolded protein state is supported by entropic and nonentropic free energies (Anfinsen, 1973).

The protein stability can be measured by classical biophysical or biochemical methods as circular dichroism (CD) (Kelly & Price, 2000), differential scanning calorimetry (DSC) (Johnson, 2013), NMR, or radioactive ligand binding assays (Bischof & He, 2005). In the case of enzymes, the enzymatic activity can be monitored as a function of temperature. Most of aforementioned assay techniques are characterized by (i) their high labor-intensive work flow, (ii) low sample throughput, and (iii) relatively high protein consumption which limit their application in screening of larger compound libraries (1000s to 100,000s).

In this section we will mainly focus on protein denaturation techniques, which can be used to screen larger compound collections to identify LMW ligands stabilizing a protein of interest. Major prerequisites for screening assays are (i) microtiter well plate compatibility (96-well microtiter plate or higher density formats), (ii) significant automation potential enabling simple workflows, (iii) low protein consumption, and (iv) high assay robustness guaranteeing good reproducibility to minimize number of false positive/negative hits. Over the last 15 years thermal denaturation-based methods have become an integral part in early lead finding projects to discover LMW compounds binding and stabilizing many different targets (Niesen, Berglund, & Vedadi, 2007; Senisterra, Chau, & Vedadi, 2012; Simeonov, 2013). Thermal denaturation of proteins can be monitored by (i) measuring the change of intrinsic fluorescence intensity of a protein, e.g., NanoDSF (Muca, Marek, Żurawski, Piątkowski, & Antos, 2017), (ii) using fluorescence dyes which are changing their optical properties during temperature-dependent protein-unfolding process (DSF) (Pantoliano et al., 2001), or (iii) following protein aggregation and particle formation with differential static light scattering (DSLS) (Senisterra et al., 2006).

## 2.1 Differential Scanning Fluorimetry

DSF, also known as ThermoFluor, fluorescent thermal shift assay (FTSA), or temperature-dependent fluorescence (TdF) is a simple, fast, cheap, and broadly applicable biophysical method that typically requires minimal assay development. DSF enables the identification and characterization of ligands binding to a protein of interest by stabilizing the target. A wide range of ligands can be analyzed by DSF, e.g., LMW organic compounds (Huang et al., 2017), carbohydrates (Hirano, Adachi, Ito, & Totani, 2015), or proteins/peptides (Kopec & Schneider, 2013). In lead finding projects, DSF is used (i) as primary screening method to identify chemical starting points for MedChem optimization, (ii) as fragment-based screening method, or (iii) as orthogonal assay format to confirm/validate hits from other sources, e.g., HTS. In addition, DSF applied to optimize protein purification (Boivin, Kozak, & Meijers, 2013; Kozak et al., 2016), protein storage/formulation (Hawe, Filipe, & Jiskoot, 2010; Wong et al., 2016), or X-ray crystallization conditions (Dupeux, Röwer, Seroul, Blot, & Márquez, 2011), e.g., pH/buffer scouting, salt/additive screens (Seabrook & Newman, 2013). DSF measures the changes of fluorescence intensity of environmental sensitive dyes in the presence of a protein as a function of temperature. DSF dyes can be grouped into the following categories by their detection principle: (i) solvatochromatic dyes (Pantoliano et al., 2001), (ii) rotor dyes (Ablinger, Leitgeb, & Zimmer, 2013; Hawe et al., 2010), and (iii) thiol-reactive dyes (Alexandrov, Mileni, Chien, Hanson, & Stevens, 2008; Bergsdorf et al., 2016). All environmental sensitive dyes have in common that their fluorescence properties are directly influenced by the thermal melting process of a protein as shown in Fig. 2. Therefore the transition from the folded to the unfolded state of a protein can be followed in real time (RT) using simple equipment as RT-PCR systems. The most commonly used DSF dyes are solvatochromatic ones as Sypro Orange or 1-anilinonaphthalene-8-sulfonic acid (1,8-ANS). These dyes have a low fluorescence in high dielectric environment, e.g., aqueous buffers and a high fluorescence in low dielectric surrounding. During the protein-unfolding process hydrophobic regions buried inside of the protein structure become solvent accessible. Dye molecules can then bind to these regions with low dielectric constant, which lead to an increase of fluorescence intensity. A drawback of these dyes is their interaction with hydrophobic components, e.g., hydrophobic patches of proteins or detergents in the buffer, leading to high starting fluorescence. Therefore, Alexandrov and colleagues introduced the use of a thiol-reactive dye, *N*-[4-(7-diethylamino-4-methyl-3-coumarinyl)phenyl]maleimide (CPM), to monitor the denaturation of a detergent-solubilized integral membrane protein which showed no interference with hydrophobic components. In the native state, the cysteine residues of the protein are located inside and not accessible by the thiol-reactive dye. During the thermal-unfolding process these cysteine residues become solvent exposed and react with the dye forming a fluorescent adduct.

### 2.1.1 Assay Development

Due to the simplistic nature DSF, assay development is a straightforward process and can typically be completed within a few days. One advantage

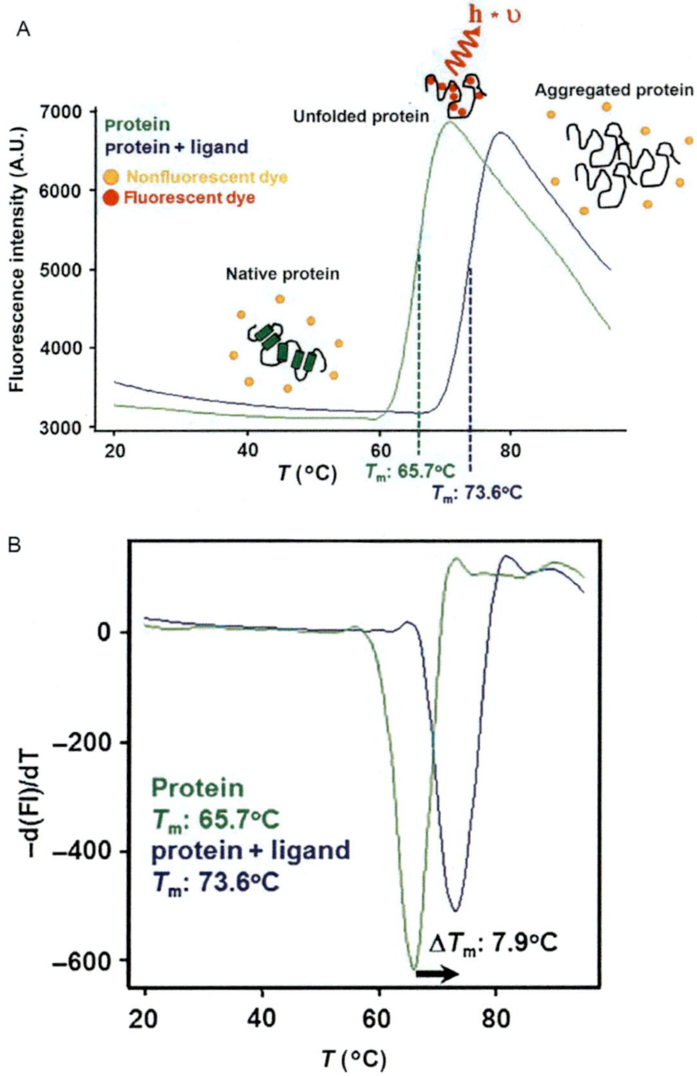


**Fig. 2** Protein unfolding profile and melting temperature determination. In (A) melting curves of a protein with (*blue*) and without (*green*) an LMW ligand and the corresponding $T_m$ values are displayed. The presence of a ligand shifts the *green melting curve* to the *right* (*blue melting curve*) resulting in a stabilization of 7.9°C. In the diagram the different phases of the protein-unfolding process and their interaction with an environmental dye are schematically shown. The nonfluorescent dye changes its fluorescence properties due to changes in the environment which are caused by unfolding the target protein. (B) The first derivative of both melting curves, which enables the determination of melting temperature $T_m$ (inflection point of graph). The degree of protein stabilization can be calculated as $\Delta T_m$.

of DSF is that tool compounds, while helpful, are not required for assay development. The major goal is to identify (i) the lowest protein concentration, (ii) the dye concentration, and (iii) a suitable assay buffer system generating highly reproducible protein melting curves to enable a sensitive test measuring protein stabilization ($\Delta T_m$) by LMW compounds.

### 2.1.2 General Considerations About Selecting Assay Buffer, DSF Dye, Protein, and Instrumentation

The buffer system should contain a minimal number of components to avoid interfering effects with DSF dye. Classical buffer ingredients like buffer agents, salts, and additives, e.g., TCEP, DTT, should be tested for their influence on the melting curve upfront. A good starting point is the protein storage buffer without additives as glycerol. Buffers such as HEPES, phosphate, or other Good buffers are a reasonable starting point. TRIS buffers are not well suited for DSF, due to their high-temperature dependency of pH (TRIS: $\Delta pK_a/10°C$: −0.28 vs phosphate $\Delta pK_a/10°C$: −0.028) (Good et al., 1966). The selection of dye type influences the buffer composition which can be applied. Solvatochromatic dyes are very sensitive to hydrophobic components, strongly increasing fluorescence intensity at room temperature, which often masks the protein melting transition. If detergents are mandatory components of assay buffer, rotor, or thiol-reactive dyes should be chosen. Especially, in the case of detergent-solubilized membrane proteins thiol-reactive dyes, e.g., CPM, BODIPY FL-cystine, have been proven as a valuable tool to monitor protein denaturation (Alexandrov et al., 2008; Bergsdorf et al., 2016; Hofmann, Gulati, Sears, Stewart, & Palczewski, 2016). For compound screening BODIPY FL-cystine with a longer wavelength excitation/emission spectrum (505/513 nm) and its broad compatibility with many RT-PCR systems in 96 up to 1536-well format is the preferred over CPM due to the unfavorable excitation/emission spectrum (387/463 nm). One should note when using rotor dyes that high concentrations of buffer components influencing the liquid density/rigidity, e.g., glycerol, sucrose, etc., should be avoided due to the potential of since high starting fluorescence intensities (Haidekker & Theodorakis, 2010) (Table 1).

DSF assays can be applied to a broad range of proteins, e.g., cytosolic, extracellular, and integral membrane proteins. As a prerequisite a protein preparation has to fulfill the following quality criteria: (i) aim for highest protein purity (at least >70%), (ii) confirmed molecular weight (LC–MS), and

**Table 1** DSF Dyes

| *Dye* | *Ex. (nm)* | *Em. (nm)* | *Dye Type* | *Interference* |
|---|---|---|---|---|
| **Sypro Orange** | 473 | 570 | Solvatochromatic dye | Detergents, hydrophobic reagents |
| **1,8-ANS** | 350–380 | 505 | | |
| **2,6-ANS** | 385–400 | 515 | | |
| **2,6-TNS** | 296 | 420 | | |
| **Nile red** | 540–580 | 660 | | |
| **CPM** | 384 | 469 | Thiol reactive dye | Reducing reagents |
| **BODIPY** | 505 | 513 | | |
| **ProteoStat** | 488 | ca. 600 | Rotor dye | Reagents changing viscosity |
| **DCVJ** | 450 | 480–505 | | |
| **CCVJ** | 430 | 480 | | |
| **Thioflavin T** | 450 | 480–490 | | |
| **DASPMI** | 475 | 605 | | |

(iii) determine dispersity of solution (DLS or SEC-MALS) to enable recording of reliable melting curves with a single transition. The apo protein $T_m$ should be in the range of 40–75°C to allow the detection of protein stabilization in a thermal gradient from 25 to 95°C. As aforementioned the protein type influences the dye selection. For most cytosolic proteins solvatochromic dyes such as Sypro Orange are well suited. In an initial experiment a cross titration of protein vs dye is performed to elucidate the optimal protein/dye concentration and to analyze their effect on the apparent $T_m$. For example, for a 60 kDa protein a concentration range between 0.2 μ*M* (12.5 μg/mL) and 6 μ*M* (400 μg/mL) is combined with dye concentrations achieving 0.5–20 μ*M* in the assay. From our experience 50–200 μg/mL protein and 2 × to 5 × Sypro Orange final assay concentration (stock 5000 × in DMSO) give acceptable results in most cases. In contrast to Sypro Orange, the apparent $T_m$ values observed with thiol-reactive dyes are very dependent on protein/dye ratio. During the unfolding process a covalent bond between protein and thiol-reactive dye is formed in a chemical reaction whose reaction speed is influenced by (i) the molar ratio of reactants and (ii) the velocity of temperature increase (reaction time). These parameters could be used to tune the assay for the most robust melting curve.

RT-PCR instruments or a temperature-controlled fluorescence spectrometer is generally used to measure DSF assays. There is a broad spectrum of RT-PCR instruments on the market, which can cope with most of the available dyes and plate formats. For dyes with a short excitation wavelength

(< 400 nm) as CPM, the number of available RT-PCR instruments is very limited. The selection of instrument is mainly driven by task which will be performed.

### 2.1.3 Setting up a Primary DSF Screen

In the following section we summarize the major steps which are performed to run a DSF primary screen. To select a suitable RT-PCR instrument, you should address the following questions: (i) compatibility of dye with available excitation/emission filters, (ii) which throughput is needed to perform the screen in a reasonable time frame, (iii) automation, and (iv) assay miniaturization potential. The relevance of the above questions is illustrated using an example of a 50K screen with 142 × 384-well or 36 × 1536-well DSF assay plates. To achieve a throughput of three to four plates per hour an automated measurement system has to be introduced in the workflow. For example, a Roche LC480, a Roche LC1536, and multiple BioRad CFX384 in combination with an automated plate feeder are possible options. Besides the throughput, the protein consumption for the whole screen has to be taken into account. A standard DSF assay with 100 μg/mL protein concentration and 10 μL assay volume point would consume 50 mg protein for a 50K screen. Therefore, during assay development miniaturization efforts should be addressed to reduce protein needs, e.g., decrease assay volume to 5 μL in 384-well format or changing to a higher plate density format (1536-well with 1–2 μL assay volumes).

### 2.1.4 Assay Protocol to Test Compounds

#### 2.1.4.1 Material

- 384-well or 1536-well RT-PCR plate
- Optical clear, thermostable, adhesive seals
- Liquid-handling devices: e.g., Giger Certus dispenser, Labcyte Echo 555
- RT-PCR instrument: Roche LightCycler480 or 1536
    - All buffer solutions are filtered by 0.2 μm filter
    - All dye solutions are protected against light

#### 2.1.4.2 Preparation of Dye/Protein Mix

- Add dye stock solution (dye dissolved in 100% DMSO) to aqueous assay buffer achieving final dye assay concentration (typically low μ*M* range, dependent on used DSF dye; SyproOrange 5 × (stock 5000 ×)). Mix dye/buffer solution and centrifuge at 10,000 × *g* for 5 min to remove particles, which can interfere in the assay

- Add sufficient volume of protein stock to dye/buffer solution (final protein concentration typically 50–200 μg/mL) and gently mix (important: test protein stability in the presence of dye during assay development)

#### 2.1.4.3 Preparation of Ready-to-Use DSF Assay Plate

- Compound stock solutions dissolved in 90%–100% DMSO (typical stock concentration 5–10 m*M*) are transferred into the wells of a PCR assay plate (384-well or 1536-well format) by an acoustic dispenser such as a Labcyte Echo555. In a standard screen a protein ligand ratio of 1:20 (e.g., 100 μg/mL (2 μ*M*) 50 kDa protein plus 40 μ*M* compound) is used to guarantee high occupancy of target protein enabling the detection of weaker binders. Each screening plate includes 16–32 control wells containing DMSO (columns 23 and 24 of 384-well PCR plate), being used to determine the apparent $T_m$ of the target protein. In an optimal case a reference binder is included as a positive control, too. The transferred compound volumes are in the range of 5–100 nL when an assay is performed in 2–10 μL total volume.

#### 2.1.4.4 Assay Assembly and Measurement

- 2–5 μL dye protein mix/well (depends on plate format) is added to the ready-to-use DSF plate by a dispenser, e.g., Giger Certus or equivalent.
- The filled plates are sealed and briefly mixed on an orbital shaker, e.g., Variomag Teleshake 1536, and then centrifuged at 2000 × *g* for 2 min.
- Afterward the assay plate is preincubated 15–30 min at room temperature in the darkness.
- For the measurement on the Roche LC480 the following protocol is used:
  - *Temperature range:* 30–95°C
  - *Temperature ramp:* 1–4.8°C/min (has to be adapted for each protein during assay development). The temperature ramp speed influences the obtained apparent $T_m$. A low scan rate can shift apparent $T_m$ to lower values (Lepock et al., 1992)
  - *Data acquisition:* 4 acquisitions per 1°C increase

### 2.1.5 DSF Data Analysis

As a result DSF screens produce a large number of protein-unfolding curves from which the melting temperature ($T_m$) of a target protein can be extracted. In general there are different methods to determine an apparent $T_m$: (i) generating the first derivative of the melting curve and the inflection point (temperature of maximal slope) represents the apparent $T_m$ and (ii) fitting the melting curve with a suitable algorithm modeling the unfolding process (commonly two-state unfolding model used). The obtained $T_m$ values are used to calculate the protein stabilization by the ligand $\Delta T_m$ in comparison to the protein in the presence of DMSO:

$$\Delta T_m = T_{m(\text{protein + ligand})} - T_{m(\text{protein + DMSO})}$$

The threshold of $\Delta T_m$ to identify ligands significantly stabilizing the protein target is set by monitoring the reproducibility and variability of controls on each plate. Therefore, each screening plate contains several negative and positive (if available) controls, e.g., 384-well microtiter plate (16 negative and 16 positive controls or 32 negative controls) to check assay quality. Usually a threshold for $\Delta T_m$ of three times the standard deviation of the negative control (protein in absence of ligand) is used to classify hits. A high assay quality enables to identify hits causing protein stabilization in the presence of ligand $\Delta T_m > 0.5$–1°C. In our case we bin the observed $\Delta T_m$ values in different categories: (i) positive shifter >1°C, (ii) weak positive shifter >0.5 and <1°C, (iii) no shifter <1 and >−1°C, (iv) negative shifter <−1°C, and (v) compounds causing curve issues, e.g., low/high fluorescence intensity, multiple transitions, etc. to further validate the identified hits. For conformation weak and positive shifters are selected and tested as 4-point concentration series (e.g., 10–80 μ*M*, primary screening concentration 40 μ*M*) in the DSF assay. Hits confirming in the concentration series experiment are subjected to further validation with a second biophysical method, e.g., SPR, NMR. From our experience weak positive shifter is often not confirmed at the primary screening concentration, since the observed $\Delta T_m$ was too close to chosen threshold of the assay. Some of these ligands are rescued since they stabilize the target at higher ligand concentrations. It is important to mention that a ligand ranking simply based on $\Delta T_m$ can be misleading, since the observed $\Delta T_m$ is not unique for a certain binding affinity. The enthalpic and entropic effects on a certain affinity $K_d$ can influence the melting process of a protein in different way. Therefore, you often find good correlation of $\Delta T_m$ values within a chemical cluster having a similar enthalpic contribution on binding, whereas between different chemical series the correlation is very limited.

In a screening setup the apparent $T_m$ values from 1000 to 100,000 melting curves have to be extracted and interpreted. Therefore, software tools were implemented to support the data analysis process as QC checks, outlier recognition, $T_m$ estimations, curve rejection, side by side curve comparison, plate normalization, connecting $\Delta T_m$ with other assays, etc. Publically available software programs such as Meltdown (Rosa et al., 2015), MTSA (Schulz, Landström, & Hubbard, 2013), or commercial products like Genedata Screener are valuable tools to speed up and enable an efficient data analysis.

## 2.2 Other Protein Denaturation Methods

The focus of this section is on additional protein denaturation methods using label-free detection readout principles, specifically highlighting the differences to the aforementioned DSF screening format.

### *2.2.1 Differential Static Light Scattering*

DSLS is an optical detection system that can be used monitor protein denaturation. The denaturation of a protein is monitored by the increase in intensity of scattered light due to aggregation induced in a temperature gradient, typically from 25 to 95°C. The aggregation temperature ($T_{agg}$) is determined by plotting the measured scattered light intensity vs the temperature. Analogous to the $T_m$ in DSF, the $T_{agg}$ of a target protein can increase with ligand interaction. The assumption behind this measurement is that protein denaturation occurs as a three-state process which combines a reversible with irreversible process. This process can be described as follows:

$$\mathrm{N} \underset{k_{-1}}{\overset{k_1}{\rightleftarrows}} \mathrm{U} \xrightarrow{k_2} \mathrm{A}$$

A native protein (N) starts the reversible process of unfolding in a temperature gradient and is converted into the unfolded state (U). This process is described with the kinetic constants $k_1$ and $k_{-1}$. Due to the unfolding of the protein, hydrophobic parts get accessible which promote intermolecular interactions between the unfolded protein molecules and result in the protein aggregation (A). This irreversible step is described with kinetic constant $k_2$, locking the protein molecules in an aggregated state. In the case where both kinetic constants $k_1$ and $k_2$ are equivalent, the three-state model can be converted into an irreversible three-state process. In this case the determination of $T_{agg}$ is a good surrogate for $T_m$ which was shown in study using 61 different proteins (Vedadi et al., 2006). $T_{agg}$ is determined by fitting an aggregation profile to a modified Boltzmann equation. The advantage of DSLS measurements is reduced compound interference and protein/dye artifacts due to long wavelength detection (reduced autofluorescence of LMW compounds) and no fluorescence dye in protein sample. Furthermore, DSLS can be applied to a broad range of targets, e.g., hydrophobic proteins, membrane proteins which cannot be accessed by solvatochromatic dyes with DSF (Bergsdorf et al., 2016; Senisterra et al., 2010). On the other hand, proteins forming no aggregates cannot be analyzed by DSLS. Furthermore, buffer components, e.g., micelle-forming agents, influence the light scattering readout and can induce a second transition. Therefore, it is mandatory to test the buffer composition without protein to evaluate this effect. Nowadays new DSLS instrumentation is on the market (Stargazer 2, Epiphyte) which addressed some of the weakness of first generation, e.g., increased temperature range 10–95°C, assay miniaturization resulting in reduced protein amount consumption, improved sensitivity, and automation potential. DSLS can be seen as a valuable alternative for fluorescence dye-based thermal shift assays in affinity screening or as an orthogonal method to validate hits.

### *2.2.2 Isothermal Chemical Denaturation*

As discussed, thermal denaturation can be used to screen for ligands that bind and alter protein stability. A positive shift in the denaturation or melting temperature ($\Delta T_m$) between the ligand-free protein and protein in the presence of the ligand is indicative of binding. Ligand ranking by the magnitude of the $\Delta T_m$ has to be done with caution since real affinities cannot be determined due to the lack of knowledge of the enthalpy and heat capacity (Schön, Brown, Hutchins, & Freire, 2013). Furthermore, thermal denaturation assay can suffer from artifacts, such as ligand–dye interaction, visualizing protein melting, and heat-induced aggregation (Mahendrarajah, Dalby, Wilkinson, Jackson, & Main, 2011).

An alternative approach to thermal denaturation is isothermal chemical denaturation (ICD), which measures the increase in the concentration of a chemical denaturant, such as urea, guanidinium hydrochloride, or pH to modulate protein stability at a constant temperature. Unlike thermal denaturation, ICD depends on the Gibbs energy of protein stability ($\Delta G$), concentration of the denaturant ($C$), and denaturant $m$ value, and the Gibbs free energy follows a linear dependence with the denaturant concentration where $\Delta G^\circ$ is the Gibbs free energy of the protein stability at zero denaturant concentration:

$$\Delta G = \Delta G^\circ - m[C]$$

$C$ is the concentration of the denaturant and the $m$ value correlates with the change in accessible surface area or change in solvent exposure upon protein unfolding (Myers, Pace, & Scholtz, 1995). The midpoint of the denaturation curve, where $\Delta G = 0$, is defined as $C_{1/2}$. A deeper insight into theory of chemical denaturation can be found in Schön et al. (2013) and Mahendrarajah et al. (2011). Chemical denaturation has long been used to interrogate the structural stabilities of proteins and to evaluate biotherapeutic formulation (Svilenov, Markoja, & Winter, 2018; Temel, Landsman, & Brader, 2016). It is only recently that ICD has been applied to characterize ligand binding:

$$\Delta G = \Delta G_{\mathrm{L}}^{\circ} - m_{\mathrm{L}}[C]$$

ICD in the presence and absence of ligand yields $\Delta G$ and $\Delta G_{\mathrm{L}}^{\circ}$ and the difference can be used to estimate the relative binding affinity (Schön et al., 2013). Ligands that bind to and stabilize the native state do so in a manner that is dependent in ligand affinity and concentration. This can result in an increase in $C_{1/2}$ and may or may not change the $m$ value. ICD can be used to measure binding affinities over a broad range from mid μ$M$ to low p$M$ (Ross et al., 2015). The recent development of automated instruments running the assay in a 96-well microtiter well format has increased the ease and throughput of ICD assays. While the throughput is less than that of thermal denaturation assays and the protein consumption is higher (~100 μg/sample), ICD provides several advantages. Real binding affinities (approximately 1 $K_{\mathrm{D}}$/h) can be determined. Affinities in the low n$M$ to p$M$ range, which are challenging to measure using standard biophysical methods such as SRP and ITC, can be determined. In addition, chemical denaturation can be used with a broad variety of proteins which are not applicable in fluorescent thermal denaturation assays due to protein dye interference. The major disadvantages of the ICD are (i) moderately high protein requirements, (ii) low sample throughput, (iii) not ideal for weak ligands found in most screening hit lists, and (iv) prone to fluorescent artifacts from compounds. Temel et al. provide an overview of ICD experimental methodology and data interpretation.

## 3. SPR IN LOW MOLECULAR WEIGHT COMPOUND SCREENING

In the last 15 years, the sensitivity and throughput of the SPR and SPR-like instruments, such as the Biacore S200, T200, Biacore 8K, Sierra Sensors MASS1 and 2, and Creoptix Wave, have advanced to the point where they can easily be used for both screening and detailed kinetic characterization of fragments and LMW compounds and thus have been widely adapted in the pharmaceutical industry.

In these systems one of the interacting molecules, often referred to as the ligand, is immobilized on a sensor surface. The other molecule, referred to as the analyte, is injected over the sensor surface. As the injected analyte binds to the ligand attached to the sensor surface, a change in SPR response is detected in real-time. The change in response level after the injection is proportional to the change in refractive index which in turn is generally proportional to mass at the surface. The kinetic rate constants ($k_a$ and $k_d$) and the equilibrium dissociation constant ($K_D$) can be derived by fitting the data to appropriate interaction models (Hämäläinen et al., 2000; Karlsson & Fält, 1997; Markgren et al., 2001, 2003).

The heart of SPR technology is the sensor chip. The sensor chip provides the physical conditions required to generate the SPR signal and provides the surface required for capture or immobilization of the ligand. Molecules are generally attached to the sensor surface using one of the three methods: (i) covalent immobilization, (ii) affinity capture, and (iii) hydrophobic adsorption. The two most common methods used for small molecule interaction analysis are the covalent immobilization and affinity capture. With the covalent modification method, the ligand is attached to the surface through a covalent chemical link. There are a variety of covalent modification chemistries available such as amine, thiol, and aldehyde coupling immobilization. With the affinity capture method, the ligand is attached or captured via a noncovalent interaction with another capturing molecule, such as Ni-NTA or streptavidin. Immobilization and capture approaches are discussed in more detail in Schasfoort (2017).

In addition to selecting the best immobilization/capture strategy, it is critical that immobilization/capture levels and the assay conditions are optimal. The surface density of the ligand needs to be optimized to provide an appropriate analyte response, while the immobilization and running buffers need to be optimized such that the surface is active and stable for the duration of the assay.

## 3.1 SPR in LMW Affinity Screening

The overall procedure for LMW affinity screening, like fragment screening, is divided into three following general steps:

**1** Clean Screen
**2** Binding Level Screen
**3** Affinity Screen

The protocols listed below are geared toward the newest generation of Biacore instruments, the Biacore 8K, utilizing a biotinylated protein captured onto a streptavidin surface. These protocols can but be applied to other instruments and immobilization strategies with minor adaptations.

### *3.1.1 Clean Screen*

The goal of an LMW Clean Screen, like a fragment Clean Screen, is to identify and remove compounds that nonspecifically bind or show residual binding to the sensor chip matrix. Because compounds are serially flowed over the SPR sensor surface, this nonspecific or residual binding can adversely affect the data quality of subsequent compound injections. It has been observed that nonspecific compound binding can also be target dependent. Therefore, it is advisable to run a clean screen in the presence of the target as well. During the Clean Screen, it is not uncommon to find compounds that have a large carryover and will effect multiple injections. In such cases, the compounds affected by the carry-over will need to be reassayed to ensure that they do not nonspecifically interact with the surface.

#### 3.1.1.1 Instrument and Sensor Chip Preparation

- A series S SA Sensor chip is warmed to room temperature and docked into the Biacore 8K instrument.
- The instrument is primed (Change Solution feature) 3 × with buffer (typically 1 × HBS-P+).
- The chip is preconditioned with four injections of 10 m*M* NaOH, 1 *M* NaCl (30 μL/min for 60 s) followed by another Change Solutions (prime) with buffer.
- Instrument is normalized with BIAnormalizing solution (70% (w/w) glycerol) followed by a final prime.
- For the Clean Screen on the target, a biotinylated protein (in this example) is captured on to flow cell two of all eight channels of the sensor chip. Biotinylated protein, at the appropriate concentration, is placed in all a row (8 wells) of a 96-well plate and then injected over the sensor surface at a flow rate of 10 μL/min until desired surface density is reached (typically between 2500 and 5000 RU). The contact time is predetermined.

All commands are queued up in the Activity Queue and are executed in order.

#### 3.1.1.2 Running Buffer Preparation

- Running Buffer (4 L) is prepared by diluting 10 × stock of HBS-P+ buffer (400 mL of 10 × stock buffer into 3600 mL of deionized water), mixed thoroughly and filtered through a 0.22 μm filter.
- 400 mL of Running Buffer (without DMSO) is set aside for compound preparation.
- DMSO will be added to remaining 3600 mL of Running Buffer to bring it to 2% DMSO. Buffer (72 mL) is removed (and can be added Running Buffer set aside for compound prep). DMSO (72 mL) is added to the remaining 3528 mL of buffer is mixed very thoroughly. DMSO-containing buffer is again filtered under vacuum.

#### 3.1.1.3 Compound Plate Preparation

- Compounds arrive in 384-well plates in columns 1–22 (columns 23 and 24 are empty) as 10 m*M* stocks in 100% DMSO. Compounds are first diluted 2 m*M*, which is 50-fold the desired final concentration (40 μ*M* in the example) in 100% DMSO in a new 384-well plate. Controls can be added to columns 23 and 24 if desired.
- 3 μL of the 2 m*M* intermediate compound dilution is added to 147 μL of Running Buffer without DMSO and mixed thoroughly in a 384-deep well plate. Final assay plates will contain solution of each compound at 40 μ*M* in Running Buffer with 2% DMSO.

#### 3.1.1.4 Instrument Operation

- Up to three assay plates are loaded into sample compartment of Biacore 8K instrument. Reagent plate (1850 μL 96-deep well) contains Running Buffer for Start-up cycles (usually 10 or more) and 50% DMSO solutions for washes between each compound. Please note that it uses alternating rows (A, C, E, G, I, K, M, and O) due to the eight needle configuration and this needs to be accounted for in plate setup). Compounds are injected at 40 μ*M*.
- Data collection rate is set to 10 Hz and High-Performance injections are used. Sample compartment and analysis temperature are set to 25°C. Contact time is set to 45 s and dissociation time is set to 45 s (contact and dissociation time can be set to be less to decrease the cycle time; however, we prefer to have the times to be the same for the Clean Screen and the Binding Level Screen). Flow rate is set to 100 μL/min.

### 3.1.1.5 Data Analysis

Data are analyzed using the Biacore 8K Evaluation software (Version 1.1.17442):

– A new Evaluation is created; the desire run and Evaluation Method are selected.
– The predefined Fragment Clean Screen method can be used to evaluate LMW clean screens. The data are plotted as the Baseline difference (RU) vs cycle number (an alternative method is to generate a user-defined method using the early stability point (relative repose) instead of the Baseline Difference (shown in Fig. 3A)).

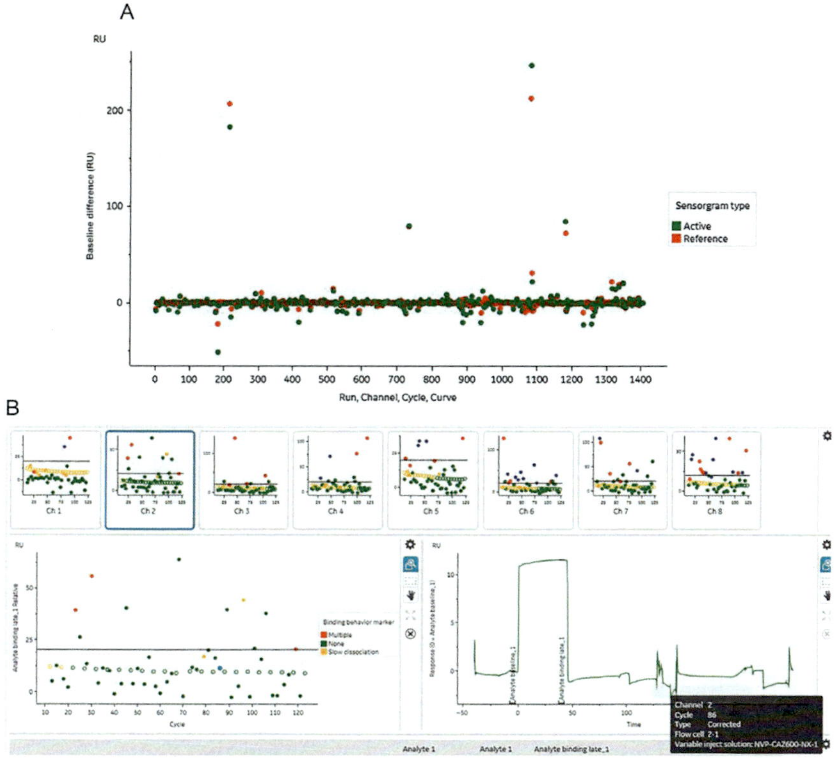


**Fig. 3** Clean Screen and Binding Level Screen using a Biacore 8K. Clean screen results are shown in (A). Reference flow cell contains only plain streptavidin surface (SA), *red dots*. Active flow contains target of interest, *green dots*. RU binding levels for each tested compound are plotted against injection cycle number. Compounds above the *blue dash line* were excluded from further analysis. The results of the single point Binding Level Screen (384-well microtiter plate) are shown in (B). Flagged compounds (*blue frame*) are visually inspected. Compounds generating appropriate sensograms are progressed to the Affinity Screen.

- The data are inspected. Compounds that residually bind to the surface at 10 RU or more are considered to be sticky and are not progressed into the Binding Level Screen. Compounds that interact with the surface between 0 and 5 RU are considered to be clean. Out of caution, compounds that interact with the surface between 5 and 10 RU are suspect and placed at the end of the Binding Level Screen.

It is important to note that compounds that fail the clean screen still have the potential to be valid binders and should be followed up by another method.

### *3.1.2 Binding Level Screen*

The goal of the Binding Level Screen is to identify compounds that bind to the target of interest and to remove the false positives originating from the screening assay. Compounds that pass the Clean Screen are next assayed for binding in the single point Binding Level Screen. In the example below is for a biotinylated protein, however, a number of capture or immobilization schemes can be utilized.

#### 3.1.2.1 Instrument and Sensor Chip Preparation

- Biacore 8K instrument and chip is prepared as described in steps 1–4 in Section 3.1.1.1 of the Clean Screen Protocol.
- Biotinylated protein is then captured on to flow cell two of all eight channels of the sensor chip as described earlier.

#### 3.1.2.2 Running Buffer Preparation

- Running Buffer is prepared as described in Section 3.1.1.2.

#### 3.1.2.3 Compound Plate and Solvent Correction Preparation

- Compounds arrive in 384-well plates in columns 1–22 (columns 23 and 24 are empty) as 10 m*M* stocks in 100% DMSO. Compounds are first diluted 2 m*M*, which is 50-fold the desired final concentration (40 μ*M* in the example) in 100% DMSO in a new 384-well plate.
- 3 μL of the 2 m*M* intermediate compound dilution is added to 147 μL of Running Buffer without DMSO and mixed thoroughly in a 384-deep well plate. Final assay plates will contain solution of each compound at 40 μ*M* in Running Buffer with 2% DMSO.
- Positive controls (and negative controls) are prepared in the same manner as the compound to be screened and in our case can be added to columns 23 and 24 or can be placed in the reagent plate.

- Solvent Correction sample (seven point) that brackets the 2% DMSO used in the assay are prepared by mixing 100% DMSO into the non-DMSO-containing Running Buffer as shown in Table 2. Each sample is thoroughly mixed by vortexing. Final volumes can be scaled as needed.

#### 3.1.2.4 Instrument Operation

- Up to three assay plates are loaded into sample compartment of Biacore 8K instrument. Reagent plate (1850 μL 96-deep well) contains Running Buffer for Start-up cycles and 50% DMSO solutions for washes between each compound and Solvent Correction samples.
- Data collection rate is set to 10 Hz and High-Performance injections are used. Sample compartment and analysis temperature is set to 25°C, although the temperature can be varied based on the needs of the protein. Contact time is set to 45 s and dissociation time is set to 45 s (dissociation time can be increased based on needs of assay or controls). Flow rate is set to 100 μL/min.
- Start-up cycles (10–15) are run to equilibrate the system. Buffer controls are typically run every 8 cycles and positive controls (preferably one with a fast off-rate) used to monitor the activity of the sensor surface are run every 24th cycle. Solvent Corrections are run every 24th cycle as well as 50% DMSO washes are run after every compound injection and are immediately followed by a buffer injection.

#### 3.1.2.5 Data Analysis

Data are analyzed using the Biacore 8K Evaluation software (Version 1.1.17442):

- As with the Clean Screen a new evaluation if created. In this example, the predefined LMW Screen method was selected.
- Solvent Correction is applied and the compound that falls outside the Solvent Correction range is not corrected and noted.

**Table 2** Solvent Correction

| Sample (μL) | 1 | 2 | 3 | 4 | 5 | 6 | 7 |
|---|---|---|---|---|---|---|---|
| Buffer | 1960 | 1960 | 1960 | 1960 | 1960 | 1960 | 1960 |
| DMSO | 25 | 30 | 35 | 40 | 45 | 50 | 55 |

- QC items are automatically generated including the Binding Level Screen plot which contains the compound response from the binding level early report point for all channels and cycles (shown in Fig. 3B).
- Adjustments for blank subtraction, molecular weight, and controls are applied as desired.
- Cutoffs are applied, and typically 3 × the standard deviation of the negative control is used for the lower level cutoff. Super stoichiometric compounds ($5 \times R_{max}$) are disregarded.
- The software automatically applies the Binding Behavior Markers (slope, slow dissociation, and multiple undesired properties) which can be used to quickly bin compounds.
- The data are then sorted in the Results Table by cutoff levels and Binding Behavior markers resulting in hits at the top of the table (some may find it more efficient to export the data and preform the sorting in a spreadsheet application). A sensorgram overview is created of the selected hits which are further inspected to ensure appropriate data quality. In addition, the sensorgrams for the nonhits are directly inspected and hits can be rescued if needed.
- Binding Level Screen hits are then progressed into the Affinity Screen.

### *3.1.3 Affinity Screen Protocol*

Compounds identified as hits in the single point Binding Level Screen are next assayed in full dose–response with the goal of determined the binding affinity. With the Biacore 8K instrument (and other eight channel instruments) dose–response data can be collected in multiple formats. Dose–response data can be collected in the traditional multicycle format or in single injection or parallel format. The traditional multicycle format each concentration of a dose–response curve is injected over the same flow cell. Using an 8-point (7 compound concentrations and 1 buffer only sample) allows for 48 compounds to be arrayed on a 384-well plate. This format is generally used when large numbers of compounds need to be assayed. In the single injection or parallel format, all eight concentrations of the dose–response experiment are injected simultaneously over the sensor surface (i.e., one compound concentration per flow cell) and a zero compound or buffer only injection is run as a separate injection. Parallel mode can be easily run in either 96-well format or and the number of compounds per plate can be maximized by utilizing 384-well plates and Reagent Bottle for zero buffer injections. The protocol below describes the traditional multicycle mode but is directly applicable to the parallel mode. Single-cycle kinetics mode is generally not used in the screening mode. It is important to note that even after the clean and single point screen that false positives and poorly behaved compounds can still be encounter in the affinity screen assays. Instrument, Sensor Chip, and Running Buffer are preparation as described earlier.

#### 3.1.3.1 Compound Plate and Solvent Correction Preparation

- Compounds arrive in 384-well plates in columns 1–22 (columns 23 and 24 are empty) as 10 m*M* stocks in 100% DMSO and are arrayed into a seven-point dose–response format bracketed by buffer injections (zeros for buffer subtraction). Compounds are first diluted 2 m*M*, which is 50-fold the desired final concentration (40 μ*M* in this example) in 100% DMSO in a new 384-well plate.
- Intermediate plate, controls, and Solvent Correction samples are prepared as described earlier.

#### 3.1.3.2 Instrument Operation

- Instrument was set up and run as previously described.

#### 3.1.3.3 Data Analysis

Data are analyzed using the Biacore 8K Evaluation software (Version 1.1.17442).

- In the 8K Evaluation software, the desired run and Evaluation method are selected. Typically the LMW multicycle affinity is used. Solvent Correction is calculated and applied.
- The software automatically generates the QC (Sensorgram, Baseline, and Binding to reference) and Kinetics/Affinity evaluation.
- Sensorgrams are inspected in the QC-Sensorgram tab and poor quality cycles are excluded from the evaluation. Binding to reference is also noted.
- The data grouping is set to the appropriate mode (serial in the example) and Affinity is selected in the Kinetics/Affinity mode (both kinetics and affinity mode can be selected in the data set which contains compounds that display kinetic behavior). The blank setting is set to use blanks from within the same series. In this example, the data are automatically fit to the 1:1 binding model for both kinetics and steady-state affinity (shown in Fig. 4).

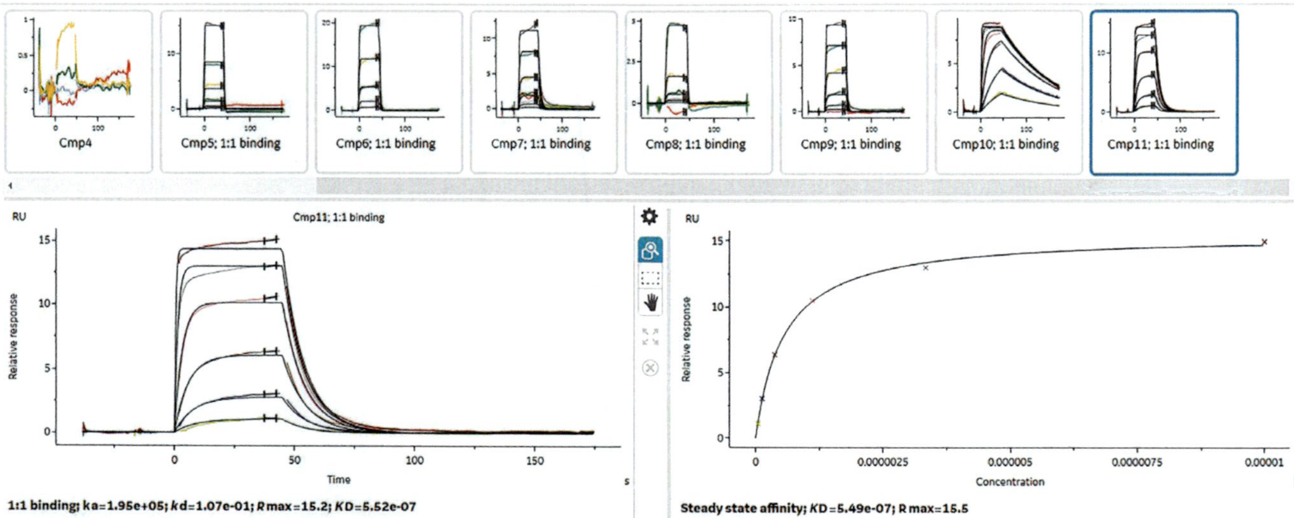


Sample table | Parameters | Residuals | Blanks

| Cycle | Channel | Sensorgram type | Cycle purpose | Injection type | Control type | Analyte 1 solution | Concentration (μM) | Immobilized ligand | Excluded | Curve markers | Blank cycles | Temperature |
|---|---|---|---|---|---|---|---|---|---|---|---|---|
| 26 | 6 | Corrected | Analysis | Analyte | Not a control | Cmp11 | 0.04115226 | | | None | [24][25] | 10 |
| 27 | 6 | Corrected | Analysis | Analyte | Not a control | Cmp11 | 0.1234568 | | | None | [24][25] | 10 |
| 28 | 6 | Corrected | Analysis | Analyte | Not a control | Cmp11 | 0.3703704 | | | None | [24][25] | 10 |
| 29 | 6 | Corrected | Analysis | Analyte | Not a control | Cmp11 | 1.111111 | | | None | [24][25] | 10 |
| 30 | 6 | Corrected | Analysis | Analyte | Not a control | Cmp11 | 3.333333 | | | None | [24][25] | 10 |
| 31 | 6 | Corrected | Analysis | Analyte | Not a control | Cmp11 | 10 | | | None | [24][25] | 10 |

**Fig. 4** Affinity Screen using a Biacore 8K. Affinity Screen results are shown. Compound 11, while not displaying ideal behavior, binds in a dose-responsive manner to the target with a $K_D$ of 0.55 μ*M*.

- The desired fit mode is selected and is applied to all compound series. The report is generally set to 5 s before the end of the injection but can be tailored to the data.
- The steady-state affinity (and/or kinetics) fits are inspected and are accepted or rejected. Results are exported from Results Table.

At this stage in the hit finding and validation process, compounds tend to be weak binders and can be binned into four categories: (1) weak binders that show dose-responsive behavior that reach saturation, (2) weak binders that show dose-responsive behavior but do not reach saturation, and (3) compounds that exhibit measurable binding kinetics, and (4) nonbinders and poorly behaved compounds. Compounds that show dose-responsive behaviors but do not reach saturation are noted as such. Because they do not reach saturation, it is not possible to determine a $K_D$. Unfortunately, it is not possible to fully distinguish these compounds from nonspecific compounds.

SPR is one of the most versatile and useful label-free biophysical methods for the characterization of biomolecular interactions. It can be used to characterize a wide range of binding interactions, from highly potent protein–protein interactions to weak fragment or LMW–protein interactions. SPR is a widely accepted method for fragment-based screening (Danielson, 2009; Erlanson, Fesik, Hubbard, Jahnke, & Jhoti, 2016; Giannetti, 2011; Hämäläinen et al., 2000; Jahnke & Erlanson, 2006; Navratilova & Hopkins, 2010; Neumann, Junker, Schmidt, & Sekul, 2007; Silvestre et al., 2013) and the process and methodologies are directly applicable to LMW affinity screening. The major difference between fragments and LMW compounds is the nature of the compounds themselves; fragments are soluble and generally well behaved and most early-stage LMW compounds are not. They tend to have limited solubility (usually less than 50 μ*M*) and are nonspecific aggregators (Coan & Shoichet, 2008) and/or are PAINS (Baell & Holloway, 2010). Because of sensitivity and throughput SPR is well suited for affinity screening and can be used to identify LMW compounds that directly interact with the target.

However, using SPR for affinity screening and/or characterization does come with its own set of challenges and well-known limitations. SPR is a serial flow technique, and although much effort is put into removing poorly behaving compounds using Clean Screens, it is still possible for compounds to damage the protein immobilized onto the sensor surface in subsequent cycles and assays. Positive control compounds are crucial ensuring that the integrity of the immobilized target on the surface can be monitor throughout the assay. The flow system also makes it challenging to work with protein complexes unless the affinity of the complex is low or the complex can be cross-linked together. The size of the target can be limiting with larger targets (>150 kDa) which become more challenging when screening LMW compounds due to the molecular weight ratio constraints inherent in SPR. This can potentially be mitigated by increasing the surface density, but care needs to be taken to ensure that the surface is active at the higher densities and that the surface is not overcrowded. However, at higher surface density, aggregation due to the high local protein concentration and mass transport must be taken into consideration. The small molecules present a wide variety of challenges including aggregation, promiscuity and nonspecific binding, solubility limitations, and weak binding interactions (small signals) and require the use of DMSO which has a very high refractive index. In addition, when in screening mode, DMSO mismatch can be challenging due to the differences in DMSO between compound samples and that used in the assay. The DMSO mismatch and void volume effects can be diminished via the use of Solvent Correction. Fortunately, the vast majority of the challenges and limitations can be overcome by careful experimental design and execution of SPR experiments.

## 4. CONCLUSION

In the last 10 years, affinity-based screening techniques have emerged as mature tools that have been successfully applied in many LMW lead finding campaigns by both academia and pharmaceutical industry. The field of biophysics has evolved rapidly in terms of increasing sensitivity, introduction of new technologies, sample throughput, degree of miniaturization, and reproducibility. These improvements have led to the integration biophysical methods early in the lead finding process strengthening the hit validation/optimization process enabling the identification of new chemical entities. In addition, targets without defined enzymatic activities or functions and challenging targets such as membrane proteins, multiprotein complexes, nucleic acids, and weak target/ligand interactions can now be characterized and/or screened using these methods. Affinity screening methods can be used to identify LWM compounds that bind to the target of interest but in most case provide no information on functional activity. Thus, affinity screening hits need to be validated in a functional assay whenever possible.

In this chapter, we described how to set up LMW compound affinity screens using DSF and SPR. The fluorescence thermal shift assay offers several advantages: (i) simple work flow, (ii) fast implementation, (iii) robust readout, (iv) modest instrumentation needs, (v) a cost-effective method to identify LMW compounds with a broad range of affinities, and (vi) the ability to work with many different target classes. Despite the simplicity of DSF, there are many details that need careful attention to successfully enable a screen. In the DSF section, we described in detail the major steps, e.g., dye selection/properties and other considerations which have to be taken into account. Furthermore, in the DSF data analysis section we highlight how to select promising compounds and do the conformation/validation. Contrary to DSF, SPR assays provide advantage of higher information content (yes/no binding, affinities ($K_D$), and kinetic parameters ($k_{on}$ and $k_{off}$ rates)) while consuming less protein. However, the sample throughput is significantly reduced compared to DSF. The newest generation of high-throughput, high-sensitivity SPR instruments has enabled the analysis of large number of compounds or multiple targets in a parallel format within a timeframe sufficient for drug discovery.

The future trends and challenges in affinity screening are (i) bridging biophysical in vitro assays to cellular environment, (ii) addressing conformational changes in a higher throughput format, and (iii) enlarging the chemical space by further implementing DNA-encoded technologies. Cellular biophysics applications such as cellular thermal shift assays (CETSA) (Martinez Molina et al., 2013) and in-cell NMR (Breindel, Burz, & Shekhtman, 2018; Lippens et al., 2018) have been successfully established. CETSA is used to analyze ligand target engagement in a complex environment offering information about compound cell permeability and target binding. This method has a potential to become an affinity screening tool and first plate-based screening assay formats have been published (Massey, 2018). The importance of conformational change in biology is well understood, and targeting new binding pockets generated by protein interactions, locking or stabilizing proteins into active or inactive conformations are all relevant and important modes of target modulation. However, monitoring conformation has been both time and labor intensive. New techniques, such as Biodesy (based on second harmonic generation), provide the means to monitor the conformational changes of a protein upon ligand binding in a rapid plated-based format binding (Butko, Moree, Mortensen, & Salafsky, 2016). The swift evolution of DNA-encoded techniques has enabled the production of compound libraries with $>10^8$ members greatly enlarging the tractable chemical space for compound screening. DNA-encoded screening workflows have been established which can identify hits from vast compound libraries via the use of affinity selection followed by the identification with PCR amplification and next-generation sequencing (reviewed in Zimmermann & Neri, 2017). The combination of the current state-of-the-art affinity selection methods with these novel techniques has the potential to revolutionize biophysical contributions of the modern drug discovery process.

## ACKNOWLEDGMENTS

We would like to thank all of our Novartis colleagues who have worked and contributed their knowledge and expertise in affinity screening.

## REFERENCES

Ablinger, E., Leitgeb, S., & Zimmer, A. (2013). Differential scanning fluorescence approach using a fluorescent molecular rotor to detect thermostability of proteins in surfactant-containing formulations. *International Journal of Pharmaceutics*, *441*, 255–260.

Alexandrov, A., Mileni, M., Chien, E. Y., Hanson, M. A., & Stevens, R. C. (2008). Microscale fluorescent thermal stability assay for membrane proteins. *Structure*, *16*(3), 351–359.

Anfinsen, C. B. (1973). Principles that govern the folding of protein chains. *Science*, *181*(4096), 223–230.

Baell, J. B., & Holloway, G. A. (2010). New substructure filters for removal of pan assay interference compounds (PAINS) from screening libraries and for their exclusion in bioassays. *Journal of Medicinal Chemistry*, *53*(7), 2719–2740.

Bergsdorf, C., Fiez-Vandal, C., Sykes, D. A., Bernet, P., Aussenac, S., Charlton, S. J., et al. (2016). An alternative thiol-reactive dye to analyze ligand interactions with the chemokine receptor CXCR2 using a new thermal shift assay format. *Journal of Biomolecular Screening*, *21*(3), 243–251.

Bergsdorf, C., & Ottl, J. (2010). Affinity-based screening techniques: Their impact and benefit to increase the number of high quality leads. *Expert Opinion on Drug Discovery*, *5*(11), 1095–1107.

Bischof, J. C., & He, X. (2005). Thermal stability of proteins. *Annals of the New York Academy of Sciences*, *1066*, 12–33.

Boivin, S., Kozak, S., & Meijers, R. (2013). Optimization of protein purification and characterization using Thermofluor screens. *Protein Expression and Purification*, *91*(2), 192–206.

Breindel, L., Burz, D. S., & Shekhtman, A. (2018). Interaction proteomics by using in-cell NMR spectroscopy. *The Journal of Proteomics*. in press. pii: S1874-3919(18)30055-1.

Butko, M. T., Moree, B., Mortensen, R. B., & Salafsky, J. (2016). Detection of ligand-induced conformational changes in oligonucleotides by second-harmonic generation at a supported lipid bilayer interface. *Analytical Chemistry*, *88*(21), 10482–10489.

Ciulli, A. (2013). Biophysical screening for the discovery of small-molecule ligands. *Methods in Molecular Biology*, *1008*, 357–388.

Coan, K. E., & Shoichet, B. K. (2008). Stoichiometry and physical chemistry of promiscuous aggregate-based inhibitors. *Journal of the American Chemical Society*, *130*(29), 9606–9612.

Danielson, U. H. (2009). Fragment library screening and lead characterization using SPR biosensors. *Current Topics in Medicinal Chemistry*, *9*(18), 1725–1735.

Dupeux, F., Röwer, M., Seroul, G., Blot, D., & Márquez, J. A. (2011). A thermal stability assay can help to estimate the crystallization likelihood of biological samples. *Acta Crystallographica. Section D, Biological Crystallography*, *67*(Pt. 11), 915–919.

Erlanson, D. A., Fesik, S. W., Hubbard, R. E., Jahnke, W., & Jhoti, H. (2016). Twenty years on: The impact of fragments on drug discovery. *Nature Reviews. Drug Discovery*, *15*(9), 605–619.

Folmer, R. H. (2016). Integrating biophysics with HTS-driven drug discovery projects. *Drug Discovery Today*, *21*(3), 491–498.

Genick, C. C., & Wright, S. K. (2017). Biophysics: For HTS hit validation, chemical lead optimization, and beyond. *Expert Opinion on Drug Discovery*, *12*(9), 897–907.

Giannetti, A. M. (2011). Chapter eight—From experimental design to validated hits: A comprehensive walk-through of fragment lead identification using surface plasmon resonance. In L. C. Kuo (Ed.), *Methods in enzymology: Vol. 493*. (pp. 169–218). Academic Press.

Good, N. E., Winget, G. D., Winter, W., Connoll, T. N., Izawa, S., & Singh, R. M. (1966). Hydrogen ion buffers for biological research. *Biochemistry*, *5*(2), 467–477.

Haidekker, M. A., & Theodorakis, E. A. (2010). Environment-sensitive behavior of fluorescent molecular rotors. *Journal of Biological Engineering*, *4*, 4–11.

Hämäläinen, M. D., Markgren, P. O., Schaal, W., Karlén, A., Classon, B., Vrang, L., et al. (2000). Characterization of a set of HIV-1 protease inhibitors using binding kinetics data from a biosensor-based screen. *Journal of Biomolecular Screening*, *5*(5), 353–360.

Hawe, A., Filipe, V., & Jiskoot, W. (2010). Fluorescent molecular rotors as dyes to characterize polysorbate-containing IgG formulations. *Pharmaceutical Research*, *27*(2), 314–326.

Hirano, M., Adachi, Y., Ito, Y., & Totani, K. (2015). Calreticulin discriminates the proximal region at the N-glycosylation site of Glc1Man9GlcNAc2 ligand. *Biochemical and Biophysical Research Communications*, *466*(3), 350–355.

Hofmann, L., Gulati, S., Sears, A., Stewart, P. L., & Palczewski, K. (2016). An effective thiol-reactive probe for differential scanning fluorimetry with a standard real-time polymerase chain reaction device. *Analytical Biochemistry*, *499*, 63–65.

Huang, R., Ayine-Tora, D. M., Muhammad Rosdi, M. N., Li, Y., Reynisson, J., & Leung, I. K. H. (2017). Virtual screening and biophysical studies lead to HSP90 inhibitors. *Bioorganic & Medicinal Chemistry Letters*, *27*(2), 277–281.

Jahnke, W., & Erlanson, D. A. (2006). Fragment-based approaches in drug discovery. *Vol. 34*. John Wiley & Sons.

Janzen, W. P. (2014). Screening technologies for small molecule discovery: The state of the art. *Chemistry & Biology*, *21*(9), 1162–1170.

Johnson, C. M. (2013). Differential scanning calorimetry as a tool for protein folding and stability. *Archives of Biochemistry and Biophysics*, *531*(1–2), 100–109.

Karlsson, R., & Fält, A. (1997). Experimental design for kinetic analysis of protein-protein interactions with surface plasmon resonance biosensors. *Journal of Immunological Methods*, *200*(1), 121–133.

Kelly, S. M., & Price, N. C. (2000). The use of circular dichroism in the investigation of protein structure and function. *Current Protein & Peptide Science*, *1*(4), 349–384.

Kopec, J., & Schneider, G. (2013). Comparison of fluorescence and light scattering based methods to assess formation and stability of protein-protein complexes. *Analytical Biochemistry*, *443*(1), 52–57.

Kozak, S., Lercher, L., Karanth, M. N., Meijers, R., Carlomagno, T., & Boivin, S. (2016). Optimization of protein samples for NMR using thermal shift assays. *Journal of Biomolecular NMR*, *64*(4), 281–289.

Lepock, J. R., Ritchie, K. P., Kolios, M. C., Rodahl, A. M., Heinz, K. A., & Kruuv, J. (1992). Influence of transition rates and scan rate on kinetic simulations of differential scanning calorimetry profiles of reversible and irreversible protein denaturation. *Biochemistry*, *31*(50), 12706–12712.

Lippens, G., Cahoreau, E., Millard, P., Charlier, C., Lopez, J., Hanoulle, X., et al. (2018). In-cell NMR: From metabolites to macromolecules. *Analyst*, *143*(3), 620–629.

Mahendrarajah, K., Dalby, P. A., Wilkinson, B., Jackson, S. E., & Main, E. R. (2011). A high-throughput fluorescence chemical denaturation assay as a general screen for protein-ligand binding. *Analytical Biochemistry*, *411*(1), 155–157.

Markgren, P. O., Lindgren, M. T., Gertow, K., Karlsson, R., Hämäläinen, M., & Danielson, U. H. (2001). Determination of interaction kinetic constants for HIV-1 protease inhibitors using optical biosensor technology. *Analytical Biochemistry*, *291*(2), 207–218.

Markgren, P.-O., Lindgren, M. T., Gertow, K., Karlsson, R., Hämäläinen, M., & Danielson, U. H. (2003). *Biacore sensor surface handbook, Version AA*. Uppsala, Sweden: Biacore.

Martinez Molina, D., Jafari, R., Ignatushchenko, M., Seki, T., Larsson, E. A., Dan, C., et al. (2013). Monitoring drug target engagement in cells and tissues using the cellular thermal shift assay. *Science*, *341*(6141), 84–87.

Massey, A. J. (2018). A high content, high throughput cellular thermal stability assay for measuring drug-target engagement in living cells. *PLoS One*, *13*(4), e0195050. https://doi.org/10.1371/journal.pone.0195050.

Mayr, L. M., & Bojanic, D. (2009). Novel trends in high-throughput screening. *Current Opinion in Pharmacology*, *9*(5), 580–588.

Muca, R., Marek, W., Żurawski, M., Piątkowski, W., & Antos, D. (2017). Effect of mass overloading on binding and elution of unstable proteins in hydrophobic interaction chromatography. *Journal of Chromatography. A*, *1492*, 79–88.

Myers, J. K., Pace, C. N., & Scholtz, J. M. (1995). Denaturant m values and heat capacity changes: Relation to changes in accessible surface areas of protein unfolding. *Protein Science*, *4*(10), 2138–2148.

Navratilova, I., & Hopkins, A. L. (2010). Fragment screening by surface plasmon resonance. *ACS Medicinal Chemistry Letters*, *1*(1), 44–48.

Neumann, T., Junker, H. D., Schmidt, K., & Sekul, R. (2007). SPR-based fragment screening: Advantages and applications. *Current Topics in Medicinal Chemistry*, 7(16), 1630–1642.

Niesen, F. H., Berglund, H., & Vedadi, M. (2007). The use of differential scanning fluorimetry to detect ligand interactions that promote protein stability. *Nature Protocols*, *2*(9), 2212–2221.

Olaru, A., Bala, C., Jaffrezic-Renault, N., & Aboul-Enein, H. Y. (2015). Surface plasmon resonance (SPR) biosensors in pharmaceutical analysis. *Critical Reviews in Analytical Chemistry*, *45*(2), 97–105.

Pantoliano, M. W., Petrella, E. C., Kwasnoski, J. D., Lobanov, V. S., Myslik, J., Graf, E., et al. (2001). High-density miniaturized thermal shift assays as a general strategy for drug discovery. *Journal of Biomolecular Screening*, *6*(6), 429–440.

Renaud, J. P., Chung, C. W., Danielson, U. H., Egner, U., Hennig, M., Hubbard, R. E., et al. (2016). Biophysics in drug discovery: Impact, challenges and opportunities. *Nature Reviews. Drug Discovery*, *15*(10), 679–698.

Rosa, N., Ristic, M., Seabrook, S. A., Lovell, D., Lucent, D., & Newman, J. (2015). Meltdown: A tool to help in the interpretation of thermal melt curves acquired by differential scanning fluorimetry. *Journal of Biomolecular Screening*, *20*(7), 898–905.

Ross, P., Weihofen, W., Siu, F., Xie, A., Katakia, H., Wright, S. K., et al. (2015). Isothermal chemical denaturation to determine binding affinity of small molecules to G-protein coupled receptors. *Analytical Biochemistry*, *473*, 41–45.

Schasfoort, R. (2017). *Handbook of surface plasmon resonance* (2nd ed.). Royal Society of Chemistry.

Schön, A., Brown, R. K., Hutchins, B. M., & Freire, E. (2013). Ligand binding analysis and screening by chemical denaturation shift. *Analytical Biochemistry*, *443*(1), 52–57.

Schulz, M. N., Landström, J., & Hubbard, R. E. (2013). MTSA—A Matlab program to fit thermal shift data. *Analytical Biochemistry*, *433*(1), 43–47.
Seabrook, S. A., & Newman, J. (2013). High-throughput thermal scanning for protein stability: Making a good technique more robust. *ACS Combinatorial Science*, *15*(8), 387–392.
Senisterra, G., Chau, I., & Vedadi, M. (2012). Thermal denaturation assays in chemical biology. *Assay and Drug Development Technologies*, *10*(2), 128–136.
Senisterra, G. A., Ghanei, H., Khutoreskaya, G., Dobrovetsky, E., Edwards, A. M., Privé, G. G., et al. (2010). Assessing the stability of membrane proteins to detect ligand binding using differential static light scattering. *Journal of Biomolecular Screening*, *15*(3), 314–320.
Senisterra, G. A., Markin, E., Yamazaki, K., Hui, R., Vedadi, M., & Awrey, D. E. (2006). Screening for ligands using a generic and high-throughput light-scattering-based assay. *Journal of Biomolecular Screening*, *11*(8), 940–948.
Silvestre, H. L., Blundell, T. L., Abell, C., & Ciulli, A. (2013). Integrated biophysical approach to fragment screening and validation for fragment-based lead discovery. *Proceedings of the National Academy of Sciences of the United States of America*, *110*(32), 12984–12989.
Simeonov, A. (2013). Recent developments in the use of differential scanning fluorometry in protein and small molecule discovery and characterization. *Expert Opinion on Drug Discovery*, *8*(9), 1071–1082.
Svilenov, H., Markoja, U., & Winter, G. (2018). Isothermal chemical denaturation as a complementary tool to overcome limitations of thermal differential scanning fluorimetry in predicting physical stability of protein formulations. *European Journal of Pharmaceutics and Biopharmaceutics*, *125*, 106–113.
Temel, D. B., Landsman, P., & Brader, M. L. (2016). Chapter fourteen—Orthogonal methods for characterizing the unfolding of therapeutic monoclonal antibodies: Differential scanning calorimetry, isothermal chemical denaturation, and intrinsic fluorescence with concomitant static light scattering. In A. L. Feig (Ed.), *Methods in enzymology: Vol. 567* (pp. 359–389). Academic Press.
Vedadi, M., Niesen, F. H., Allali-Hassani, A., Fedorov, O. Y., Finerty, P. J., Wasney, G. A., et al. (2006). Chemical screening methods to identify ligands that promote protein stability, protein crystallization, and structure determination. *Proceedings of the National Academy of Sciences of the United States of America*, *103*(43), 15835–15840.
Wong, J. J., Wright, S. K., Ghozalli, I., Mehra, R., Furuya, K., & Katayama, D. S. (2016). Screening using a fluorescent molecular rotor dye, 4-(4-(dimethylamino)styryl)-N-methylpyridinium iodide (DASPMI). *Journal of Biomolecular Screening*, *21*(8), 842–850.
Zimmermann, G., & Neri, D. (2017). DNA-encoded chemical libraries: Foundations and applications in lead discovery. *Drug Discovery Today*, *21*(11), 1828–1834.

CHAPTER EIGHT

# Second-Harmonic Generation (SHG) for Conformational Measurements: Assay Development, Optimization, and Screening

**Tracy A. Young*, Ben Moree*, Margaret T. Butko*, Bason Clancy*, Mary Geck Do†, Tarun Gheyi‡, John Strelow§, Juan J. Carrillo‡, Joshua Salafsky*[,1]**

*Biodesy, Inc., South San Francisco, CA, United States
†The University of Texas MD Anderson Cancer Center, Institute for Applied Cancer Science, Houston, TX, United States
‡Lilly Biotechnology Center, San Diego, CA, United States
§Lilly Research Laboratories, Eli Lilly and Company, Lilly Corporate Center, Indianapolis, IN, United States
[1]Corresponding author: e-mail address: joshua.salafsky@biodesy.com

## Contents



## Abstract

Second-harmonic generation (SHG) has recently emerged as a biophysical tool for conformational sensing of a target biomolecule upon binding to ligands such as small molecules, fragments, proteins, peptides, and oligonucleotides. To date, SHG has been used

*Methods in Enzymology*, Volume 610
ISSN 0076-6879
https://doi.org/10.1016/bs.mie.2018.09.017

to measure conformational changes of targets such as soluble proteins, protein complexes, intrinsically disordered proteins, peripheral and integral membrane proteins, peptides, and oligonucleotides upon binding of ligands over a wide range of affinities. In this chapter, we will provide a technology overview, detailed protocols for optimizing assays and screening, practical considerations, and an example case study to guide the reader in developing robust and informative measurements using the Biodesy Delta SHG platform.

## ABBREVIATIONS

| | |
|---|---|
| **CMC** | critical micelle concentration |
| **CRC** | concentration response curve |
| **DoL** | degree of labeling |
| **DOPC** | dioleoyl phosphatidylcholine |
| **His** | histidine |
| **Mal** | maleimide |
| **Ni-NTA** | nickel-nitrilotriacetic acid |
| **PS** | phosphatidylserine |
| **SE** | succinimidyl ester |
| **SH-active** | second-harmonic active |
| **SHG** | second-harmonic generation |
| **SLB** | supported lipid bilayer |
| **SUVs** | small unilamellar vesicles |
| **TIR** | total internal reflection |

## 1. INTRODUCTION

Second-harmonic generation (SHG) has recently emerged as a biophysical tool for conformational sensing of a target biomolecule upon binding to ligands such as small molecules, fragments, proteins, peptides, and oligonucleotides. To date, SHG has been used to measure conformational changes of targets such as soluble proteins, protein complexes, intrinsically disordered proteins, peripheral and integral membrane proteins, peptides, and oligonucleotides upon binding of ligands over a wide range of affinities (Butko, Moree, Mortensen, & Salafsky, 2016; Ma et al., 2016; Moree et al., 2015; Moree et al., 2015; Rizvi et al., 2018; Salafsky, 2003, 2006, 2007; Wong et al., 2017). In this chapter, we will provide a technology overview, detailed protocols for optimizing assays and screening, practical considerations, and an example case study to guide the reader in developing robust and informative measurements using the Biodesy Delta SHG platform.

## 2. TECHNOLOGY BACKGROUND

The Biodesy Delta is a high-throughput screening instrument for measuring conformational changes of biomolecules rendered second-harmonic active (SH-active) by the use of dye probes. Conformational changes are detected spectroscopically using SHG, a two-photon process involving the conversion of two photons from an incident laser into a single photon of twice the energy (Tran, Sly, & Conboy, 2017). In the Biodesy Delta, light from an 800 nm pulsed laser is converted into 400 nm with an efficiency that depends very sensitively on the angular orientation of SH-active dyes with respect to the surface to which the labeled biomolecules are tethered (Fig. 1). When a SH-active dye is conjugated to a biomolecule of interest and the conjugate is tethered to a surface, ligand-induced conformational changes of the biomolecule that result in net dye movement are observed by a change of the SHG signal (Moree, Connell, et al., 2015). Because this process is independent of the mass of the biomolecules or the ligands, a wide variety of interactions can be studied using this technique.

The Biodesy Delta is a well-based medium- to high-throughput instrument using 384- or 1536-well microtiter plates, respectively. Each well within the plate is coupled to a custom optical prism that enables the focusing of the incident laser into the well. The prisms are engineered for total internal reflection (TIR) of the light from the glass-buffer interface, creating

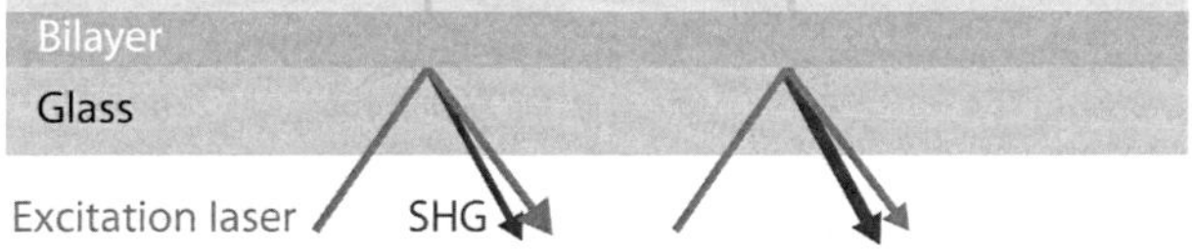


**Fig. 1** *Cartoon representation* of the experimental setup. Proteins conjugated with a SHG-active dye are attached to a supported lipid bilayer via a His-tag. The SHG intensity depends upon the angle $\theta$, which is the angle between the dye transition dipole moment and the surface normal ($z$-axis). Conformational changes are detected upon ligand binding as the dye orientation changes relative to the orientation before adding ligand.

an exponentially decaying electric field over a distance of tens of nanometers from the interface into the buffer. The intrinsic surface selectivity of SHG allows for discrimination between surface-bound SH-active molecules, which possess a net, average orientation with respect to the surface, and molecules present in the bulk; molecules randomly oriented in bulk produce no SHG signal because second-harmonic waves originating from different molecules in a given volume interfere completely destructively. Many different assay formats can be developed using SHG. In the simplest assay, the bottom of the well is covered with a Ni-NTA (nickel-nitrilotriacetic acid) supported lipid bilayer (SLB), which serves to passivate the glass surface and allows biomolecules to bind to a biomimetic membrane via a His-tag. Incubation of biomolecules in the well after bilayer formation creates a molecular monolayer with an average density of thousands of molecules per square micron (Moree, Connell, et al., 2015). By conjugating SH-active dyes to the biomolecule at one or multiple site(s), the SHG intensity, $I_{SHG}$, can be used to monitor the conformational change at each site of interest where the signal is proportional to the square of the molecular surface density, $N$, and the orientational average of the dye relative to the surface normal according to:

$$I_{SHG} \propto N^2 \left\langle \cos^3\theta \right\rangle^2 \tag{1}$$

where $\theta$ is the angle between the dye transition dipole moment and the surface normal (Fig. 1) and the brackets denote an average over all orientations including the conformational landscape of the biomolecule and flexibility of the dye-biomolecule bond.

In general, higher SHG signals will be produced as the dye becomes more aligned with the surface normal (i.e., more vertical, $\theta \rightarrow 0$), as demonstrated by Eq. (1). Changes in the conformation and thus the dye orientation at a given site on the biomolecule are registered instantaneously as a change in the SHG intensity. The high sensitivity of SHG to probe orientation (and thus biomolecule conformation) originates from the nonlinear photophysics that govern SHG. SHG is therefore physically distinct from and more sensitive than linear techniques such as fluorescence (Brasselet, 2011; Lakowicz, 2006). When two incident photons are converted by the dye probe into a single SHG photon, no energy is lost and the SHG beam maintains characteristics of the incident laser such as directionality and coherence. In essence, a second, less powerful laser beam is produced—the second harmonic—at half the original wavelength. The directionality of SHG greatly simplifies detection, and the signal is easily collected and spatially and spectrally separated from the incident laser.

# 3. ASSAY DEVELOPMENT

SHG assays have been developed for a wide variety of biomolecules (Butko et al., 2016; Ma et al., 2016; Moree, Connell, et al., 2015; Moree, Yin, et al., 2015; Rizvi et al., 2018; Wong et al., 2017). This chapter will focus on the two main applications of SHG that are most widely used today. The two applications differ from each other mainly in how the target biomolecule is labeled. In the first application, small molecule or fragment libraries are screened to discover new chemical matter that targets a specific biomolecule. In these assays, it is important to label the target so that the conjugate sensitively detects conformational changes associated with binding of chemical entities to sites throughout the target molecule. Therefore, assay development is simplified by not requiring precise placement of the SH-active dye on the target. As the SHG signal depends on the net, average orientation of probes, native residue labeling of a biomolecule fractionally labeled at multiple sites is a powerful means to sense conformation at each of these sites simultaneously.

In the second application, specific conformational states are resolved with high resolution to classify or bin compounds by mechanism of action or the conformation induced upon binding. Such information may be correlated with any functional data available, such as pharmacological activity in cells and enzymatic activity, thus providing different and unique signatures of the conformations that underlie biological activity and function. For example, assays can be built that distinguish antagonists from agonists, inhibitors from activators, and orthosteric ligands from allosteric ones. This application typically requires precise placement of the dye and fine-tuning of the assay conditions, and is used to best advantage when existing structural or functional information is available. It is important to note that regardless of the goal of the experiment, the following steps should be considered a general starting point for developing the desired assay; assay development may require several iterations to achieve the proper dye placement and final optimized assay readout for discrimination of distinct conformational states. Alternatively, engineered residue labeling can be achieved, for example, with a cysteine mutation at the desired site of labeling.

Developing a SHG assay consists of six main steps as outlined in Fig. 2. For simplicity, the example here describes a His-tagged soluble protein tethered to a Ni-NTA SLB. Briefly, a purified, His-tagged protein is labeled with a SH-active dye (Fig. 2A). The labeled protein is then deposited into selected wells of a 384-well Biodesy Delta plate (Fig. 2B) that have been

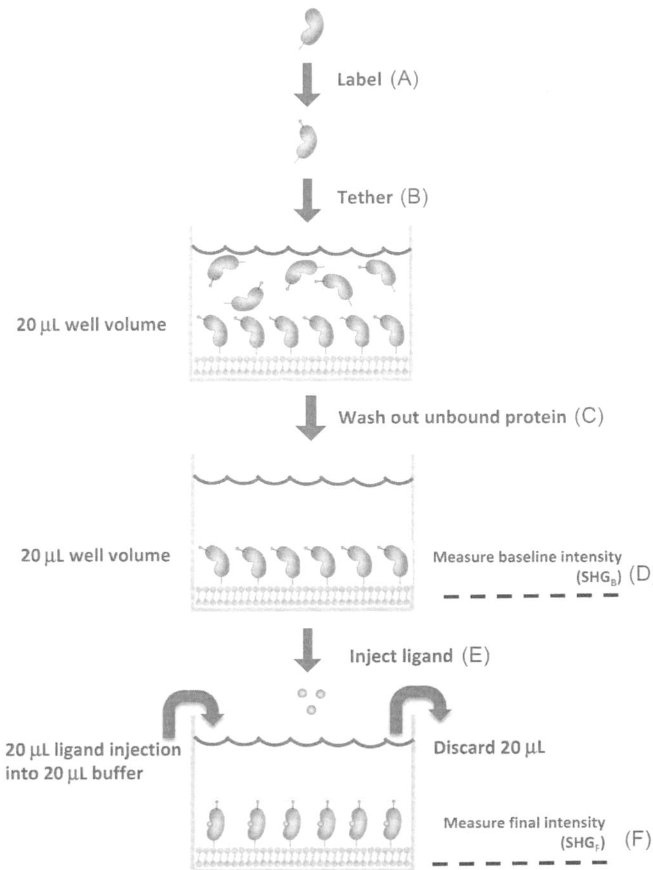


**Fig. 2** Schematic of SHG assay workflow. (A) His-tagged, unlabeled protein is labeled with SH-active dye to produce the conjugate. (B) The conjugate is tethered to the Biodesy Ni-NTA surface via His-tag. (C) Unbound conjugate is washed out. (D) Baseline SHG intensity is measured. (E) Ligand is injected. (F) Final SHG intensity is measured.

precoated with the Ni-NTA SLB. The sample is incubated in the well to allow the protein sufficient time to tether to the SLB surface. After tethering, unbound protein is washed out (Fig. 2C) and the baseline SHG signal ($SHG_B$) is read (Fig. 2d). After reading the baseline, a ligand is injected (Fig. 2E), and the SHG signal is read afterward ($SHG_F$) (Fig. 2F).

## 3.1 Before Starting

### *3.1.1 Choosing the Surface*

Biodesy offers a number of different tethering surfaces. Three are commercially available SLB surfaces, and two are nonlipid-based surfaces currently in development but available for testing. The available SLB preparations include one with Ni-NTA head groups for capture of His-tagged biomolecules and another with biotin head groups for capture of avidin followed by secondary capture of biotinylated biomolecules. For oligonucleotides, the biotin SLB surface is recommended because biotin is easily added to oligonucleotides during synthesis (Butko et al., 2016; Rizvi et al., 2018). To use the biotin SLB surface with proteins, an AviTag™ is recommended for biotinylation to ensure a single site-specific biotinylation site per protein, allowing the protein to be uniformly oriented on the surface. The third available SLB preparation is based on phosphatidylserine (PS) lipid. This surface allows tethering of peripheral membrane proteins without the need for an affinity tag through native interactions with the PS lipid. The two nonlipid-based surfaces currently in development are compatible with His-tagged or biotinylated molecules, and these surfaces are resistant to detergents and other buffer conditions that are incompatible with SLB surfaces. An in-depth protocol for setting up all of the available surfaces is beyond the scope of this chapter. Instead, this chapter will describe assay development and case studies on the most commonly used surface, the Biodesy Ni-NTA bilayer.

### 3.1.2 SHG Assay-Compatible Buffers

The Biodesy Delta is compatible with a wide range of buffer conditions and most buffers used in other biophysical methods and biochemical assays are compatible with the SHG assay. However, it is important to consider the type of SLB surface and the interactions necessary to tether a molecule to the surface when choosing an assay buffer. The primary consideration is the use of detergents in the buffer. In most cases, buffers containing nonionic detergents below the critical micelle concentration (CMC) preserve SLB integrity. In the case of integral membrane proteins or other proteins that require detergent concentrations above the CMC, a nonlipid-based surface is required. For the Ni-NTA SLB, it is important to avoid conditions and reagents that disrupt the Ni-NTA:His-tag interaction. These include chelating agents such as EDTA, high concentrations of reducing agents, and buffers at $pH < 6.0$.

Another practical consideration in the selection of a buffer is the refractive index of the solution. High concentrations of any buffer components, including but not limited to salt ($>1$ *M*), glycerol ($>10\%$), and DMSO ($>5\%$), lead to a change in the refractive index of the solution, which alters the path of the incident laser beam in the well and leads to a loss of SHG signal in the system. Therefore, high concentrations of these components should be avoided or washed out prior to SHG measurements.

### 3.1.3 Choosing SHG Assay Controls

SHG assays can be developed both with and without specific control ligands. If an assay is developed with the goal of discovering new chemical matter without regard to the specific conformational change induced in the target, small molecule controls may not be available and are not necessary. Furthermore, any molecule that binds to the target of interest and produces a SHG signal change may serve as a positive control for developing a small molecule or fragment screening assay. A wide range of generic SHG assays for screening have been developed with ligands including antibodies, DNA, RNA, protein-binding partners, peptides, and nucleotides when small molecule controls were not available. If, on the other hand, the aim is to build an assay that resolves specific conformational states, then well-characterized control ligands with orthogonal data (e.g., X-ray structures, NMR, cellular assays, etc.) are necessary. For example, if an assay is desired to discriminate between and classify unknown small molecules—for example, orthosteric vs allosteric inhibitors or agonists vs antagonists—then at least one and ideally two or more control ligands from each class should guide assay development.

## 3.2 Labeling

The most critical step in developing an informative SHG assay is dye placement on the target biomolecule. There are currently two commercially available dyes probes with distinct conjugation chemistries. One dye, SHG1-SE (Biodesy, Inc.), covalently reacts with surface-exposed primary amine groups (e.g., lysine residues) through succinimidyl ester (SE) chemistry. The other dye probe, SHG2-Mal (Biodesy, Inc.), covalently reacts with surface-exposed free thiol groups (e.g., reduced cysteine residues) through maleimide (Mal) chemistry. Typically, both chemistries are employed during the first step of assay development because optimal dye placement may be difficult to predict a priori. Generally, multiple assays with different conjugation chemistries and reaction conditions are developed and tested in parallel. Employing both chemistries at the outset allows for a diverse number of regions of the biomolecule to be sampled, thereby allowing a wide range of assays for resolving desired conformational states and maximizing assay windows to be tested. For most conjugations, regardless of dye chemistry, a degree of labeling (DoL) of 0.5–2 dyes per protein is found to be minimally perturbative to protein activity, although successful assays have been developed outside this range. The labeling protocols described in detail below are recommended for proteins. Oligonucleotide conjugation is described in a separate section below.

### 3.2.1 General Considerations for Labeling: SE Chemistry

Before providing a detailed protocol for labeling a target protein with the SHG1-SE dye, several important considerations regarding lysine residues should be discussed. Lysine residues are usually abundantly represented in the primary protein sequence and solvent-exposed due to the presence of a primary, charged amine on the ε-carbon of the amino acid side chain. The typical p$K_a$ of this primary amine is ~10.5, although the actual p$K_a$ of any specific lysine residue may vary with the local protein microenvironment. These local effects, in turn, affect the reactivity of the primary amine, which serves as a nucleophile in the reaction with SE dyes (Hermanson, 2013). The combination of the relatively high lysine abundance and the sensitivity to local environmental effects makes it difficult to predict a priori which residue(s) will be targeted in a given labeling reaction. However, dye: protein conjugates of interest can be characterized after the reaction by mass spectrometry to determine the label site(s).

A common labeling buffer for the lysine-SE reaction is 150 m*M* NaCl, 50 m*M* sodium bicarbonate buffer at pH 8.3, although the optimal buffer composition will depend on protein-specific requirements. In general, it is best to avoid buffering reagents or other components that introduce primary amines into the reaction solution, as they will compete with the lysines on the target protein, decreasing the efficiency of the labeling reaction. The optimal pH range for SE conjugation on lysine residues is between pH 7 and 9 (Hermanson, 2013), but can be adjusted outside this range when necessary. Varying the pH of this reaction alters the reactivity of surface-exposed lysine residues, resulting in unique distributions of lysine residues labeled at different pH levels (Moree, Connell, et al., 2015). Accordingly, conjugates produced by labeling at different pH levels typically exhibit different sensitivities to a given ligand-induced conformational change.

Most SE labeling reactions produce a heterogeneous population of modified lysine residues as determined by mass spectrometry (Ma et al., 2016; Moree, Connell, et al., 2015; Wong et al., 2017). While we have observed highly specific targeting of a single lysine residue, this is protein-specific and generally not the rule (unpublished data). However, despite any observed heterogeneity of labeled lysines, the sites labeled and the total DoL is highly reproducible when conjugation reactions are performed under identical conditions as confirmed by mass spectrometry (data not shown).

### 3.2.2 General Considerations for Labeling: Maleimide Chemistry

Maleimide labeling chemistry has a distinct set of requirements for the labeling reaction that must be considered prior to conjugation. Maleimide groups typically react with the thiol side chain of cysteine residues, which exhibit a p$K_a$ ~8.4. To maximize the specificity and reactivity of the cysteine–maleimide reaction, conjugations are typically carried out between pH 6.5 and 7.5 (Hermanson, 2013).

A common labeling buffer for this reaction is 150 m*M* NaCl, 50 m*M* sodium phosphate buffer at pH 6.5, which minimizes off-target reactivity, but the optimal buffer depends on any protein-specific requirements. In addition, it is important to not use reducing agents such as DTT or BME in the labeling buffer, due to their free thiols, as they will generally compete with the cysteine–maleimide reaction. In the event that reducing agents are required, TCEP is preferred because it does not compete as effectively as DTT or BME (Getz, Xiao, Chakrabarty, Cooke, & Selvin, 1999).

Another important consideration for maleimide labeling is the presence of free surface-exposed cysteine residues. On average, cysteine residues are less abundant in the primary amino acid sequence of a protein than lysines, and they may be solvent-exposed, buried, or paired in disulfide bonds. Therefore, the total available reactive cysteine residues in a target protein are often limited, so maleimide reactions generally produce more selective and less heterogeneous conjugates than SE reactions (Moree, Yin, et al., 2015).

### 3.2.3 Labeling Protocol

1. Incubate all buffers and protein on ice until cold, set up all reactions on ice.
2. Buffer exchange 100 μL of stock protein into desired labeling buffer with a 0.5 mL Zeba™ Spin Desalting Column according to the manufacturer's instructions (ThermoFisher). Repeat for each labeling condition to be tested.
3. Determine the buffer-exchanged protein concentration using a UV–vis spectrometer such as a NanoDrop (ThermoFisher Scientific).
4. Dilute the protein to 25 μ*M* in labeling buffer for a final volume of 100 μL. Repeat for each labeling condition to be tested.
5. Solubilize the lyophilized dye in 100% anhydrous DMSO to a stock concentration of up to 20 m*M*.
   - Note: Lyophilized dye should be diluted to a concentration that will result in a final DMSO concentration of 1%–3% in the reaction.
6. Add 1 μL of 5 m*M* SHG dye in DMSO to the inside of the cap of the tube containing 100 μL of 25 μ*M* protein.

- Note: The above reaction conditions result in a 2:1 dye to protein ratio. If the available protein amount allows, perform 2:1 and 5:1 (dye:protein) reaction ratios for both SHG1-SE and SHG2-Mal dyes as a starting point.

7. Close the cap and mix gently by inversion to ensure immediate and even distribution of the dye in the protein solution.
8. Incubate the reaction on ice for 10 min (SHG1-SE) or 90 min (SHG2-Mal).
9. Centrifuge at 16,000 × *g* for 20 min at 4°C.
   - Note: If there is precipitate at this step, decreasing the dye and/or protein concentration, or reaction time, can be attempted in order to increase yield. Glycerol may also be added to increase yield.
10. Buffer exchange the supernatant into storage buffer and remove unreacted dye with a 0.5 mL Zeba™ Spin Desalting Column according to the manufacturer's instructions (ThermoFisher).
11. Aliquot and flash-freeze any conjugate that will not be used the same day in liquid $N_2$ and store at −80°C until use.

### *3.2.4 Conjugate Characterization: Determining DoL*

After conjugation, the average number of dye molecules per protein, or DoL, can be determined by measuring absorbance at 280 and 410 nm with a UV–vis spectrometer such as a Nanodrop and the following equations:

$$[\text{Protein}] = (A_{280} - (A_{410}{}^{*}0.65))/\text{Extinction Coefficient}_{\text{protein}}(M^{-1}\,\text{cm}^{-1})$$
$$[\text{Dye}] = A_{410}/\text{Extinction Coefficient}_{\text{dye}}(M^{-1}\,\text{cm}^{-1})$$
$$\text{DoL} = [\text{Dye}]/[\text{Protein}]$$

Note that the peak absorption of the SH-active dye is at 410 nm, but the dye also contributes to absorption at 280 nm; an empirically determined correction factor of 0.65 is used to correct for the dye contribution at 280 nm to accurately determine the protein concentration after conjugation. The extinction coefficients for the SHG1-SE and SHG2-Mal dyes are 25,020 ($M^{-1}$ cm$^{-1}$) and 23,000 ($M^{-1}$ cm$^{-1}$), respectively. In order to determine the exact conjugation site, peptide mapping by mass spectrometry analysis is required, but in most cases this is not necessary for successful SHG assay development.

### 3.2.5 Fine-Tuning the Native Residue Labeling Reaction

The above labeling conditions are meant as starting points for new assays. After the resulting DoL is known, and the conjugates have gone through the following assay development steps to assess the SHG signal change upon ligand binding, the conjugation reactions can be fine-tuned for further assay optimization, if necessary. If the conjugates from the starting reactions are not optimal, decreasing or increasing the DoL and labeling with SE chemistry at multiple pHs in order to label new sites should be tested. The following list provides the main conditions that can be varied to alter the DoL and labeling site(s): pH, dye to protein ratio, time and temperature of incubation, buffer composition, and protein concentration.

### 3.2.6 Engineered Residue Labeling vs Native Residue Labeling

If the goal is to build an assay that reports on a specific conformational change, site-specific mutagenesis can be employed to engineer nonnative cysteine residues for precision labeling. This process can decrease the amount of time for assay development and guarantees that the dye is placed on the protein at the desired location. Engineered residue labeling requires the introduction of a cysteine residue at a predetermined site and concomitant removal of any solvent-exposed native cysteine residues from the protein. Cysteine residues that are buried or paired in disulfide bonds can be retained in the native protein sequence. We recommend test labeling the wild-type protein prior to mutagenesis to confirm that no reactive cysteines are present, even in cases in which the X-ray structures indicate the cysteines are buried, as the dynamic nature of proteins can lead to their exposure in solvent.

### 3.2.7 Labeling Oligonucleotides

DNA- and RNA-based oligonucleotides shorter than ~100 bases can be custom synthesized by commercial vendors with primary amine groups placed on the 5′ or 3′ end, on all four deoxyribonucleotides, and on uridine ribonucleotides. These amine groups can be conjugated with the SHG1-SE dye as described above for proteins, and engineering the position of the dye at multiple sites around the oligonucleotide is recommended to determine the optimal dye location. Alternatively, TriLink Biosciences (San Diego, CA) offers conjugation of the SHG1-SE dye (Biodesy1 in their catalog) with custom synthesis. Note that in all of these scenarios, there is a C6 linker between the dye and the nucleotide. It is also possible to order a C3 linker at the 3′ end of the oligo if desired, and changing the linker length at this terminus can result in different assay readouts. Therefore, both linker lengths should be tested during assay development, if possible.

## 3.3 Tethering

After labeling, conjugates are tethered to the Ni-NTA SLB coating the wells of the Biodesy Delta plate to determine the baseline SHG signal and the signal change upon ligand injection. All assay development is generally done using 384-well plates as follows:

### *3.3.1 Preparing the Ni-NTA SLB Surface*

1. Resuspend a vial of lyophilized Ni-NTA bilayer (Biodesy, Inc.) in purified water to a final concentration of 0.5 mg/mL.
2. Vortex the mixture for 5 min.
3. Transfer 2 mL aliquots of the sonicated sample into 6.5 mL (85 × 13 mm) polystyrene tubes on ice.
4. Sonicate on ice with a microtip at 25% power until 800 J of energy is transferred to each tube.
   - Note: The relative transparency of a sonicated sample provides a simple test of adequate solubilization, as the fully sonicated sample will be completely clear while the nonsonicated sample is turbid.
5. Transfer the sonicated samples to microcentrifuge tubes and centrifuge at 16,000 × *g* for 30 min at 4°C.
6. Pool the supernatant and store at 4°C until use.
   - Note: The prepared small unilamellar vesicles (SUVs) are stable at 4°C for 1 month.
7. Dilute the SUVs 5 × in TBS.
8. Add 20 μL of the diluted SUVs to each well of a 384-well Biodesy Delta plate and seal.
   - Note: Separately seal all unused wells for later use.
9. Incubate the SUVs in the wells for at least 30 min at room temperature to form the SLB.
   - Note: SUVs can be incubated for longer times if convenient; if overnight incubation is desired, the plate should be transferred to 4°C and used within 1 week.
10. Wash out excess SUVs with 10 50 μL washes into the desired assay buffer.
    - Note: 20 μL of buffer is always present in the well to keep the bilayer from being exposed to air, which will destroy it. During this and all

subsequent wash steps, it is imperative that the bilayer remain fully hydrated at all times. Disruption of the bilayer is most commonly caused by inadvertent exposure to air, harsh mixing, or physical contact of the pipet tip with the bilayer during mixing. Disruption of the bilayer can cause errors in the SHG signal and response, which can lead to incorrect data interpretation. If bilayer disruption is suspected, visualization of the SLB with an inverted fluorescence microscope with 4–20 × magnification is recommended. The Biodesy Ni/NTA bilayer preparation is manufactured with a fluorescent dye that allows visualization and easy detection of bilayer disruptions with excitation and emission wavelengths of 560 and 583 nm, respectively. If the SLB has been disrupted, those wells should not be used in experiments.

11. Immediately proceed to the conjugate deposition step.

### 3.3.2 Conjugate Deposition

When testing conjugates for the first time, it is recommended to test assay buffer with and without Tween-20, as Tween-20 has been shown to have a positive impact in some assays and is deleterious in others. The SHG assay is typically carried out anywhere between 125 n*M* and 1 μ*M* final conjugate concentration, but the optimal conjugate concentration should be empirically determined for each assay. To deposit conjugates:

1. First remove 10 μL of the buffer volume, next add 10 μL of conjugate diluted in assay buffer to each well, and mix gently.
   - Note: Conjugate should be added to wells at 2 × the desired final conjugate concentration because it is diluted 1:1 upon deposition (10 μL of conjugate is added to 10 μL of the assay buffer in the well for a final assay volume of 20 μL).
2. Incubate the conjugate in the well overnight at 4°C to allow for complete tethering of the conjugate to the surface.
   - Note: The incubation time/temperature of this step can be varied during assay optimization. Assays have been developed with tethering times/temperatures ranging from 30 min at room temperature to 4°C overnight. Generally, tethering at 4°C overnight works well but shorter times may be required for unstable biomolecules or complexes.
   - Note: We recommend testing for specific tethering of the protein to the surface as a routine part of assay development. Specific tethering of the protein to the surface through the His-tag:Ni-NTA

interaction can be tested by depositing conjugate onto a DOPC SLB (which lacks Ni-NTA) or by protein tethering in the presence of EDTA or imidazole.

## 4. RUNNING AN SHG ASSAY

To determine the optimal conjugate(s), the next step is to measure conformational change upon ligand injection for all conjugates. After overnight conjugate incubation on the plate proceed as follows:

1. Equilibrate the plate to room temperature for 30 min.
2. Wash out unbound conjugate with five 50 μL washes into assay buffer containing DMSO at the concentration used for compound preparation (step 4 below).
   - Note: This protocol describes setting up a small molecule compound plate but can be applied to any ligand in the presence or absence of DMSO.
3. Allow the plate to equilibrate at room temperature for 20 min in buffer containing DMSO.
4. Prepare the compound plate by diluting compounds to be tested into assay buffer at 2 × the desired well concentration.
5. Transfer compounds (minimum of 30 μL/well) to the wells of the source plate (Greiner Bio-One 781280) for injection on the Biodesy Delta.
   - Note: Compounds are injected into user-defined wells of the Biodesy Delta plate by the liquid handling robot in the Biodesy Delta. Each compound, as well as a negative control buffer injection, should be tested in triplicate at each concentration or buffer condition being tested. To begin, compounds or ligands should be injected at 10 × the $K_D$ or at least 10 μ*M* for nanomolar and higher affinity binders.

SHG assays have been successfully developed with baseline signals ($SHG_B$) across a wide range of magnitudes, including those which are lower than the background signal level in the absence of protein (surface alone). Therefore, rather than maximizing for SHG baseline signal, the sign and magnitude of the SHG signal change upon ligand addition should be optimized when determining the best assay conditions. For a screening assay, the optimal assay will exhibit the largest assay window with the available controls. For an assay aimed at discriminating specific conformations, the optimal one will exhibit the largest signal change difference between the two types of control ligands. If the conditions tested at this point do not result in an appropriate assay, then the steps of assay development should be modified and repeated.

## 5. SHG ASSAY DEVELOPMENT: EXAMPLE

Fig. 3 shows an example of the data output from assay development on the protein tyrosine phosphatase Shp2 supplied by The University of Texas MD Anderson Cancer Center. Shp2 was conjugated with SHG1-SE at pH 7.5 and 8.2, resulting in DoLs of 1.8 and 2.5, respectively. Cysteine labeling was not attempted due to the presence of an active-site cysteine residue. Both conjugates were tested at 500 n*M* tethering concentration in both buffer with and without 0.005% Tween-20. The SHG baseline signal of these conjugates was higher without Tween (Fig. 3A). However, in buffer containing Tween-20, both conjugates exhibited a positive SHG signal change in response to allosteric compounds and a negative SHG signal change in response to active-site compounds, resulting in good discrimination between the two ligand-bound conformations, while the conjugates without Tween-20 in the buffer did not respond to the active-site compounds (Fig. 3B). Based on these data, an assay buffer with Tween-20 was chosen as the best condition for discrimination of allosteric and active-site inhibitors and this buffer was used in all subsequent steps of assay development. Of the two conjugates, the DoL 2.5 conjugate offered a slightly better assay window due to the larger signal change for the compound designated Active site #1. However, because both conjugates produced similar results, each was tested in the next step of assay development.

To optimize the assay while minimizing protein consumption, 500, 250 and 125 n*M* protein tethering concentrations were tested in assay buffer with Tween-20. Fig. 3C shows that the baseline signal decreases with lower protein concentrations. Upon injection of the control compounds, the magnitude of SHG signal change was similar for the allosteric compounds at 500 and 250 n*M* protein tethering concentration, but was lower at 125 n*M* for both conjugates (Fig. 3D). The active-site compounds induced a negative signal change at most concentrations, but the largest signal change, and the least amount of noise in the signal change, was achieved at 500 n*M* of Shp2-SE (DoL 2.5). This conjugate at this protein tethering concentration produced an assay with robust signal change and discriminated between allosteric and active-site binders by the sign (direction) of the SHG signal change (positive vs negative signal change relative to baseline), and thus it was selected as the best conjugate with which to move forward for screening and binning of compounds by mechanism of action.

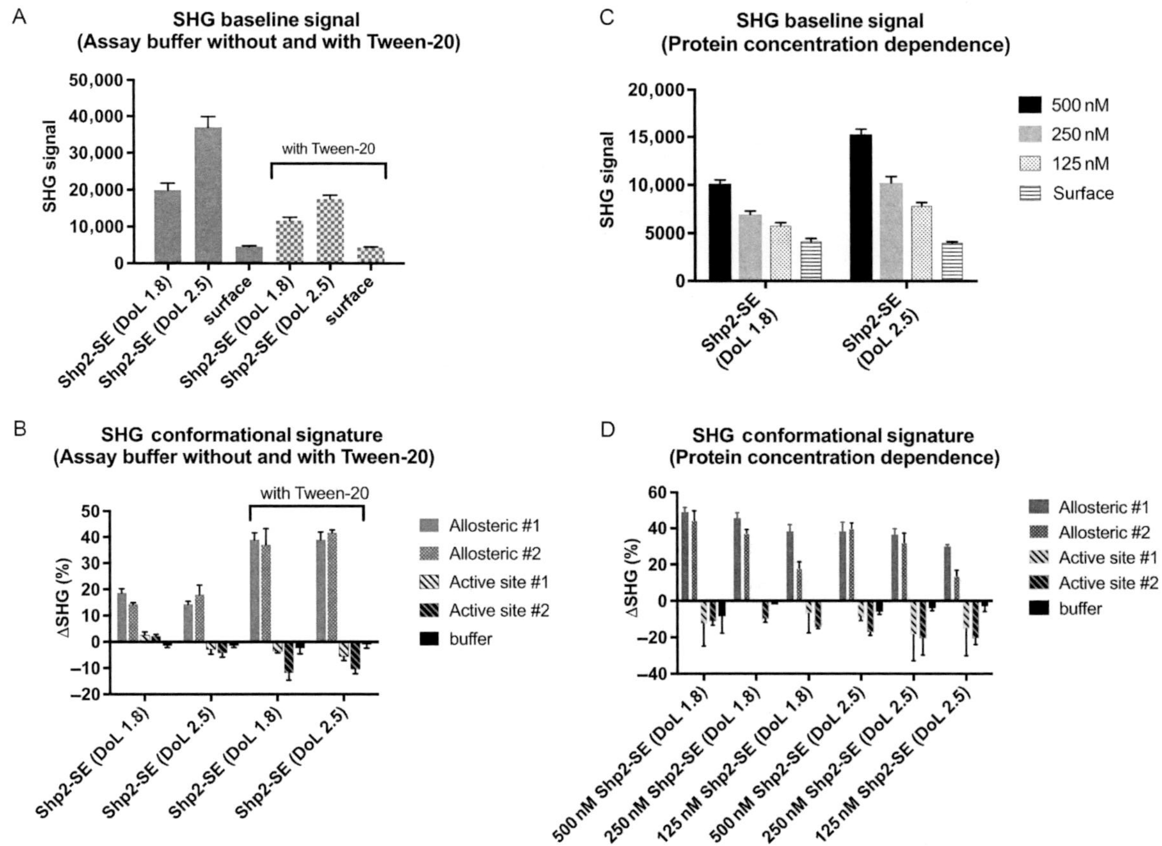


**Fig. 3** Shp2 assay development example data set (data presented as the mean ± standard deviation of triplicate data points). (A) Baseline SHG intensity for two Shp2 conjugates was measured in assay buffer without and with Tween-20. (B) SHG signal change was monitored in response to two control allosteric inhibitors and two active-site inhibitors of Shp2. The SHG assay discriminated active-site and allosteric binding modes by the sign of the SHG signal change relative to baseline (positive or negative changes, respectively). (C) SHG baseline signal as a function of protein tethering concentration. (D) SHG signal change for control compounds was similar for 500 and 250 n*M* protein tethering concentrations, but decreased in magnitude and increased in noise (standard error) at 125 n*M*.

# 6. SCREENING

After developing a SHG assay for a target of interest the next step is to optimize the assay for a screening campaign. Because the initiation of a screening campaign can be resource- and time-intensive, it is best to optimize the screening assay at the outset. This process involves several steps, outlined below for a 384-well plate format. A 1536-well format is also available from Biodesy for higher-throughput screening but will not be discussed in detail here.

## 6.1 Scale to Screen

The goal of this process is to convert the assay from one performed on a fraction of wells in the 384-well Biodesy Delta plate during assay development to one that performs robustly across the entire plate. First, run assay controls on the Biodesy Delta using the optimized small-scale assay conditions across at least 96 wells that sample all regions of the 384-well plate, including the first and last columns of the plate, to test the stability of the assay over the experimental time frame required for running the assay on a full plate. An example of this uniformity test is shown in Fig. 4, in which 352 wells were

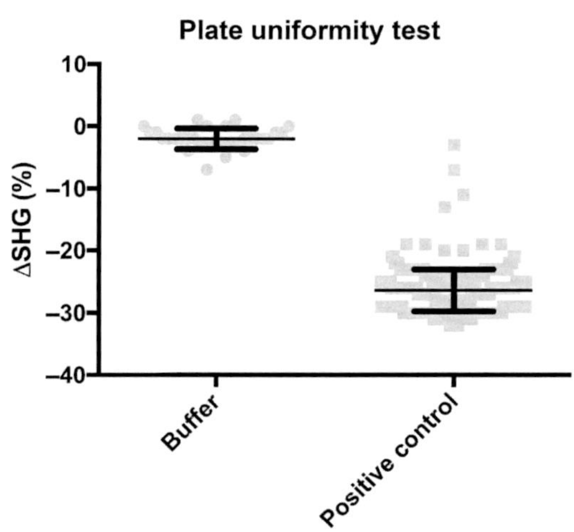


**Fig. 4** Plate uniformity test. Injection of the positive control compound across 22 of the available 24 columns ($n = 352$) of the 384-well plate (the remaining columns were injected with buffer, $n = 32$) resulted in an average SHG signal change of negative 26% (−26%) and a coefficient of variation of 13% for this signal change.

injected with positive control and the remaining 32 wells were injected with negative control, resulting in a coefficient of variation (CV) of the positive control of 13%. In this assay, these results were acceptable to move the current assay conditions forward to a full plate format. If the CV of either the baseline SHG signal or the signal change magnitude is larger than that shown in the small-scale assay, further optimization of the assay conditions will be required to reduce the variability before screening. Further optimizations to scale the assay may include, but are not limited to, the following strategies: reoptimization of conjugation conditions to produce the same DoL and conjugation pattern at higher reaction volumes, adjusting plate preparation protocols for full plate formats, altering buffer components to keep the protein stable on the surface for longer experimental run times, and changing data collection time intervals and the frequency of reads to decrease the experimental run time.

## 6.2 Small Molecule/Fragment Screening Concentrations

Before starting a full screen, one 384-well plate from the library to be screened should be tested at the desired concentration and assay conditions. To ensure the assay performs as expected over the entire run time, positive and negative control compounds (or buffer alone instead of a negative control) should be placed in columns 1 and 24, which are read by the Biodesy Delta at the beginning and end of each experiment, respectively. This step allows for a rough estimate of hit rate and also tests the full plate preparation protocol before performing the full screen. For small molecules, screens have been run at 5–20 μ*M* with a typical hit rate of 1%–2%. For fragments, screens have been run at 200–500 μ*M* with a typical hit rate of 10%–15%.

## 6.3 Running the Screen

After scaling the assay and determining the appropriate compound concentration, running the screening campaign is straightforward. Most screens take about 40 min per plate on the Biodesy Delta. The plate read time includes injection and mixing of compound by the on-board liquid-handling robot as well as reading all 384 wells both pre- and 5 min post-compound injection.

## 6.4 Hit Calling

After completing the screen, hits from the data set can be assigned using constant or empirically derived hit-calling thresholds. Typically, one computes the mean and standard deviation of all negative control compound responses across the entire screen (in cases where negative controls are not available, we use buffer injections). In this case, compounds that induce SHG responses which are at least 3 standard deviations away from the mean negative control response would be considered hits. Because SHG responses can be either an increase or decrease in signal relative to baseline, compounds that are 3 standard deviations above or below the mean negative control response ($>+3$ sigma or $<-3$ sigma) should be considered hits in the analysis. However, while small SHG signal change magnitudes and low noise are typical for negative controls, using the mean and standard deviation of these controls to set hit-calling thresholds may be too stringent, in which case thresholds of $\pm 10\%$ are recommended. An alternative approach is to compute the global mean and standard deviation of SHG responses from the full screening library data set. In this case, compounds that induce SHG responses at least 3 standard deviations above or below the global mean are considered hits.

## 6.5 Nuisance Compounds

A common concern in most screening campaigns is false positives caused by the presence of nuisance compounds in the library. An experiment designed by scientists at Eli Lilly directly assessed the susceptibility of a typical SHG assay to produce a false-positive response from nuisance compounds, including many well-known redox compounds, compounds contaminated by heavy metals, aggregators, pan-assay interfering compounds (PAINS), and compounds purified on strong cation exchange columns which have been shown to cause artifacts in other assays. Eighty-seven percent of these nuisance compounds did not show a significant change in the SHG response ($>10\%$ or $\leq 10\%$) relative to baseline, and thus would not have been called as hits in the screen run at $10\,\mu M$ compound concentration (Fig. 5). The remaining 13% of hits were distributed equally across all nuisance mechanisms listed above. In some cases, nuisance mechanisms can be flagged in SHG follow-up validation studies as described below.

## 6.6 Hit Validation

After identifying hit compounds from the screen, all hits should be validated with additional follow-up SHG measurements. First, all the putative hits from the screen should be retested in triplicate, ideally by producing new

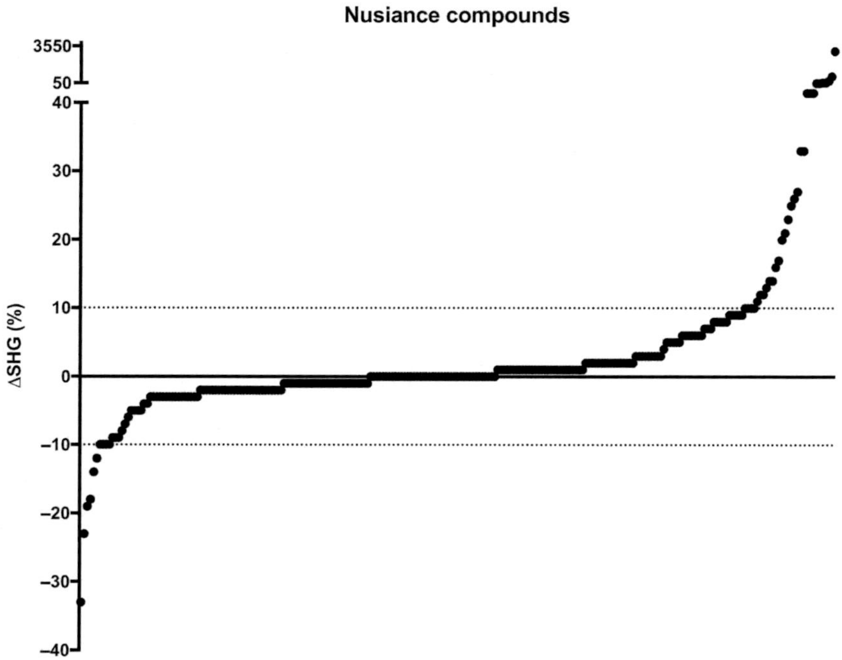


**Fig. 5** Nuisance compounds in the SHG assay. Eighty-seven percent (87%) of all nuisance compounds tested are silent in the SHG assay.

DMSO stocks of each hit from fresh powder whenever possible. In many cases, libraries composed of previously plated DMSO stock compounds will degrade due to age and shipping and/or storage conditions. Plating compounds from fresh powder can be omitted if the library is well curated to ensure that only high fidelity compounds are present. After preparing fresh stocks, each compound should be tested at the screening concentration, and all compounds that reproduce the original screening response should advance to the next phase of testing.

All hits that confirm in the triplicate measurements should be further validated by performing concentration response curves (CRCs) and measuring the SHG response over time. In this follow-up step, a dilution series for each compound is prepared in the source plate and the SHG response at each concentration is measured in real time using the timecourse mode (at least three different time points after injection) over a predetermined time interval, typically 5–10 min. The total number of concentrations needed to provide the best fit for the CRC should be empirically determined; at least 10 points per curve is recommended. Please note that as SHG is a nonlinear process the standard equations to calculate the $K_D$ typically result in a $K_D$ ~2× higher than the actual values, as these equations assume the signal response varies linearly with the number of molecules that bind ligand. However, the SHG signal change is quadratically dependent on the number of molecules that bind ligand and undergo conformational change (Eq. 1). To account for this quadratic dependence, Biodesy has derived an equation to correctly fit the SHG data as:

$$\% \text{ Change} = 100\left(\left(1 + C\frac{[\text{L}]}{[\text{L}] + K_D}\right)^2 - 1\right) \tag{2}$$

where $C$ is a fitting parameter proportional to the overall response, [L] is the ligand concentration, $K_D$ is the equilibrium dissociation constant, and % Change is 100% × (SHG final signal-SHG baseline signal)/SHG baseline signal. If there is an overall offset in the data (the % change does not asymptote to zero at low ligand concentration), a second fit parameter can be added to the end of Eq. (2) to account for the offset. Eq. (2) will be valid in most cases where effects due to depletion of compound from solution are negligible. In cases where depletion effects are nonnegligible the above equation will return a $K_D$ which is an overestimate of the true value.

The goal of this step is to further validate true hits and distinguish them from false positives. Fig. 6A (top) is an example of a SHG signal change timecourse taken at multiple ligand concentrations for a small molecule compound known to bind this target specifically with a 1:1 stoichiometry. The SHG signal change shows one transition and is stable over the course of the data collection indicative of a true binder. Fig. 6A (bottom) shows the SHG signal change at an endpoint 5 min postcompound injection plotted as a function of the concentration of the compound. True single-site binders will show a sigmoidal curve that can be fit with Eq. (2) for $K_D$ determination. Comparatively, nuisance compounds can exhibit very different responses, an example of which is in Fig. 6B where SHG signal change over time shows unstable behavior (top) and the CRC is nonsigmoidal (bottom). Only compounds with normal, monotonically changing timecourses that reach a plateau and acceptable CRC profiles should be advanced to additional validation and follow-up studies with orthogonal techniques.

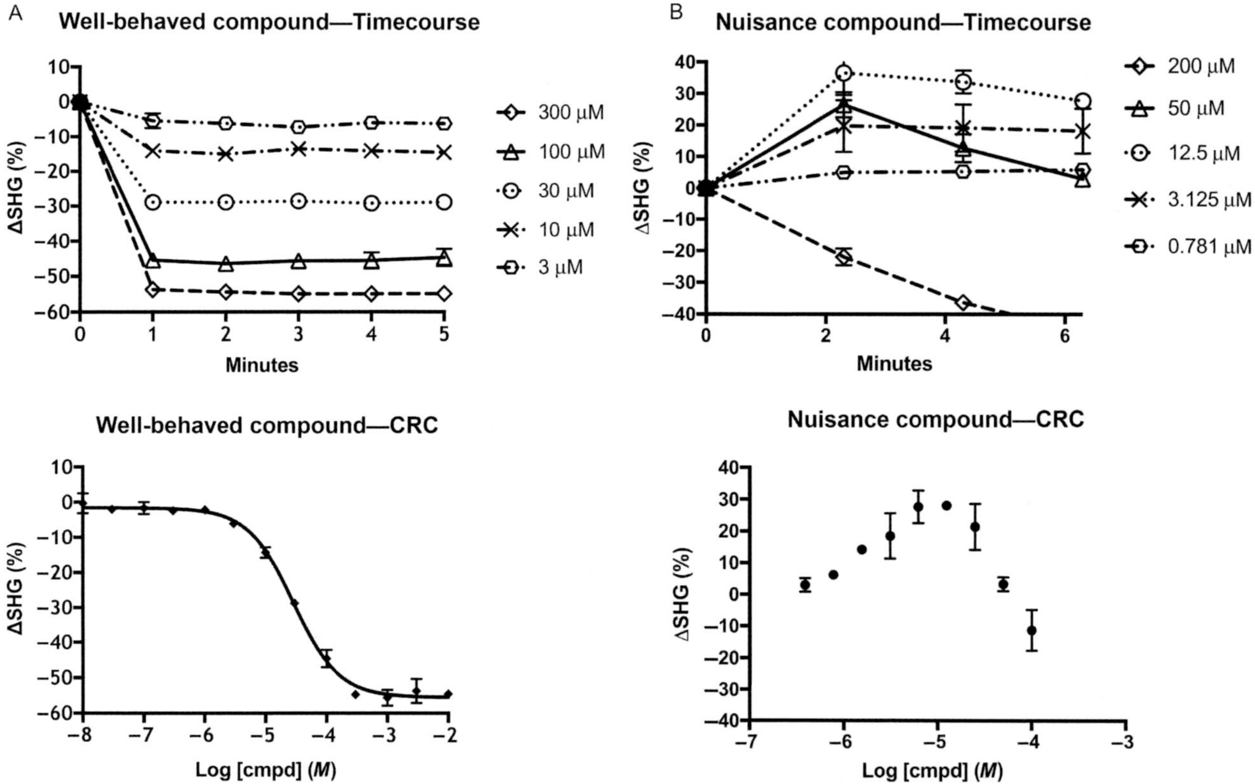


**Fig. 6** SHG signal timecourse measurements (*top*) and SHG concentration response curves (CRCs) (*bottom*) for well-behaved responses of a specific ligand (A) and ill-behaved responses of a nuisance compound (B) after ligand injection ($t = 0$ min).

## REFERENCES


Brasselet, S. (2011). Polarization-resolved nonlinear microscopy: Application to structural molecular and biological imaging. *Advances in Optics and Photonics*, *3*(3), 205. https://doi.org/10.1364/AOP.3.000205.

Butko, M. T., Moree, B., Mortensen, R. B., & Salafsky, J. (2016). Detection of ligand-induced conformational changes in oligonucleotides by second-harmonic generation at a supported lipid bilayer interface. *Analytical Chemistry*, *88*(21), 10482–10489. https://doi.org/10.1021/acs.analchem.6b02498.

Getz, E. B., Xiao, M., Chakrabarty, T., Cooke, R., & Selvin, P. R. (1999). A comparison between the sulfhydryl reductants Tris(2-carboxyethyl)phosphine and dithiothreitol for use in protein biochemistry. *Analytical Biochemistry*, *273*(1), 73–80.

Hermanson, G. T. (2013). *Bioconjugate techniques* (3rd ed.). Academic Press.

Lakowicz, J. R. (2006). *Principles of fluorescence spectroscopy* (3rd ed.). Springer.

Ma, B., Marcotte, D., Paramasivam, M., Michelsen, K., Wang, T., Bertolotti-Ciarlet, A., et al. (2016). ATP-competitive MLKL binders have no functional impact on necroptosis. *PLoS One*, *11*(11), e0165983.

Moree, B., Connell, K., Mortensen, R. B., Liu, C. T., Benkovic, S. J., & Salafsky, J. (2015). Protein conformational changes are detected and resolved site specifically by second-harmonic generation. *Biophysical Journal*, *109*(4), 806–815. https://doi.org/10.1016/j.bpj.2015.07.016.

Moree, B., Yin, G., Lazaro, D. F., Munari, F., Strohaker, T., Giller, K., et al. (2015). Small molecules detected by second-harmonic generation modulate the conformation of monomeric α-synuclein and reduce its aggregation in cells. *The Journal of Biological Chemistry*, *290*, 27582–27593.

Rizvi, N. F., Howe, J. A., Nahvi, A., Klein, D. J., Fischmann, T. O., Kim, H.-Y., et al. (2018). Discovery of selective RNA-binding small molecules by affinity-selection mass spectrometry. *ACS Chemical Biology*, *13*(3), 820–831. https://doi.org/10.1021/acschembio.7b01013.

Salafsky, J. S. (2003). Second-harmonic generation as a probe of conformational change in molecules. *Chemical Physics Letters*, *381*(5–6), 705–709.

Salafsky, J. S. (2006). Detection of protein conformational change by optical second-harmonic generation. *The Journal of Chemical Physics*, *125*(7), 074701.

Salafsky, J. S. (2007). Second-harmonic generation for studying structural motion of biological molecules in real time and space. *Physical Chemistry Chemical Physics*, *9*(42), 5704–5711. https://doi.org/10.1039/B710505C.

Tran, R. J., Sly, K. L., & Conboy, J. C. (2017). Applications of surface second harmonic generation in biological sensing. *Annual Review of Analytical Chemistry*, *10*(1), 387–414. https://doi.org/10.1146/annurev-anchem-071015-041453.

Wong, J. J. W., Young, T. A., Zhang, J., Liu, S., Leser, G. P., Komives, E. A., et al. (2017). Monomeric ephrinB2 binding induces allosteric changes in Nipah virus G that precede its full activation. *Nature Communications*, *8*(1), 781. https://doi.org/10.1038/s41467-017-00863-3.

CHAPTER NINE

# An Array-Based Ligand Discovery Platform for Proteins With Short Half-Lives

**Becky S. Leifer**[*,†,‡,2], **Shelby K. Doyle**[*,†,‡,§,2], **André Richters**[*,†,‡], **Helen L. Evans**[*,†,‡], **Angela N. Koehler**[*,†,‡,§,1]

[*]David H. Koch Institute for Integrative Cancer Research, Massachusetts Institute of Technology, Cambridge, MA, United States
[†]Center for Precision Cancer Medicine, Massachusetts Institute of Technology, Cambridge, MA, United States
[‡]The Broad Institute of MIT and Harvard, Cambridge, MA, United States
[§]Department of Biological Engineering, Massachusetts Institute of Technology, Cambridge, MA, United States
[1]Corresponding author: e-mail address: koehler@mit.edu

## Contents



## Abstract

Many promising therapeutic protein targets were previously considered "undruggable" due to a deficit in structural information to guide drug design and/or a lack of an obvious binding pocket. Fortunately, array-based methods for evaluating protein binding against large chemical libraries, such as small-molecule microarray screening, have

[2] These authors contributed equally to this work.

*Methods in Enzymology*, Volume 610
ISSN 0076-6879
https://doi.org/10.1016/bs.mie.2018.09.019

provided one of several emerging inroads to ligand discovery for these elusive targets. Despite the advance in the area of ligand discovery for poorly structured and intrinsically disordered proteins provided by array-based technologies involving cell lysates, the extension of this technology for screening proteins with short half-lives in physiologically relevant conformations has been technically challenging. In this chapter we present a protocol for leveraging in vitro translation strategies to enable array-based screening of short-lived proteins against large small-molecule libraries for ligand discovery.

## ABBREVIATIONS

| | |
|---|---|
| **BCA** | bicinchoninic acid |
| **BSA** | bovine serum albumin |
| **CV** | coefficient of variation |
| **EDTA** | ethylenediaminetetraacetic acid |
| **EGTA** | ethylene glycol-bis(β-aminoethyl ether)-*N*,*N*,*N*′,*N*′-tetraacetic acid |
| **HTS** | high-throughput screening |
| **IVT** | in vitro translation |
| **PBS** | phosphate-buffered saline |
| **PMT** | photomultiplier tube |
| **RT** | room temperature |
| **SDS-PAGE** | sodium dodecyl sulfate polyacrylamide gel electrophoresis |
| **SMM** | small-molecule microarray |
| **SNR** | signal-to-noise ratio |
| **SPR** | surface plasmon resonance |
| **TBS-T** | Tris-buffered saline and Tween 20 |

## 1. INTRODUCTION

Spatially addressable chemical screening approaches (e.g., small-molecule microarrays (SMMs), surface plasmon resonance (SPR)-coupled fragment arrays, spatially resolved DNA-encoded libraries) enable the rapid evaluation of potential binding interactions of a molecular therapeutic target of interest (e.g., protein, RNA) and a broad swath of chemotypes (Neumann, Junker, Schmidt, & Sekul, 2007; Vegas, Fuller, & Koehler, 2008; Zambaldo, Barluenga, & Winssinger, 2015).

There is a renewed interest in the use of these spatially addressable screening systems in the discovery path for therapeutics because of the growing interest in the development of heterobifunctional molecules that can degrade target proteins catalytically (Crews, 2018). These emerging modalities are comprised of (i) a target-specific chemotype capable of binding a therapeutic target that is covalently tethered to (ii) a second chemotype that can recruit intracellular proteasomal machinery to degrade the target

(Bondeson et al., 2015; Winter et al., 2015). Because spatially addressable screening systems can enable identification of putative binders to a target in a manner that is agnostic of inhibitory function, they are potentially more attractive than high-throughput functional assays (e.g., enzymatic activity, DNA binding) early on in the development of these chimeric molecules. Additionally, because the assay positives from these assays are immobilized and/or encoded with nucleic acid tags, the immobilization or encoding strategy can readily suggest a vector on the scaffold for building out the linker and degrading moiety, giving these screening approaches additional advantage over alternative high-throughput binding assays (e.g., SPR, isothermal calorimetry).

The SMM platform, specifically, has proven to be a robust technology for discovering protein–small-molecule interactions for a variety of different target classes for both human and nonhuman therapeutic targets (Daguer et al., 2014; Felsenstein et al., 2016; Hong, Neel, Wassaf, Caballero, & Koehler, 2014; Pop et al., 2014; Sadaghiani et al., 2014; Sztuba-Solinska et al., 2014; Velagapudi & Disney, 2014).

In order to systematically immobilize diverse screening libraries in an array format, several covalent attachment approaches have been developed to anchor small molecules on solid substrates (e.g., glass, silicon). Reported methodologies typically leverage either a selective attachment chemistry, relying on specific moieties introduced to each screening molecule to react with a chemically modified surface, or a more permissive attachment chemistry, where a latent chemotype at the substrate surface is able to react with a number of different functional groups presumed to be present in the library. Either of these approaches can result in homogenous (Koehler, Shamji, & Schreiber, 2003; Kohn et al., 2003; Urbina et al., 2006; Vegas et al., 2007) or heterogeneous (Bradner, McPherson, & Koehler, 2006; Bradner, McPherson, Mazitschek, et al., 2006; Miyazaki et al., 2009) display of small molecules, depending on the contents of library being employed, but are enabling toward the screening of biological targets for putative binding interactions.

Previously, we have reported on the use of isocyanate-mediate capture of small molecules on glass surfaces to enable array-based chemical screening of protein targets, where the highly reactive isocyanate capture increases the diversity of compounds compatible with the screening system (Bradner, McPherson, & Koehler, 2006) and is stable in the presence of cellular esterases to enable screens for nonpurified proteins from cell lysates (Bradner, McPherson, Mazitschek, et al., 2006).

We have in previous years shared detailed protocols for leveraging these arrays for the screening of both purified protein targets and targets residing in cell lysates. But, while the use of SMM screening has proved fruitful in uncovering small-molecule binders of a variety of recalcitrant targets through pure protein and lysate-based screening (Boskovic et al., 2016; Pop et al., 2014; Stanton et al., 2009; Yang, Koehler, & Wang, 2016; Young et al., 2015), in our hands the lysate-based approach has proved challenging for proteins with short half-lives, as it is difficult to achieve a sufficiently high relative expression level in live cells for SMM screening.

Overcoming this challenge is important for drug discovery, as many oncogenic transcriptional regulators (e.g., c-MYB, c-MYC) and oncogenic fusion proteins (e.g., EWS-FLI1, MYB-NFIB) are not only difficult to purify in physiologically relevant confirmations for SMM screening but are short lived in cells, with short (<4 h) to very short (<30 min) half-lives (Eden et al., 2011; Schwanhäusser et al., 2011; Yen, Xu, Chou, Zhao, & Elledge, 2008).

In our development of assays for short-lived proteins, attempts to overcome degradation of the target protein have not been sufficient to enable screening by SMM. Exogenous overexpression coupled to universal proteasomal inhibition can successfully boost target protein concentration, but not without global accumulation of protein and an increase in misfolded proteins. This sharp increase in hydrophobic protein surfaces on the array resulted in high background signal from the detection antibody, making it difficult to discern meaningful small-molecule interactions with the target of interest. More target-specific approaches to increasing target protein concentration include the modulation of genes known to regulate target stability and/or mutation of amino acids predicted to be phosphorylated or ubiquitinated to signal the degradation pathway of the target. These approaches are labor intensive, as they require generation and optimization of target-specific reagents, and, importantly, have not been universally successful in our hands, especially for proteins with half-lives on the order of minutes (30–120 min).

Additionally, we have explored immunoprecipitation approaches to increase relative protein concentration by reducing sample complexity. While this approach is more systematically applicable by precipitating protein through the same epitope tag used for SMM detection, this method is also labor intensive in addition to being somewhat resource intensive. The recovery of target protein through nondenaturing conditions—such as competition with peptide epitope tag—is very low yielding and can produce high false positives through detection of the peptide. Reducing this complexity further by size-exclusion filtration is again low yielding and therefore resource intensive as an approach for high-throughput screening (HTS).

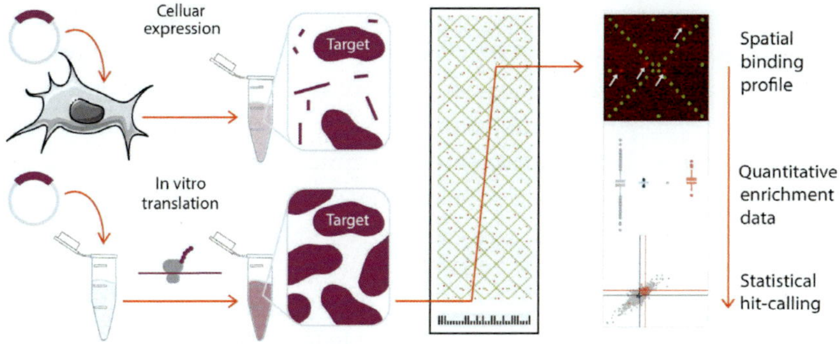


**Fig. 1** Overview of the small molecule microarray (SMM) screening process using in vitro translation (IVT) technology for target protein expression. Through IVT, target proteins with short half-lives can be expressed at concentrations suitable for SMM screening, enabling the identification potential small molecule binders of the target protein.

Here, we describe a robust, straightforward solution to this ligand discovery challenge that is readily extensible to any short-lived protein of interest. This chapter describes a protocol that leverages in vitro translation (IVT) to enable SMM screening for protein targets with short half-lives. We detail: (i) the expression of a representative screening target by IVT; (ii) modifications to published SMM workflows from our laboratory for use with this method; (iii) the data analysis we currently employ for such screens; and (iv) the recommended course for optimizing such an approach on a target-by-target basis (Fig. 1).

## 2. EXPRESSION OF PROTEIN BY IVT

A SMM screen can be conceptually described as a biological sample, containing a target of interest, incubated with a glass slide presenting an arrayed library of small molecules, and the quantification of the amount of the molecular target bound to each displayed small molecule via fluorescently labeled antibody. This section will detail the leveraging of IVT to prepare a biological sample in such a way that screening-relevant concentrations of short-lived targets of interest.

In order to demonstrate the utility of this method, we will detail the screening of both FKBP12, a well-studied protein with known interactions that we recommend as a positive control throughout the screening process, and ODC1, a therapeutic target of interest in a variety of oncology settings.

ODC1, or ornithine decarboxylase, is frequently upregulated in cancer as a direct target of Myc-driven transcription, is a promoter of polyamine synthesis-dependent proliferation in a variety of settings, and is one of the most tightly regulated enzymes in the human proteome (Agarwal, Katiyar, Zaidi, & Mukhtar, 1992; Funakoshi-Tago, Sumi, Kasahara, & Tago, 2013; He et al., 2017; Hogarty et al., 2008; Jang et al., 2017; Kim et al., 2017; Mangold & Leberer, 2005; Mohammed et al., 2014; Patel, Loboda, Showe, Showe, & McMahon, 2004; Pegg, 2006; Singh et al., 2018). While there are long-standing, low-throughput assays for endogenous ODC1 activity (Gupta, Ahmad, Mohan, Husain, & Mukhtar, 1999; Verma & Boutwell, 1977; Verma, Shapas, Rice, & Boutwell, 1979), more recent approaches for ligand discovery in the literature focus on either the active site of the enzyme (Smithson, Shelat, Baldwin, Phillips, & Guy, 2010) or on molecular docking studies in silico (Kim et al., 2016). Given the attractiveness of array-based platforms for novel ligand discovery mentioned earlier and the short half-life of this therapeutically relevant protein, we present to the reader a protocol for screening ODC1 by SMM as an illustrative screen for the method.

## 2.1 IVT Protein Expression

To express full-length FKBP12 and ODC1, we utilized the expression vector included with the IVT kit used to demonstrate this approach, the 1-Step Human Coupled IVT Kit-DNA (Thermo Scientific™, #88881). The pT7CFE1-CHis plasmid (Thermo Scientific™, #88860) contains: (i) a T7 promoter; (ii) an EMCV IRES sequence; (iii) a multicloning site; (iv) a 6 × His affinity tag; and (v) a T7 terminator.

Ultimately, the protein will be detected in the SMM screen with a primary antibody conjugated to a fluorophore. This can be an antibody raised against the protein itself or against an affinity tag. As high-quality, fluorescent antibodies can be expensive and are not available for all proteins, we routinely use an affinity tag for detection. While the platform is compatible with nearly any tag/detection system, we recommend the use of an HA tag over the His tag that is present in the vector included with the kit (pT7CFE1-CHis) because in our experience anti-HA antibodies give a cleaner screening profile in the absence of target protein for lysate-based and IVT-based screening. To insert this tag, an N- or C-terminal HA tag sequence (YPYDVPDYA) can be added to the target protein-coding sequence during PCR amplification prior to cloning into the pT7CFE1-CHis vector. Alternatively, the pT7CFE1-CHA plasmid (Thermo Scientific™, #88862), which is an identical vector that contains a C-terminal HA tag, can be purchased and utilized in the same manner.

Once the desired protein-coding sequence is amplified, it can be cloned into the pT7CFE1 plasmid using restriction enzyme cloning. When devising a cloning scheme, there are several restriction enzyme cloning sites in the multicloning site of the pT7CFE1-CHis vector. We found that maximal expression occurred when using *Nde*I as the 5′ restriction site, as it preserves the Kozak sequence (5′-(GCC)GCCACCATGG-3′) and places the beginning of the target sequence adjacent to the start codon. If *Nde*I is not compatible with the coding sequence for the target protein of interest, you may either mutate out restriction enzyme sites within the gene of interest for a synonymous sequence or reintroduce the Kozak sequence upstream of the start codon to boost expression.

For FKBP12, we utilized *Nde*I and *Xho*I (NEB), ligated with T4 DNA ligase (NEB), and confirmed the product by sequencing (F: 5′-GGAATTCCATATGATGTACCCCTACGACGTGCCCGACTACGCCGGTGGAGGAGTGCAGGTGGAAACC-3′; R: 5′-GGCCTCGAGTCATTCCAGTTTTAGAAGCTCCACATC-3′).

For ODC1, we utilized *Msc*I and *Xho*I (NEB), ligated with T4 DNA ligase (NEB), and confirmed the product by sequencing (F: 5′-GGGATCCATGTACCCCTACGACGTGCCCGACTACGCCGGTGGAATGAACAACTTTGGTAATGAAGAG-TTTG-3′; R: 5′-GGCCTCGAGTCACACATTAATACTAGCCGAAGCACAG-3′). Once we had the desired clone, the Kozak sequence/*Nde*I sequence was reintroduced by site-directed mutagenesis using the Q5 kit (NEB) using primers (F: 5′-CATATGATGTACCCCTACGACGTG-3′; R: 5′-GGTGGTGGCCATATTATCATCGTG-3′).

The final expression vectors should be purified and concentrated for successful use with the IVT kit. A protocol for DNA precipitation is provided in the IVT kit manual. Briefly, samples were precipitated with 1/10 volume of 3 *M* sodium acetate (pH 5.5) and two volumes of 200 proof ethanol (−20°C, 15 min), centrifuged at 14,000 × *g* (4°C, 15 min), washed with 70% ethanol, centrifuged once more at 14,000 × *g* (4°C, 5 min), and then left to air-dry for (5 min) before resuspension in nuclease-free water for quantitation and use. Ideally, the template vector concentration should be 500–1000 ng/μL to minimize residual contaminants from the DNA preparation.

### 2.1.1 Reagents

- Nuclease-free pipette tips
- Nuclease-free qPCR tubes
- RNA-Zap

- 1-Step Human Coupled IVT Kit-DNA (Thermo Fisher catalog #88881)
- Gel
- Running buffer
- Transfer buffer
- Loading buffer
- REVERT Total Protein Stain (LI-COR #926-11010)
- Bovine serum albumin (BSA)
- TBS-T (50 m*M* Tris–Cl, pH 7.6, 150 m*M* NaCl, 0.05% Tween 20)
- Primary antibody for western blotting (e.g., anti-HA.11 BioLegend #MMS-101P)
- Detection antibody for western blotting (e.g., LI-COR #925-32210)

### *2.1.2 Equipment*

- Thermocycler (e.g., Bio-Rad T100™ Thermal Cycler #1861096)
- Table-top mini centrifuge that can accommodate 0.2 mL PCR tubes (e.g., Denville #C0812)
- SDS-PAGE System (e.g., Bio-Rad Criterion™ Cell and PowerPac™ Basic Power Supply #1656019)
- Western Blotting System (e.g., Bio-Rad Trans-Blot® Turbo™ Transfer System #1704150)

### *2.1.3 Protocol*

1. Clean working area, pipets, and gloves with RNAse Zap. Use only RNAse-free pipet tips and PCR tubes. Set thermocycler to an infinite hold at 30°C with lid no higher than 50°C.
2. Defrost contents of 1-Step Human Coupled IVT Kit on ice. Aliquot each of the non-DNA kit components into RNAse-free PCR tubes: HeLa lysate (red component), $8 \times 12.5\,\mu L$; accessory proteins (green), $8 \times 2.75\,\mu L$; reaction mix (yellow component), $8 \times 5\,\mu L$; and nuclease-free water (blue component), $8 \times 15\,\mu L$. Quickly return all aliquots not intended for immediate use to −80°C for storage.

   *Note*: Each IVT reaction will require one full aliquot of each component, with the exception of the nuclease-free water. In addition to running one reaction per target expression vector, we recommend running a control reaction with the pT7CFE1-GFP control plasmid provided in the kit.
3. For each reaction, combine 12.5 μL HeLa lysate (red component) and 2.5 μL accessory proteins (green component) in a single RNAse-free

PCR tube. Lightly flick (1–2 times) securely capped tubes to mix contents. Spin briefly using a table top centrifuge. Incubate at 30°C on thermocycler for 10 min.

4. Next, move PCR tubes from thermocycler to nuclease-free workspace and add to each reaction 5 μL reaction mix (yellow component), 1 μg of template DNA (target protein or GFP control vector), and nuclease-free water to 5 μL total volume (blue component), e.g., addition of 2.5 μL template DNA would require addition of 2.5 μL nuclease-free water.
5. Again, lightly flick and briefly spin tubes before returning to thermocycler for 1–4 h at 30°C in thermocycler. For a first assessment of protein expression, we recommend an incubation time of 1 h. For screening, the optimal incubation time should be assessed as described in Section 2.2.
6. After incubation is complete, samples should be analyzed for protein expression or put on ice for use in screening. To prepare reaction mixtures for analysis of target protein expression via western blot, add an equal volume of 2 × Laemmli loading buffer (5% beta-mercaptoethanol, freshly added) and heat to 90°C for 5 min for use in subsequent steps 7–11.
7. Resolve samples alongside a protein ladder by SDS-PAGE and transfer polyvinylidene difluoride (PVDF) membrane. We use Precision Plus™ Dual Color Protein Standards, 4%–15% Criterion™ TGX™ Precast gels, and the Trans-Blot® Turbo™ Transfer System, all from Bio-Rad with associated equipment and reagents.

   *Note*: Typically, no more than 10 μL of the 1:1 (IVT:buffer) sample should be loaded to avoid overloading the gel lane. It is ideal to run at least two replicate lanes for the ladder and all samples to enable a separate western blot for detection of the target protein by both an epitope tag antibody and an antibody specific to the protein of interest.
8. Next, rinse the membrane twice with water and stain for 5 min at room temperature (RT) with REVERT Total Protein Stain, by adding 5 mL of protein stain supplemented with methanol according to the manufacturers instruction. This step is included to monitor equal loading of IVT reaction mix for each experimental lane.

   *Note*: Depending on the membrane and dish size, 5 mL may be insufficient to cover the membrane evenly. We routinely dilute the stain working solution by half with ultrapure water with no effect on staining efficiency to achieve a uniform staining of the membrane.

9. Decant the stain off of the membrane and wash 2 × 5 mL with the included wash solution (6.7% (v/v) glacial acetic acid, 30% (v/v) methanol, in water). Lastly, rinse briefly with ultrapure water before imaging in the 700 nm channel with an Odyssey® CLx Imaging System from LI-COR.
10. Incubate the membrane with blocking buffer (5% (w/v) BSA in Tris-buffered saline, with 0.1% Tween 20 (TBS-T)) overnight at 4°C to prevent nonspecific binding to the membrane.
11. Incubate membrane with the primary antibody (either anti-His for GFP control, anti-HA, or an antibody for the target protein) diluted in 1% (w/v) BSA in TBS-T. We used anti-HA.11 (BioLegend #MMS-101P) at a 1:1000 dilution, incubated at 90 min at RT.
12. Wash membrane 3 × 5 min with TBS-T. Then, incubate with a secondary antibody for detection, wash, and image. We used IRDye® 800CW Goat anti-Mouse IgG (LI-COR P/n 925-32210) diluted 1:15,000 in 5% (w/v) BSA in TBS-T, incubated for 90 min at RT. After washing 3 × 5 min with TBS-T and once with water, the membrane was imaged using the Odyssey® CLx Imaging System from LI-COR (Fig. 2).

## 2.2 Target Protein Quantification and Stability Assessment

Prior to screening, we recommend assess the time to maximum protein expression as well as the stability of target protein over the anticipated screening period by western blot.

### *2.2.1 Target Protein Quantification*

To monitor the accumulation of target protein over time, use the same workflow for running the IVT reaction as outlined in steps 1 through 5 of Section 2.1, but shorten the initial incubation time to the desired time interval of protein assessment (e.g., use a 30-min incubation to measure expression at 30-min intervals).

After the time interval has elapsed, move PCR tubes from thermocycler to nuclease-free workspace and remove 2 μL of the reaction to a fresh PCR tube prefilled with 2 μL phosphate-buffered saline (PBS) and 4 μL 2 × Laemmli loading buffer (5% beta-mercaptoethanol, freshly added). Then, lightly flick and briefly spin tubes containing remaining IVT reaction and return to thermocycler to incubate at 30°C for the next time interval.

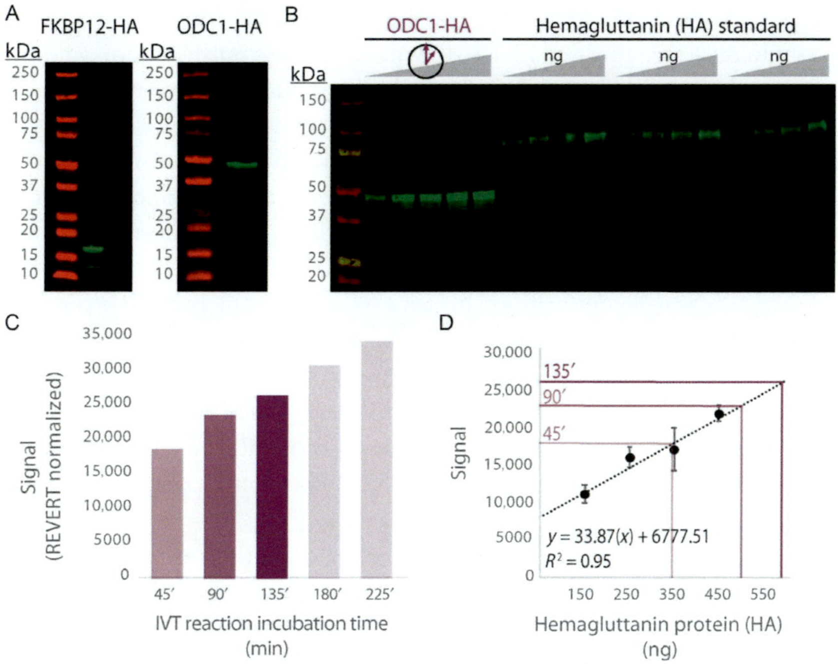


**Fig. 2** Expression of target protein by IVT. (A) Western blot for FKBP12 and ODC1 from IVT reactions using primary antibodies specific to FKBP12 (Abcam #ab2918) and ODC1 (Santa Cruz #sc-390366), respectively. (B) Western blot comparing ODC1 accumulation over time (45′ intervals) in the IVT reaction to a hemagglutinin (HA) protein standard using an anti-HA primary antibody (anti-HA.11 BioLegend #MMS-101P). (C) Quantification of the ODC1 signal accumulation over time (panel B) normalized to the REVERT Total Protein Stain signal (not shown) using the LI-COR Image Studio™ software according to the manufacturers documentation. (D) Quantification of the HA protein standard signal (panel B) and the corresponding standard curve from the linear fit of this data. A visualization of the intersection of the ODC1 signal accumulation illustrates that by 135 min, the ODC1 signal correlates with ~580 ng of HA signal. Based on the volume of sample loaded on the gel and an adjustment for differences in molecular weight of ODC1 and HA, the concentration of ODC1 was determined to be ~360 ng/μL.

Repeat this cycle until the final desired time point, so long as the remaining IVT reaction maintains a volume above 5 μL. In this manner, one can assess the accumulation of protein over time with the minimum consumption of IVT kit components without considerably compromising the IVT reaction performance.

Once each sample has been collected and diluted in Laemmli buffer, proceed with steps 6 through 11 as described in Section 2.1, but resolving the samples by SDS-PAGE alongside a purified protein that contains the HA epitope sequence (e.g., purified FKBP12 with an HA tag; purified hemagglutinin protein, such as Abcam #ab217652). This pure protein standard should be loaded on the gel at multiple (three or more) concentrations in triplicate to generate a standard curve (AU Eaton et al., 2014; Janes, 2015). In this manner, quantitation of signal for the target protein relative to the quantitation of the signal for the protein standard curve will yield the concentration of the target protein present in the IVT reaction over time.

From this result, determine the time to maximum protein expression without considerable degradation products. Degradation products often appear as a diffuse trail of signal above or below the punctate band at the expected height of the protein, where diffuse signal above can be protein tagged for degradation (e.g., ubiquitinated) and diffuse signal below can indicate either partially degraded protein or incompletely translated protein products.

Ideally, target protein expression should be on the order of hundreds of nanograms per microliter but should be no lower than 30 ng/μL. This lower limit equates to roughly 0.2% target protein in the IVT reaction mixture based on quantification of the IVT reaction mixture by BCA (15 μg/μL).

### 2.2.2 Target Stability Assessment

To monitor the accumulation of target protein over time, use the same workflow for running the IVT reaction as outlined in steps 1 through 5 of Section 2.1, using the incubation time determined from the target accumulation studies just described.

Following completion of the IVT reaction, move the reaction to the benchtop and remove 2 μL to a fresh PCR tube prefilled with 2 μL PBS and 4 μL 2 × Laemmli loading buffer. Over a period of 3 h, allow the sample to remain at RT, removing 2 μL from the reaction to a separate PCR tube prefilled with 2 μL PBS and 4 μL 2 × Laemmli loading buffer every hour. Once these time points have been collected and diluted in Laemmli buffer, proceed with steps 6 through 11 as described in Section 2.1. In this manner, the stability of the target protein over can be assessed.

If there is a marked decrease in target protein or a marked increase in degradation products over this time period, the reader may repeat this exercise, keeping the samples at 4°C and collecting samples every 2 h over an 8-h period, as SMM screening can performed at 4°C with longer incubation times.

# 3. SMM SCREENING

## 3.1 SMM Manufacture

The manufacture of isocyanate-based SMMs has been previously reported by our laboratory, with the most recent report outlining the method of manufacture as it is currently employed by our group (Casalena, Wassaf, & Koehler, 2012; Sections 2.1 and 3.1). Briefly, an activated isocyanate surface is generated from commercially available amine-functionalized glass slides. Screening compounds—including negative controls (e.g., DMSO), any available positive controls, and an amine-functionalized fluorescent dye for grid alignment—are printed to the surface using a 2470 model microarrayer from Aushon Biosystems. Finally, unreacted isocyanate groups are quenched to present a terminal alcohol and prevent covalent binding of proteins to the array surface. If following our method of manufacture, the resulting array will contain 48 subarrays, each comprising a 16 × 16 matrix of printed features.

Before outlining the use of these arrays for screening from IVT mixtures, we wish to highlight the following considerations.

### 3.1.1 Selection of Glass Substrate

Previously, we have recommended the use of GAPSII-Coated Slides (Corning Life Sciences, Inc.) as an ideal substrate for manufacturing isocyanate-based SMMs. While our laboratory still utilizes this product routinely, including in the studies presented here to demonstrate this method, Corning has recently discontinued the production of this product. The reader may, instead, employ any number of amine-functionalized glass slides commercially available, but should be cautioned that amine-functionalization achieved through adsorption (e.g., polyamino surfaces) or three-dimensional amine surfaces (e.g., matrix display) are substantially differentiated from the covalent, two-dimensional display of amines with a gamma-aminopropylsilane (GAPS) surface.

### 3.1.2 Inclusion of Small-Molecule Controls

In any high-throughput screen, positive and negative controls are critical for establishing assay conditions and confirming the validity of assay results. Both types of controls should be included in each of 48 subarrays of the microarray as a quality control for individual subarrays. Negative controls, as previously reported, should include DMSO alone and may also include "empty wells," where the microarrayer deposits no solvent to a particular

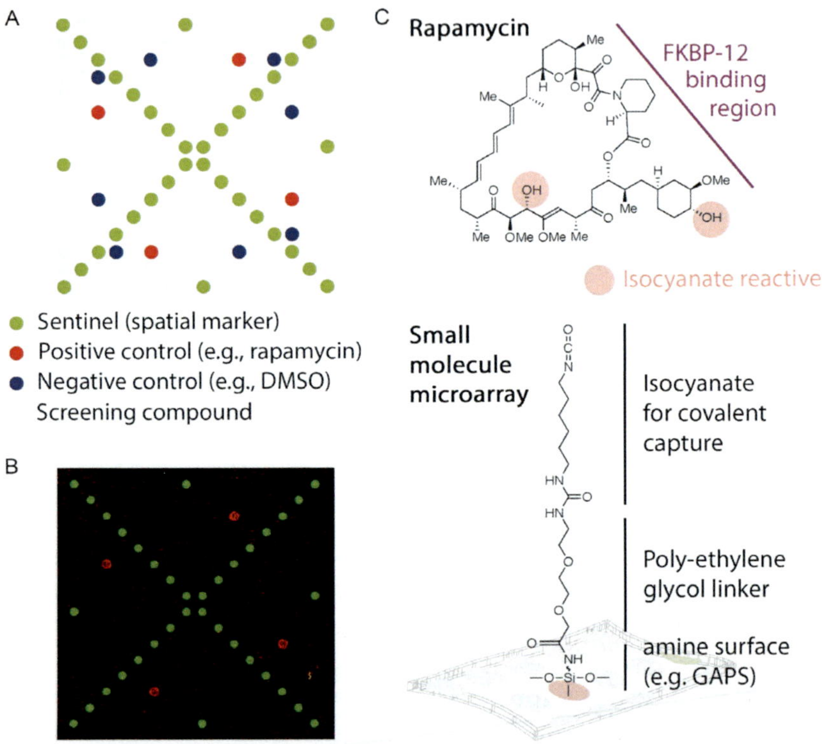


**Fig. 3** Considerations for isocyanate-based SMM manufacturing. (A) An example layout for each type of feature desired in a single subarray. The isocyanate-reactive fluorescent dye sentinels (spatial markers) mark the edge of the subarray and uniformity of the feature spacing, any negative and positives controls should be distributed throughout the subarray, and the remaining subarray positions should display screening compounds, with a minimum of two features per compound tested per subarray. (B) A representative subarray from an IVT-based screen of FKBP12 to illustrate the screening result from the type of layout described in panel A, where the positive control is rapamycin. *Note*: Rapamycin binds FKBP12 with nanomolar affinity. Not all positive controls will produce signals of this magnitude. (C) An illustration of the molecular considerations that should be made when selecting a positive control for isocyanate-based SMM screening. Rapamycin is a known FKBP12 binder, with multiple sites of isocyanate reactivity in a region of the molecule unlikely to inhibit FKBP12 binding when the compound is displayed on the subarray through the multiatom linker.

microarray feature position. If there is a known ligand of the protein of interest, we recommend its inclusion in the printing library as a positive control, but with several caveats that the reader should consider (Fig. 3).

First, not all small molecules annotated as modulators of a protein in the public domain are robustly characterized for direct interaction with the target.

Because of this, we encourage the reader to investigate the supporting data in the public domain surrounding a small molecule and its molecular target. If these data are primarily based on cellular and molecular phenotypes—without either sufficient molecular profiling and mechanism of action data in vitro (Frye, 2010) and/or demonstrated target engagement and exposure at the site of action in cells (Bunnage, Chekler, & Jones, 2013)—it may not be a true positive control. Additionally, because the isocyanate surface will preferentially react with primary amines and alcohols (Bradner, McPherson, Mazitschek, et al., 2006), the reader may find that the likely orientation of immobilized positive control on the array will directly block or sterically hinder the recognition of the molecule by the target. In this case, installation of a short, amine-terminated polyethylene glycol (PEG) or alkane linker at a site on the molecule permissive to substitution can increase the likelihood of detecting the interaction by SMM.

## 3.2 Screening of Target Protein

Screening of the IVT-expressed target protein involves diluting the IVT reaction in MIPP buffer before incubation with the microarray. In the development of the SMM screening conditions, we recommend testing a range of concentrations against the identical microarrays (i.e., same compound set). In an optimized screen, we recommend preparing a single working solution of target protein to test against the screening library, screening at least two replicate microarrays for each unique microarray required to print the full screening library.

### *3.2.1 Reagents*

- Target proteins expressed by IVT, as in Section 2.1 of this protocol (steps 1–5)
- Printed SMM slides as referenced in Section 3.1 of this protocol
- 4-well dishes (Thermo Fisher Scientific #267061)
- MIPP buffer (20 m*M* $NaH_2PO_4$ (pH 7.2), 25 m*M* β-glycerophosphate, 2 m*M* EGTA, 2 m*M* EDTA, 0.5% (v/v) Triton X-100)
- Protease inhibitor cocktail (e.g., Roche #04693159001)
- TBS-T (Tris-buffered saline + Tween 20)
- Fluorophore-labeled primary antibody (e.g., Thermo Fisher Scientific #26183-D650)
- TBS (Tris-buffered saline)
- Milli-Q water or equivalent

### 3.2.2 Equipment

- Wafer handling tweezers (e.g., VWR #63042-358)
- Rocking shaker (e.g., Denville #S7100)
- Glass reservoir, 2 (e.g., Corning #3140190)
- Laboratory slide spinner (e.g., Labnet, cat no C1303T)
- GenePix® 4300 microarray scanner (Molecular Devices)
- GenePix® Pro software (Molecular Devices)

### 3.2.3 Protocol

1. If SMM slides have been previously manufactured and stored at −20°C, let slides thaw completely before opening slide package.
2. Meanwhile, prepare protein working solutions for screening by diluting the samples into cold MIPP buffer supplemented with 1 × protease inhibitor cocktail in an appropriately sized microcentrifuge or conical tube, making one working solution per screening concentration. For well-expressed proteins, we routinely screen from IVT reactions at a dilution in the range of 1:250–1:500 (i.e., 10 μL IVT reaction into 2.5 mL MIPP buffer for 1:250 dilution) as a first assessment of target protein binding profiles when screening by IVT. Depending on the amount of target protein expression, any dilution of the reaction that yields a final target protein concentration in the range of 500 pg/μL–5 ng/μL is reasonable starting point.

   *Note*: Each well in the recommended dish holds a maximum of 4 mL, but 2.5 mL of protein working solution is sufficient to coat the slide and our recommendations for working solution concentrations and protein loading are based on this volume.

   *Note*: If the target protein of interest requires small molecule or peptide cofactors to function in a mode of interest (e.g., to bind a positive control, to engage with a protein–protein interaction partner), small-molecule cofactors may be added to the protein working solution, as well as to washing buffers and antibody working solution if necessary.
3. Place each slide in a well of a 4-well dish using wafer handling tweezers on the barcoded end of the slide to manipulate the arrays. Be sure to record the barcodes and associated metadata for the intended screening conditions for each.
4. Add protein working solution to the microarray by applying the solution on the barcoded region of the slide with a serological pipet or pipetman. Use good pipetting technique to avoid generating bubbles.

*Note*: Do not touch the slide with pipet tip at any point throughout the protocol, as contact with the array portion of the slide can disrupt transient binding interactions.

5. Immediately cover the lidded dish with aluminum foil to shield from light and incubate at RT on rocker for 1 h.

   *Note*: The rocker should be on a slow speed—about one rocking cycle every 5 s—as faster settings may introduce currents to the system that disturb weak binding interactions expected from screening compounds.

6. Remove cover and wash the slides $3 \times 2$ min with 3 mL of TBS-T per slide.

   *Note*: Larger molecular weight proteins (>90 kDa) may require longer wash times. This may be determined in the assay development phase and is discussed in Section 5.

7. After the last wash, carefully move each slide to a new well in a fresh 4-well dish using the wafer tweezers to reduce persistence of unbound target protein and IVT reaction components.

8. Prepare a working solution of fluorescently labeled antibody in cold TBS-T. Prepare a sufficient quantity to apply 2.5 mL to each slide.

   *Note*: We recommend using a working concentration 200 μg/mL as an initial working concentration, which is a 1:5000 dilution for antibodies stored as a 1-mg/mL solution. This may be modified as further discussed in Section 5.

9. Add antibody working solution to the microarray by applying the solution on the barcoded region of the slide with a serological pipet or pipetman. Use good pipetting technique to avoid generating bubbles.

10. Immediately cover the lidded dish with aluminum foil to shield from light and incubate at RT on rocker for 1 h.

11. Remove cover and wash the slides $3 \times 2$ min with 3 mL of TBS-T per slide, followed by $1 \times 2$ min with 3 mL of TBS to remove excess detergent.

12. After the last wash, fill two glass reservoirs with clean Milli-Q water. One by one, carefully remove each slide from 4-well dish by inserting the slide tongs underneath the barcode and gently lifting upward until the tweezers have a secure hold on the slide. Immerse the slide at hand in Milli-Q water to wash, immersing the slide three consecutive times in the same reservoir, and spin dry using table top slide spinner (~1 min). Repeat the same repeated immersion of the slide in the

second reservoir and then spin a final time. Slides should be completely dry at this point.

*Note*: Slides are immersed serially in the two reservoirs to remove excess salt and detergent. Reservoirs can be used to wash multiple slides without refreshing the water, but we recommend designating one reservoir for the first wash and one for the final wash to keep detergent and salt in the final reservoir to a minimum.

13. Scan each slide using the GenePix 4300 microarray scanner to generate multilayer TIFF images, one layer for the sentinel features and a second for the detection antibody (e.g., 532 nm scan to detect rhodamine cadaverine sentinel features, 635 nm scan to detect target protein via anti-HA DyLight® 650). Particularly for the detection channel scan, the optimal photomultiplier tube (PMT) setting should be determined to yield the maximum signal-to-noise ratio (SNR), which can be assessed visually at this stage of analysis.

    *Note*: Slides should be scanned as soon as possible after assay is complete for best results, but slides can also be stored for a short time in a closed slide mailer in the dark for additional scans if necessary.

    *Note*: Once the optimal PMT level is found, all subsequent slides should be scanned using the detection settings in order to reasonable compare visual and numerical data from individual slides.

# 4. ANALYSIS OF SMM SCREENING DATA

## 4.1 Defining SNR Values for Each Array Feature

To extract numerical data on a feature-by-feature basis from fluorescent slide scan images, open each multilayer TIFF and load the corresponding GAL file for the microarray generated by the microarrayer during the print run in into the GenePix software. The GAL file contains the $(x, y)$ spatial information for the features on the array and the metadata for the compound stock plate well from which the feature was printed at the time of manufacture.

Use the adjustment tools in the workspace of the software to align the GAL file grid to the array, using the printed fluorescent sentinels (e.g., 532 nm) are used to align the grid to the printed region of the microarray. Then, use the automated fit function in the software workspace to fit the diameter of all grid features to the fluorescent features.

Before generating numerical data from the array through the calculate function, manually inspect the fitted GAL file grid to be sure that the software has sized the features appropriately, correcting any incorrect alignment settings and adjusting individual feature diameters for any compound features showing a well-defined feature border that is larger or smaller than the GAL file feature diameter. You may also wish to use the software's function for annotating features that are either clearly showing a positive signal (e.g., bright, dense signal with a well-defined border) or clearly a fluorescent artifact (e.g., bright streak or scratch). Because the software simply calculates fluorescent intensities within the GAL file feature boundaries and within a small area surrounding the feature, visually inspecting the array data is important to reduce false negatives and false positives.

Once the GAL file grid has been properly aligned, use the calculate function in the GenePix software to generate a GPR file that contains tabular, numerical data for each feature (fluorescent intensities, feature diameters and coordinates, percentage of skewed pixels, etc.) under a header containing relevant metadata to the manufacturing and scanning of the array. To define a SNR for each feature, we recommend taking the ratio of the median fluorescence intensity of the feature to the median fluorescence intensity of the background. This data should then be analyzed to assess the validity of the screen and identify assay positives for further study.

## 4.2 Computing and Interpreting Screening Statistics

Typically, the distribution of SNR values from an SMM screen is skewed right due to the nature of the detection mode. For a given screen, we expect there will be many more outliers with higher feature intensity compared to background (i.e., a visible fluorescent spot) than outliers with background intensity greater than the feature intensity (i.e., a spot on the array darker than background). For this reason, we routinely use a robust $z$-score calculation to determine the compounds with significant performance in the assay. Because the robust $z$-score calculation relies on median-based metrics rather than mean-based metrics, it is less biased by outliers or high-performing compounds in the right-hand tail of the distribution. The robust $z$-score ($z_i$) can be calculated for each feature ($i$) by the following equation (Fig. 4):

$$z_i = \frac{\mathrm{SNR}_i - Mdn(\mathrm{SNR})}{MAD(\mathrm{SNR}) * 1.486}$$

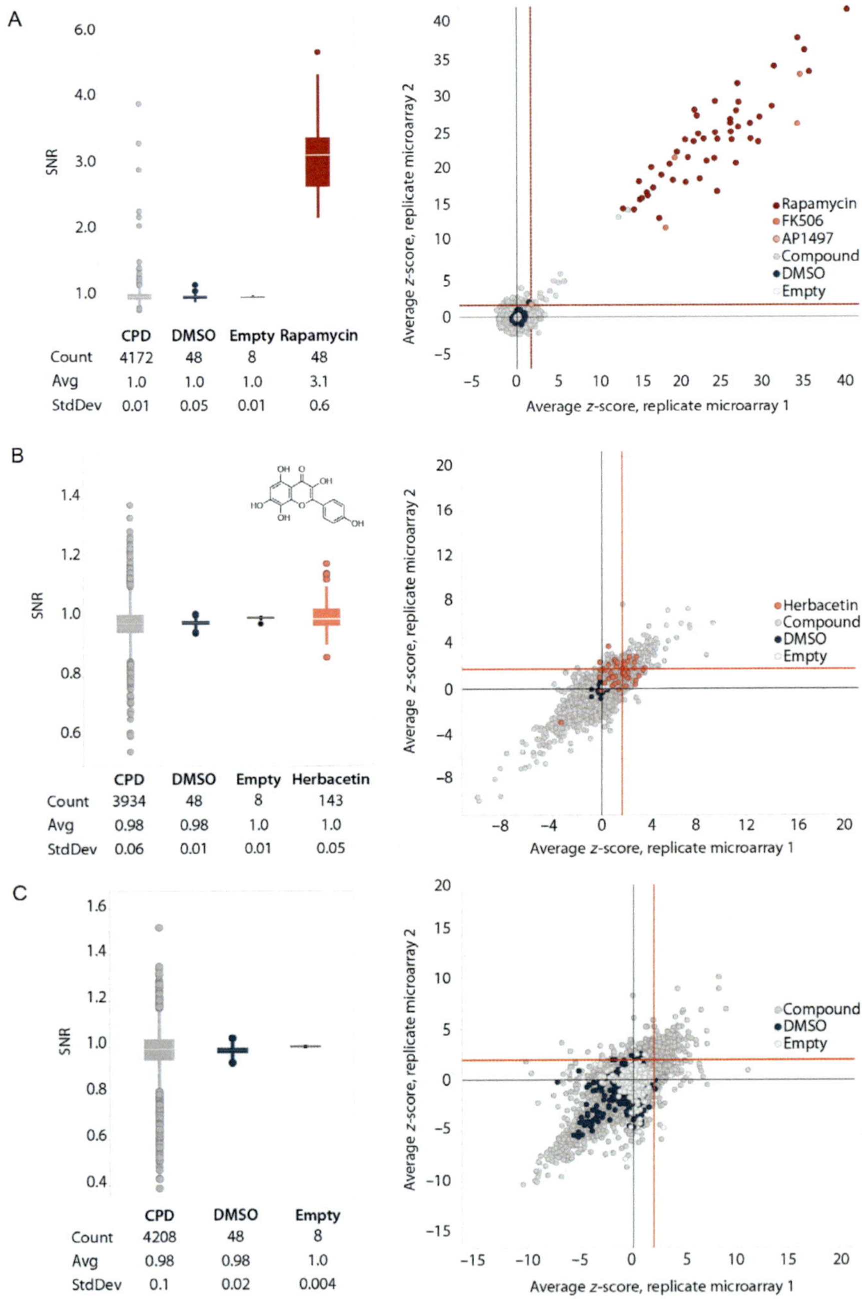


**Fig. 4** Representative IVT-based SMM screening data sets for three different protein targets. (A) FKBP12 was expressed by IVT with an incubation time of 45 min and diluted 1:100 in MIPP buffer. The data for a representative microarray are shown. Rapamycin,

where $SNR_i$ is the SNR value for a given feature, $Mdn$(SNR) is the median of the SNR values for all features in the subarray, and the $MAD$(SNR) is the median absolute deviation of the SNR values for all features in the subarray. The reader may adjust the robust $z$-score threshold for nominating assay positives from the screen based on the desired confidence interval.

Lastly, before moving forward with a given assay positive list, it is important to compute statistical measures of assay performance to assess the validity of the screen, specifically the coefficient of variation (CV) values for the negative control feature populations and the performance of any positive control compounds. We do appreciate that for many novel therapeutic targets, positive controls for direct binding may not exist. In this case, the CV value for the negative control population and a hit rate (percentage of assay positives identified from the library) within the range of 0.1%–1% may be sufficient for promoting compounds for downstream evaluation.

These calculations for identifying assay positives and quality metrics are routinely used for other types of high-throughput screens, and, alternatively, other statistical methods of normalization and quality control for HTS may also be employed if the reader so chooses (Birmingham et al., 2009), keeping in mind that the distribution of SNR for positive controls is frequently skewed right, as well.

**Fig. 4—cont'd** which binds FKBP12 with 0.5 n*M* affinity, is included as a true positive control. Additionally, other known FKBP12 ligands were included in the screening library, specifically FK506 (3 n*M* affinity) and AP1497 (18 n*M* affinity) (Keenan et al., 1998), which are both shown above a *z*-score threshold of 1.64 (90% confidence interval). (B) ODC1 was expressed by IVT with an incubation time of 2 h and 15 min and diluted 1:250 in MIPP buffer supplemented with 20 μ*M* pyridoxal phosphate (a cofactor) in the presence of DTT. We included in the screening library, herbacetin, a small molecule with isocyanate reactivity discovered as an allosteric site binder of ODC1 through molecular docking studies and confirmed to engage ODC1 by compound-anchored pulldown (Kim et al., 2016). The microarray features for this compound perform distinctly from the screening library population, but not all subarrays show herbacetin performance above a *z*-score threshold of 1.64 (90% confidence interval). Confirming the ability of immobilized herbacetin to bind ODC1 in a pulldown assay with density of compound display and immobilization strategy that more closely mimics the SMM would suggest further optimization of this screen prior to defining assay positives. (C) MYB was expressed by IVT with an incubation time of 2 h and diluted 1:250 in MIPP buffer. In the absence of positive controls, a more stringent *z*-score threshold of 1.96 (95%) is shown.

# 5. ITERATIVE OPTIMIZATION OF SMM SCREENING CONDITIONS

There are several parameters that may be optimized through iterative assay development screens in order to deduce the screening conditions which yield maximal SNR. This iterative process is crucial for maximizing ligand discovery potential and improving downstream translation rates.

## 5.1 Counter-Screening

Any detection antibody used for screening will interact with some small percentage of small molecules printed on the array. In order to identify small molecules that interact with the target itself, a counter screen for the detection antibody should be conducted, as well. An appropriate counter screen should be run with the exact same conditions (i.e., amount of IVT reaction per slide, amount of antibody per slide, incubation times, washing protocols, and scanning parameters) as the target screen, but where the IVT reaction is run without the target DNA template. For this activity, an IVT reaction run in the absence of any DNA template is sufficient for either epitope-tag or direct target detection strategies, but the use of the IVT control GFP vector is not recommended due to the fluorescent nature of the protein and its absence in the target screen. We routinely use the same data analyses described in Section 4 to interpret counter-screening data and deprioritize any compounds scoring as in both the target screen and the counter screen in downstream assays.

## 5.2 Optimization of Protein Concentration and Antibody-Based Detection

When determining the amount of target protein to apply to each microarray, keep in mind that by this method any change in target protein loading will be accompanied by a change in accessory/lysate protein loading, which will act like a blocking agent on the array. Consequently, for any target protein sample at a given IVT incubation time (i.e., target protein concentration), we recommend testing at least three different total protein loading amounts to determine the ideal screening condition (i.e., maximum SNR for positive controls, near 1.0 signal for negative controls, and uniform, low background signal).

In the absence of positive controls, the reader may also use the hit rate as a metric to monitor the performance of the screen. As stated in Section 4, hit rates for these screens, after the exclusion of antibody counter-screening positives, should be 0.1%–1% of the size of the library screened. A hit rate below or above this range may indicate a high rate of false negatives or false positives, respectively, and necessitate further assay development. If using this metric for optimization, we stress the importance of simultaneously monitoring the performance of the negative control features (e.g., DMSO) on the array. A reasonable hit rate without a narrowly distributed negative control population centered at (or very near) an SNR of 1.0 is insufficiently optimized and likely harbors an exceedingly high degree of false positives and/or false negatives.

When surveying screening conditions, we recommended selecting several concentrations of the protein working solution (Fig. 5). With this method, we typically screen target proteins in the range of 500 pg/μL–5 ng/μL of the target protein in the protein working solution. For example, a protein expressed at 300 ng/μL in the IVT reaction by quantitative western blot could be tested anywhere within the range of 1:60–1:600 dilution. Because the IVT reaction kit contains approximately 15 μg/μL of HeLa lysate and accessory proteins, this dilution range will also result in 25–250 μg/mL additional protein on the microarray.

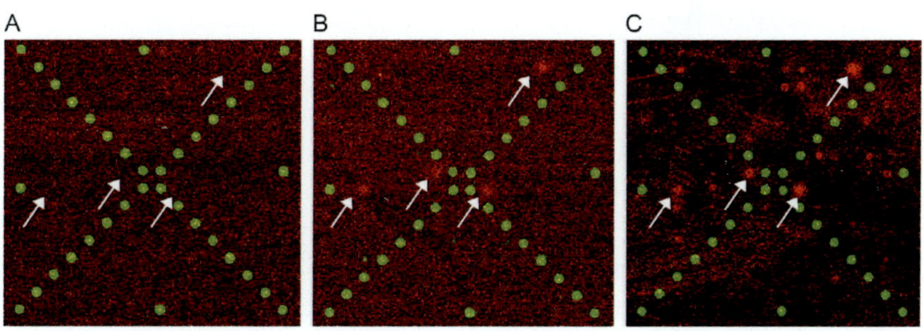


**Fig. 5** Representative subarray images from screening optimization. This example shows the result of diluting the same protein expression sample diluted to three different protein working solution concentrations, with *white arrow* indicating the features for two compounds, in duplicate, that scored as assay positives in an optimized screen. (A) Uniform background but with little to no signal is observed with a protein working solution at the lower limit of dilution recommended in the text. (B) Uniform background with pronounced, well-defined features is observed when the protein working solution is 1.6 × more concentrated than the recommended lower limit of dilution (panel A). This was empirically determined ideal concentration for the screen. (C) Nonuniform background and a high number of features with a variety of morphologies is observed with a protein working solution that is 2.5 × more concentrated than the recommended lower limit of dilution (panel A).

Insufficient signal can be improved by increasing protein concentration, but, if an increase in signal is needed beyond this range, we recommend preincubation of the antibody with the IVT reaction prior to screening to reduce signal loss due to washing rather than increasing total protein loading. For this approach, add detection antibody and IVT sample to the same working solution of MIPP and add to a single well of the 4-well dish prior to adding the SMM slide to the well. Allow this to incubate for 1 h, rock gently, shielded from light, before proceeding with steps 11–14 of Section 3.2.

In the alternative scenario, if the signal is too high, resulting in high global background, the reader should decrease the protein applied to the slide. However, if high global background persists despite decreases in total protein, we recommend testing a lower gain setting (PMT value) on the microarray scanner and/or decreasing the concentration of the antibody working solution. The counter-screening condition for these adjustments should be helpful in determining whether this issue is intrinsic to some component of the detection system (i.e., antibody selection, antibody working concentration, scanning parameters) or an issue with the target protein working solution in combination with the detection system.

## 6. SUMMARY

In summary, commercially available IVT kits can be leveraged to expand the compatibility of array-based chemical screening approaches for proteins with short half-lives. We have detailed a method for screening short-lived proteins expressed by IVT against a small-molecule library through the use of isocyanate-based microarrays, using FKBP12 and ODC1 as example proteins to demonstrate the method. In our own hands, this method has allowed for the nomination of putative binders of both naturally short-lived proteins and short-lived oncogenic protein fusions. As with other high-throughput assays, particularly with fluorescent readouts, these data are assumed to present a high false-positive rate. The assay positives for such a screen should be further evaluated for on-target cellular and molecule phenotypes, as well as for direct target binding through biophysical methods where pure protein is attainable, or through less quantitative in vitro and cellular methods. Executing the assay positives from an IVT-based SMM screen into a target-specific critical path of evaluation and prioritization, this array-based system can enable ligand discovery for recalcitrant therapeutic targets with short half-lives.

## ACKNOWLEDGMENTS

This material is based upon work supported by the Adenoid Cystic Carcinoma Research Foundation, the MIT Center for Precision Cancer Medicine, the Royal G. and Mae H. Westaway Family Memorial Fund, the National Science Foundation Graduate Research Fellowship under Grant No. 1122374 (S.K.D.), and a fellowship from the Ludwig Center for Molecular Oncology (S.K.D.). The authors thank Jesse Vigoda Kurland for consultation related to in vitro translation protocols, Dr. Amedeo Vetere, and Dr. Nicole Spardy-Burr at the AACRF. Additional thanks to Dr. Bert van de Kooij for helpful discussion related to global protein stability.

## REFERENCES

Agarwal, R., Katiyar, S. K., Zaidi, S. I., & Mukhtar, H. (1992). Inhibition of skin tumor promoter-caused induction of epidermal ornithine decarboxylase in SENCAR mice by polyphenolic fraction isolated from green tea and its individual epicatechin derivatives. *Cancer Research*, *52*(13), 3582–3588.

AU Eaton, S. L., AU Hurtado, M. L., AU Oldknow, K. J., AU Graham, L. C., AU Marchant, T. W., AU Gillingwater, T. H., et al. (2014). A guide to modern quantitative fluorescent Western blotting with troubleshooting strategies. *JoVE*, *93*, e52099.

Birmingham, A., Selfors, L. M., Forster, T., Wrobel, D., Kennedy, C. J., Shanks, E., et al. (2009). Statistical methods for analysis of high-throughput RNA interference screens. *Nature Methods*, *6*(8), 569–575. https://doi.org/10.1038/nmeth.1351.

Bondeson, D. P., Mares, A., Smith, I. E. D., Ko, E., Campos, S., Miah, A. H., et al. (2015). Catalytic in vivo protein knockdown by small-molecule PROTACs. *Nature Chemical Biology*, *11*(8), 611–617. https://doi.org/10.1038/nchembio.1858.

Boskovic, Z. V., Kemp, M. M., Freedy, A. M., Viswanathan, V. S., Pop, M. S., Fuller, J. H., et al. (2016). Inhibition of zinc-dependent histone deacetylases with a chemically triggered electrophile. *ACS Chemical Biology*, *11*(7), 1844–1851. https://doi.org/10.1021/acschembio.6b00012.

Bradner, J. E., McPherson, O. M., & Koehler, A. N. (2006a). A method for the covalent capture and screening of diverse small molecules in a microarray format. *Nature Protocols*, *1*(5), 2344–2352. https://doi.org/10.1038/nprot.2006.282.

Bradner, J. E., McPherson, O. M., Mazitschek, R., Barnes-Seeman, D., Shen, J. P., Dhaliwal, J., et al. (2006b). A robust small-molecule microarray platform for screening cell lysates. *Chemistry & Biology*, *13*(5), 493–504. https://doi.org/10.1016/j.chembiol.2006.03.004.

Bunnage, M. E., Chekler, E. L. P., & Jones, L. H. (2013). Target validation using chemical probes. *Nature Publishing Group*, *9*(4), 195–199. https://doi.org/10.1038/nchembio.1197.

Casalena, D. E., Wassaf, D., & Koehler, A. N. (2012). Ligand discovery using small-molecule microarrays. *Methods in Molecular Biology (Clifton, N.J.)*, *803*, 249–263. https://doi.org/10.1007/978-1-61779-364-6_17(Chapter 17).

Crews, C. M. (2018). Inducing protein degradation as a therapeutic strategy. *Journal of Medicinal Chemistry*, *61*(2), 403–404. https://doi.org/10.1021/acs.jmedchem.7b01333.

Daguer, J. P., Zambaldo, C., Ciobanu, M., Morieux, P., Barluenga, S., & Winssinger, N. (2014). DNA display of fragment pairs as a tool for the discovery of novel biologically active small molecules. *Chemical Science*, *6*, 739–744. https://doi.org/10.1039/C4SC01654H.

Eden, E., Geva-Zatorsky, N., Issaeva, I., Cohen, A., Dekel, E., Danon, T., et al. (2011). Proteome half-life dynamics in living human cells. *Science (New York)*, *331*(6018), 764–768. https://doi.org/10.1126/science.1199784.

Felsenstein, K. M., Saunders, L. B., Simmons, J. K., Leon, E., Calabrese, D. R., Zhang, S., et al. (2016). Small molecule microarrays enable the identification of a selective, quadruplex-binding inhibitor of MYC expression. *ACS Chemical Biology*, *11*(1), 139–148. https://doi.org/10.1021/acschembio.5b00577.

Frye, S. V. (2010). The art of the chemical probe. *Nature Publishing Group*, *6*(3), 159–161. https://doi.org/10.1038/nchembio.296.

Funakoshi-Tago, M., Sumi, K., Kasahara, T., & Tago, K. (2013). Critical roles of Myc-ODC axis in the cellular transformation induced by myeloproliferative neoplasm-associated JAK2 V617F mutant. *PLoS One*, *8*(1), e52844. https://doi.org/10.1371/journal.pone.0052844.

Gupta, S., Ahmad, N., Mohan, R. R., Husain, M. M., & Mukhtar, H. (1999). Prostate cancer chemoprevention by green tea: In vitro and in vivo inhibition of testosterone-mediated induction of ornithine decarboxylase. *Cancer Research*, *59*(9), 2115–2120.

He, W., Roh, E., Yao, K., Liu, K., Meng, X., Liu, F., et al. (2017). Targeting ornithine decarboxylase (ODC) inhibits esophageal squamous cell carcinoma progression. *Npj Precision Oncology*, *1*(1), E359. https://doi.org/10.1038/s41698-017-0014-1.

Hogarty, M. D., Norris, M. D., Davis, K., Liu, X., Evageliou, N. F., Hayes, C. S., et al. (2008). ODC1 is a critical determinant of MYCN oncogenesis and a therapeutic target in neuroblastoma. *Cancer Research*, *68*(23), 9735–9745. https://doi.org/10.1158/0008-5472.CAN-07-6866.

Hong, J. A., Neel, D. V., Wassaf, D., Caballero, F., & Koehler, A. N. (2014). Recent discoveries and applications involving small-molecule microarrays. *Current Opinion in Chemical Biology*, *18*, 21–28. https://doi.org/10.1016/j.cbpa.2013.09.020.

Janes, K. A. (2015). An analysis of critical factors for quantitative immunoblotting. *Science Signaling*, *8*(371), rs2. https://doi.org/10.1126/scisignal.2005966.

Jang, W.-J., Choi, B., Song, S.-H., Lee, N., Kim, D.-J., Lee, S., et al. (2017). Multi-omics analysis reveals that ornithine decarboxylase contributes to erlotinib resistance in pancreatic cancer cells. *Oncotarget*, *8*(54), 92727–92742. https://doi.org/10.18632/oncotarget.21572.

Keenan, T., Yaeger, D. R., Courage, N. L., Rollins, C. T., Pavone, M. E., Rivera, V. M., et al. (1998). Synthesis and activity of bivalent FKBP12 ligands for the regulated dimerization of proteins. *Bioorganic & Medicinal Chemistry*, *6*(8), 1309–1335.

Kim, D. J., Roh, E., Lee, M. H., Oi, N., Lim, D. Y., Kim, M. O., et al. (2016). Herbacetin is a novel allosteric inhibitor of ornithine decarboxylase with antitumor activity. *Cancer Research*, *76*(5), 1146–1157. https://doi.org/10.1158/0008-5472.CAN-15-0442.

Kim, H. I., Schultz, C. R., Buras, A. L., Friedman, E., Fedorko, A., Seamon, L., et al. (2017). Ornithine decarboxylase as a therapeutic target for endometrial cancer. *PLoS One*, *12*(12). https://doi.org/10.1371/journal.pone.0189044e0189044–18.

Koehler, A. N., Shamji, A. F., & Schreiber, S. L. (2003). Discovery of an inhibitor of a transcription factor using small molecule microarrays and diversity-oriented synthesis. *Journal of the American Chemical Society*, *125*(28), 8420–8421. https://doi.org/10.1021/ja0352698.

Kohn, M., Wacker, R., Peters, C., Schroder, H., Soulere, L., Breinbauer, R., et al. (2003). Staudinger ligation: A new immobilization strategy for the preparation of small-molecule arrays. *Angewandte Chemie (International Ed. in English)*, *42*(47), 5830–5834. https://doi.org/10.1002/anie.200352877.

Mangold, U., & Leberer, E. (2005). Regulation of all members of the antizyme family by antizyme inhibitor. *The Biochemical Journal*, *385*(Pt. 1), 21–28. https://doi.org/10.1042/BJ20040547.

Miyazaki, I., Okumura, H., Simizu, S., Takahashi, Y., Kanoh, N., Muraoka, Y., et al. (2009). Structure-affinity relationship study of bleomycins and Shble protein by use of a chemical array. *Chembiochem: A European Journal of Chemical Biology*, *10*(5), 845–852. https://doi.org/10.1002/cbic.200800728.

Mohammed, A., Janakiram, N. B., Madka, V., Ritchie, R. L., Brewer, M., Biddick, L., et al. (2014). Eflornithine (DFMO) prevents progression of pancreatic cancer by modulating ornithine decarboxylase signaling. *Cancer Prevention Research (Philadelphia, Pa.)*, 7(12), 1198–1209. https://doi.org/10.1158/1940-6207.CAPR-14-0176.

Neumann, T., Junker, H.-D., Schmidt, K., & Sekul, R. (2007). SPR-based fragment screening: Advantages and applications. *Current Topics in Medicinal Chemistry*, 7(16), 1630–1642.

Patel, J. H., Loboda, A. P., Showe, M. K., Showe, L. C., & McMahon, S. B. (2004). Analysis of genomic targets reveals complex functions of MYC. *Nature Reviews Cancer*, *4*(7), 562–568. https://doi.org/10.1038/nrc1393.

Pegg, A. E. (2006). Regulation of ornithine decarboxylase. *Journal of Biological Chemistry*, *281*(21), 14529–14532. https://doi.org/10.1074/jbc.R500031200.

Pop, M. S., Stransky, N., Garvie, C. W., Theurillat, J.-P., Hartman, E. C., Lewis, T. A., et al. (2014). A small molecule that binds and inhibits the ETV1 transcription factor oncoprotein. *Molecular Cancer Therapeutics*, *13*(6), 1492–1502. https://doi.org/10.1158/1535-7163.MCT-13-0689.

Sadaghiani, A. M., Lee, S. M., Odegaard, J. I., Leveson-Gower, D. B., McPherson, O. M., Novick, P., et al. (2014). Identification of Orai1 channel inhibitors by using minimal functional domains to screen small molecule microarrays. *Chemistry & Biology*, *21*(10), 1278–1292. https://doi.org/10.1016/j.chembiol.2014.08.016.

Schwanhäusser, B., Busse, D., Li, N., Dittmar, G., Schuchhardt, J., Wolf, J., et al. (2011). Global quantification of mammalian gene expression control. *Nature*, *473*(7347), 337–342. https://doi.org/10.1038/nature10098.

Singh, K., Coburn, L. A., Asim, M., Barry, D. P., Allaman, M. M., Shi, C., et al. (2018). Ornithine decarboxylase in macrophages exacerbates colitis and promotes colitis-associated Colon carcinogenesis by impairing M1 immune responses. *Cancer Research*, *78*(15), 4303–4315. https://doi.org/10.1158/0008-5472.CAN-18-0116.

Smithson, D. C., Shelat, A. A., Baldwin, J., Phillips, M. A., & Guy, R. K. (2010). Optimization of a non-radioactive high-throughput assay for decarboxylase enzymes. *Assay and Drug Development Technologies*, *8*(2), 175–185. https://doi.org/10.1089/adt.2009.0249.

Stanton, B. Z., Peng, L. F., Maloof, N., Nakai, K., Wang, X., Duffner, J. L., et al. (2009). A small molecule that binds Hedgehog and blocks its signaling in human cells. *Nature Chemical Biology*, *5*(3), 154–156. https://doi.org/10.1038/nchembio.142.

Sztuba-Solinska, J., Shenoy, S. R., Gareiss, P., Krumpe, L. R. H., Le Grice, S. F. J., O'Keefe, B. R., et al. (2014). Identification of biologically active, HIV TAR RNA-binding small molecules using small molecule microarrays. *Journal of the American Chemical Society*, *136*(23), 8402–8410. https://doi.org/10.1021/ja502754f.

Urbina, H. D., Debaene, F., Jost, B., Bole-Feysot, C., Mason, D. E., Kuzmic, P., et al. (2006). Self-assembled small-molecule microarrays for protease screening and profiling. *Chembiochem: A European Journal of Chemical Biology*, 7(11), 1790–1797. https://doi.org/10.1002/cbic.200600242.

Vegas, A. J., Bradner, J. E., Tang, W., McPherson, O. M., Greenberg, E. F., Koehler, A. N., et al. (2007). Fluorous-based small-molecule microarrays for the discovery of histone deacetylase inhibitors. *Angewandte Chemie (International Ed. in English)*, *46*(42), 7960–7964. https://doi.org/10.1002/anie.200703198.

Vegas, A. J., Fuller, J. H., & Koehler, A. N. (2008). Small-molecule microarrays as tools in ligand discovery. *Chemical Society Reviews*, *37*(7), 1385–1394. https://doi.org/10.1039/b703568n.

Velagapudi, S. P., & Disney, M. D. (2014). Two-dimensional combinatorial screening enables the bottom-up design of a microRNA-10b inhibitor. *Chemical Communications*, *50*(23), 3027–3029. https://doi.org/10.1039/c3cc00173c.

Verma, A. K., & Boutwell, R. K. (1977). Vitamin A acid (retinoic acid), a potent inhibitor of 12-O-tetradecanoyl-phorbol-13-acetate-induced ornithine decarboxylase activity in mouse epidermis. *Cancer Research*, *37*(7 Pt. 1), 2196–2201.

Verma, A. K., Shapas, B. G., Rice, H. M., & Boutwell, R. K. (1979). Correlation of the inhibition by retinoids of tumor promoter-induced mouse epidermal ornithine decarboxylase activity and of skin tumor promotion. *Cancer Research*, *39*(2 Pt. 1), 419–425.

Winter, G. E., Buckley, D. L., Paulk, J., Roberts, J. M., Souza, A., Dhe-Paganon, S., et al. (2015). Phthalimide conjugation as a strategy for in vivo target protein degradation. *Science (New York, N.Y.)*, *348*(6241), 1376–1381. https://doi.org/10.1126/science.aab1433.

Yang, Z., Koehler, A. N., & Wang, L. (2016). A novel small molecule activator of nuclear receptor SHP inhibits HCC cell migration via suppressing Ccl2. *Molecular Cancer Therapeutics*, *15*(10), 2294–2301. https://doi.org/10.1158/1535-7163.MCT-16-0153.

Yen, H.-C. S., Xu, Q., Chou, D. M., Zhao, Z., & Elledge, S. J. (2008). Global protein stability profiling in mammalian cells. *Science (New York, NY)*, *322*(5903), 918–923. https://doi.org/10.1126/science.1160489.

Young, L. M., Saunders, J. C., Mahood, R. A., Revill, C. H., Foster, R. J., Tu, L.-H., et al. (2015). Screening and classifying small-molecule inhibitors of amyloid formation using ion mobility spectrometry-mass spectrometry. *Nature Chemistry*, 7(1), 73–81. https://doi.org/10.1038/nchem.2129.

Zambaldo, C., Barluenga, S., & Winssinger, N. (2015). PNA-encoded chemical libraries. *Current Opinion in Chemical Biology*, *26*, 8–15. https://doi.org/10.1016/j.cbpa.2015.01.005.

CHAPTER TEN

# High Content, Phenotypic Assays and Screens for Compounds Modulating Cellular Processes in Primary Neurons

**Miklos Kepiro, Boglarka H. Varkuti, Ronald L. Davis[1]**
Department of Neuroscience, The Scripps Research Institute Florida, Jupiter, FL, United States
[1]Corresponding author: e-mail address: rdavis@scripps.edu

## Contents



## Abstract

High content, phenotypic screens offer a powerful approach to systems biology at the cellular level. The approach employs cells carrying fluorescently labeled molecules or organelles in 384- or 1536-well microplates, and an automated confocal screening microscope for capturing images from each well. Although some specifics vary according to the assay type, each will apply some degree of image processing and feature extraction followed by a data analysis pipeline to identify the perturbations (small molecules, etc.) of interest. We describe and discuss the advantages and limitations of high content assays and screens using the specific example of assaying mitochondrial dynamics in primary neurons. We provide a detailed description of our culturing methods, imaging and data analysis techniques and provide an open source, ready to use CellProfiler pipeline for high-throughput image segmentation and quantification tool for mitochondrial parameters.

*Methods in Enzymology*, Volume 610
ISSN 0076-6879
https://doi.org/10.1016/bs.mie.2018.09.021

## 1. INTRODUCTION

High content, phenotypic assays and screens offer powerful platforms for identifying and characterizing small molecules, genes, or other agents that modulate cellular processes (Singh, Carpenter, & Genovesio, 2014; Taylor, Haskins, & Giuliano, 2007; Zanella, Lorens, & Link, 2010). There are many advantages offered from high content, or image-based assays. Scientific data come in many different forms but no type seems more powerful than a robust image for providing confidence in the conclusions. Image-based assays also can provide both temporal and spatial information at the subcellular level about the activities of organelles or other targets being monitored. Moreover, the investigator can design such assays to be multiplexed, revealing relationships between several cellular processes or targets being monitored. The latter advantage has been driven by the discovery and generation over the last two decades of an extraordinary number of new fluorescent proteins and dyes that provide the flexibility to label most cellular targets with fluorescent reporters with different spectral properties. Furthermore, imaging technology, including both hardware and software tools, has improved dramatically allowing the investigator to rapidly capture thousands of images from cells arrayed in either two- or three-dimensions. Phenotypic screens, which monitor cellular processes rather than a specific molecular target, cast a larger net and can offer increased value in the search for agents that modulate a cellular process or interest (Swinney & Anthony, 2011).

Nevertheless, high content, phenotypic assays do offer challenges compared to target-based assays. These assays require advanced imaging instrumentation that calls for sophisticated knowledge of imaging techniques and software algorithms for processing the images. The advanced imaging instrumentation can increase costs and equipment breakdowns that create delays in obtaining experimental results. Moreover, the depth and rate of image capture demand external storage devices for terabytes of digital information. Although phenotypic screens cast a broader net in searches for agents of interest, they do not reveal the mechanism of action of such agents and this requires further studies compared to target-based screens based on a known mechanism of action.

Although such high content, phenotypic assays and screens are extraordinarily flexible and can be designed to query a myriad of cellular processes, providing essentially a systems-level snapshot of cellular processes, we focus here on techniques we have employed to monitor mitochondrial dynamics in primary neuronal cultures. Nevertheless, the principles employed in our studies are applicable to many other high content, phenotypic assays and screens that have or could be designed and implemented. Our focus on this particular organelle and its dynamics in neurons is driven by two fundamental points. First, mitochondrial loss and dysfunction are common hallmarks of neurodegenerative diseases such as Alzheimer's disease, Parkinson's disease, Huntington's disease, or Amyotrophic Lateral Sclerosis, which are characterized by a gradual, progressive, and selective loss of certain neuronal systems in each disease (Nunnari & Suomalainen, 2012). The lack of biologically relevant screening platforms, validated targets, and the high failure rate of clinical trials present an urgent need for novel screening technologies, new drugs, and druggable targets (Cummings, 2018; Hawkes, 2018). Since the pathophysiology of neurodegenerative diseases is complex and not fully understood, the implementation of phenotypic screens for successful drug development is preferable over target-based approaches. Second, given the unique physiology of neurons, screening directly in the neuronal environment holds more promise toward the goal to identify novel CNS active compounds, than screening in nonneuronal cell types.

Neurons are highly dependent on energy, supplied predominantly by the mitochondrial system. Neuronal mitochondria are highly dynamic, continuously being born through biogenesis and turned over through mitophagy, being transported from the soma to distal cellular compartments in axons and dendrites, and fusing with each other and dividing (fission) in order to adapt to the local energy requirements. These mechanisms provide homeostatic control over intracellular calcium levels and help with quality control over the mitochondrial system (Chen & Chan, 2009; Misgeld & Schwarz, 2017). Mixing mitochondrial content by fusion and isolating damaged parts by fission maintains healthy mitochondria and the effectiveness of the mitochondrial system. Excessive fission, which produces small, round mitochondria and can lead to cell death, has been reported in several neurodegenerative diseases (Johri & Beal, 2012). In contrast, elongation of mitochondria either by excessive fusion or the inhibition of fission increases efficiency and is protective under stressed conditions (Wang et al., 2012; Westermann, 2012). Mitochondrial mass has also been reported to decrease in neurodegenerative diseases (Sheng et al., 2012), either by the decrease of mitochondrial biogenesis and/or increased elimination by mitophagy. To capture the complexity of neuronal mitochondrial dynamics and screen for effective neuronal mitotherapeutics, we have employed high-content, phenotypic assays and screens (Daub, Sharma, & Finkbeiner, 2009).

# 2. HCS FOR COMPOUNDS INCREASING MITOCHONDRIAL CONTENT, ELONGATION, AND IMPROVING HEALTH

## 2.1 Labeling of Mitochondria in Primary Neurons

High content assays and screening are centered on fluorescence microscopy, where a wide variety of new generation fluorescent probes, markers, and dyes are available for visualizing whole cells, subcellular compartments, organelles, or proteins. Live-cell fluorescence imaging enables the examination of the anatomy of cellular networks, subcellular structures, their dynamics, localization, and even the mechanism of action of proteins. The cell type to be used, the specific phenotypic target, and the fluorescent reporter are dictated by the goal of assay development or screen.

### *2.1.1 Fluorescent Dyes*

Several cell-permeable mitochondrion-selective fluorescent dyes exist that allow live cell imaging among which many are also retained in mitochondria after fixation and therefore compatible with antibody-based imaging. The mitochondrial accumulation of lipophilic cationic dyes is dependent on the inner mitochondrial membrane potential and has been used widely to stain healthy mitochondria ($DiOC_1(3)$ (Korchak, Rich, Wilkenfeld, Rutherford, & Weissmann, 1982), Rhodamine 123 (Emaus, Grunwald, & Lemasters, 1986), TMRM (Ehrenberg, Montana, Wei, Wuskell, & Loew, 1988), TMRE (Ehrenberg et al., 1988), DASPMI (Bereiter-Hahn, 1976), DASPEI (Rafael, 1980)). Actively respiring mitochondria with high membrane potential will accumulate more of the fluorescent dye, therefore changes in mitochondrial membrane potential—the driving force for ATP synthase—allows for following alterations in mitochondrial function. Besides the classic cationic dyes, a newer generation of probes are also available that exhibit potential-dependent accumulation in mitochondria. For example, the cationic JC-1 (Smiley et al., 1991) carbocyanine dye exists as a monomer at low concentrations emitting green fluorescence in depolarized mitochondria, while when it is accumulated at high concentrations in hyperpolarized mitochondria, it aggregates and emits orange fluorescence. This provides the possibility of determining the ratio of polarized/depolarized mitochondria within a cell. A wide variety of MitoTracker dyes (Chazotte, 2011) with different fluorophores exist whose accumulation is dependent on or independent of the inner mitochondrial membrane potential, some of which are retained after fixation as well as after permeabilization of the membranes, giving researchers a wide range of dyes to choose from.

Staining of mitochondria is usually rapid (within an hour) compared to probes that require new protein expression, although dye concentration and incubation time should be optimized by the user for each cell type. The use of fluorescent dyes in HCS may be limited by increased variance occurring from differences in staining efficiency from experiment to experiment. The efficiency of the staining itself may be influenced by the state of the cell cultures as well and other factors. When applying dyes, their high cell permeability, specificity, and nontoxicity to cells are crucial for the development of successful and robust assays. However, the main drawback of dyes in terms of high content assays and screens for neuronal mitochondria stems from the high density of mitochondria in neurons. In such cultures where all mitochondria are stained the high number of overlapping mitochondria hinders their individual identification for further analysis. On the other hand, primary neurons fail to thrive when plated at lower densities.

### 2.1.2 Expression of Fluorescent Reporters

Transient or stable expression of fluorescent reporter proteins in primary neurons usually provides better signal-to-noise ratio than dyes, persistent expression, and the flexibility of performing cell-type or region-specific assays. Plasmid transfection of primary neurons has been a challenge in the past because of toxicity, low and variable efficiency across experiments. However, efficient viral transduction systems for neurons are available now with boosted uptake and expression such as modified baculoviral systems (Li, Yang, & Wang, 2005; Mansouri et al., 2016) or adeno-associated viral vectors (AAV) which can be used in vitro and in vivo as well. The different serotypes of AAV vectors offer a cell type-specific infection of mixed cultures, and additional specificity can be obtained by using cell type-specific promoters to express the fluorescent reporter protein. Neurons can be infected with different AAV serotypes including AAV1, AAV2, AAV5, AAV6, AAV8, or AAV9 (Aschauer, Kreuz, & Rumpel, 2013; Hammond, Leek, Richman, & Tjalkens, 2017; Michelfelder & Trepel, 2009; Murlidharan, Samulski, & Asokan, 2014; Wu, Asokan, & Samulski, 2006). The different glycan architectures of the AAV capsids contribute to the variations of efficiency of the gene transfer, as well as how they manage endosomal escape, genome release, or transcription (Murlidharan et al., 2014).

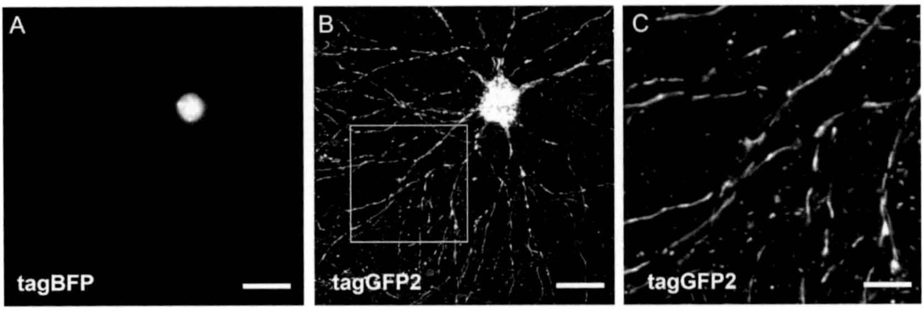


**Fig. 1** Neuron expressing a mitochondrial-localized tagGFP2 and a nuclear-localized tagBFP. Independent labeling of the nucleus by tagBFP (A) and a representative display of several hundred neuritic mitochondria and a large, essentially contiguous mitochondrion in the soma labeled by tagGFP2 (B). The *inset* bordered with a *solid line* shows a magnified image of the neuritic mitochondria (C). Scalebars are 20 μm (A and B) and 8 μm (C).

We have used a mouse line expressing a Cre-dependent dual reporter of mitochondrial tagGFP2 and nuclear tagBFP to visualize mitochondria and nuclei (Fig. 1). The mitochondria targeted GFP2 (Rizzuto, Brini, Pizzo, Murgia, & Pozzan, 1995) is an improved mutant variant of GFP that matures faster, is more pH stable than EGFP, shows improved performance in protein fusions, and is optimized for expression at 37°C. To induce the expression of the reporters specifically in neurons obtained from P0 mouse forebrains, we used a replication-deficient AAV9 packaged with a plasmid containing a CaMKII neuronal-specific promoter upstream of the coding sequences for a mammalian codon optimized (improved) Cre-recombinase (iCRE).

## 2.2 Culturing of Primary Neurons

Primary neuronal cultures are powerful and relatively simple model systems to perform various screening projects toward the goal of finding effective compounds for the treatment of brain disorders. Homogenous cultures of neurons and/or glia can be maintained in 384- or 1536-well plates, in a highly controllable environment suitable for HTS and HCS.

Primary neuronal cultures are obtained by dissecting the brains of mice or rats—the two most commonly used mammals in research—obtained from adults, newborn pups, or embryos. Using newborn pups or embryos is preferable, since their brain is immature and less susceptible to damage. Different brain regions can be isolated and cultured, the cerebellum, cortex, or hippocampus as examples, as well as the components of the peripheral system such as the dorsal root ganglia. Dissection should occur in a semi-sterile environment, with the freshly isolated tissue placed immediately in ice-cold buffered saline. Dissociation of the tissue into individual cells may involve both proteolytic (papain, trypsin) as well as mechanic procedures (mincing, trituration). Plating and maintenance include special supplements that support neuronal growth including vitamins, amino acids, or glucose (insulin, transferrin, putrescine, progesterone, etc.) and protective supplements (antioxidants such as catalase, glutathione or superoxide dismutase, and L-carnitine for oxygen–glucose deprivation). Chemicals that prevent the growth of nonneuronal cells can also be added to the culture (e.g., 5-Fluoro-2′-deoxyuridine), or if astrocyte cultures are needed, special media and supplementation can be used. Neurons can be plated onto different dishes, plates, or cover glasses placed into dishes, but surfaces need to be pretreated with some extracellular matrix protein (poly-D-lysine, poly-amine, laminin) for proper attachment and to support growth. Neuronal processes of cells begin to extend a few hours after plating, but fully mature neurons are only developed after culturing at least 10–11 days in vitro. Cultures can be maintained for approximately a month, with a gradual decrease of the health of the neurons until the majority of cells die (Lesuisse & Martin, 2002).

Selecting the proper plate (material, coating, vendor) is an important decision for a successful HCS project. Primary neuronal cultures do not grow well on glass, so polystyrene or cycloolefin plates are used; the latter has better optical properties (Niles & Coassin, 2008). Poly-D-lysine or poly-amine both work well with primary forebrain neurons. If cells clump together after a few days of plating, or do not attach to the surface of the plate within the first 4 h, the coating may be at fault. Fig. 2 shows the same batch of primary neurons plated on the same cycloolefin plates but coated with different substrates. Unfortunately, the quality of plate coating from the same vendor can vary across different batches of plates.

In the following, we describe the maintenance of neuronal forebrain cultures isolated from newborn pups (P0). The expression of mitochondrial GFP2 was induced using optimized cell and viral particle numbers (Fig. 3).

### 2.2.1 Equipment and Tools

- Incubator (37°C, 5% $CO_2$), microscope for dissection (Zeiss Stereoscope Discovery V8 or equivalent)
- Forceps and scissors
- Centrifuge, cell counter, 40 μm mesh-sized cell strainer

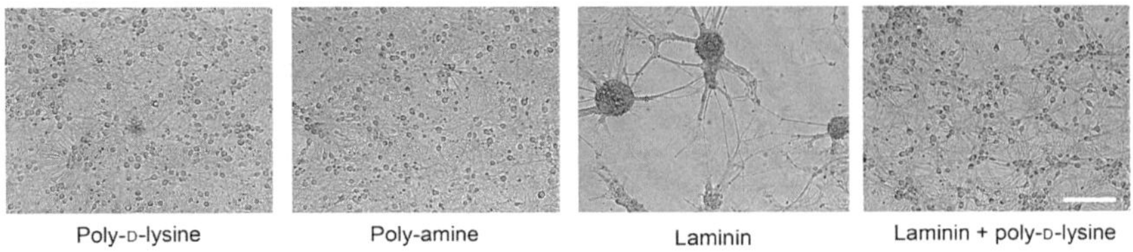


**Fig. 2** Culturing of neurons on surfaces with different coatings. Bright-field images of DIV 14 (14 days in vitro) primary forebrain neurons in 384-well plates plated from the same P0 pups at a seeding density of 15,000 cells/well. Poly-D-lysine and poly-amine treated surfaces create a homogeneous layer of neurites with evenly scattered cell bodies, showing that neurons are healthy and not stressed. The stressed neurons plated on laminin form clumps of cell bodies with neurites that bundle together to form bridges between cell body clumps. Adding poly-D-lysine to laminin resolves the clumping, though these cell cultures are still less homogeneous than treating surfaces with solely poly-D-lysine or poly-amine. Scalebar is 200 μm.

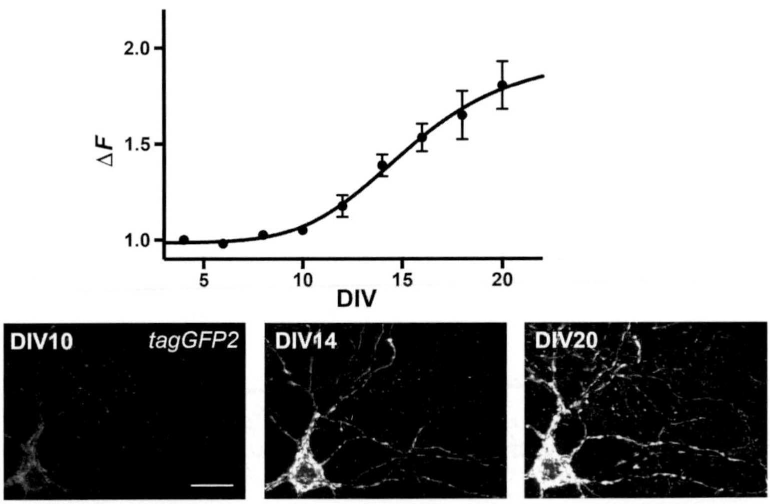


**Fig. 3** Expression of mitochondrial tagGFP2. Expression of mitochondrial tagGFP2 in neurites over time shows that cultured neurons on DIV14 have an average 1.33 more intense signal than at DIV2, which signal plateaus at 1.97 (*black curve*, sigmoid fit). At DIV14 the signal is sufficiently strong to be analyzed and neurons are mature but still in good health. By DIV20, mitochondria show fragmentation and swelling, signs of mitochondrial degradation. Data are mean ± SD from three independent measurements. Scalebar is 20 μm.

- Biomek FX$^P$ liquid handling workstation or equivalent
- 45 μL Beckman Biomek XL filter tips (BA-0051-3FC)
- Poly-D-lysine precoated Aurora 384-well plates with evaporation barrier (ABE1-01201-B), Aurora 1536-well plates (ABI2-01000-B) or equivalent

### 2.2.2 Buffers and Reagents

- Dissection buffer ($Ca^{2+}$ and $Mg^{2+}$ free HBSS supplemented with 2 μg/mL gentamicin, 25 m*M* D-glucose, 1 m*M* pyruvate, and 20 m*M* HEPES)
- Plating media (Neurobasal supplemented with 0.75% FBS, 4 m*M* GlutaMAX and 2 μg/mL gentamicin, 2% B27, 16 μ*M* 5-Fluoro-2′-deoxyuridine)
- Feeding media (Neurobasal-A supplemented with 4 m*M* GlutaMAX, 2 μg/mL gentamicin, 2% B27)
- Papain

### 2.2.3 Procedure

1. Collect forebrains of P0 pups in ice-cold dissection buffer and remove olfactory bulbs and meninges. Put 2–3 forebrains in 4 mL of ice-cold dissection buffer.
2. Digest tissue with 0.6 mg/mL prewarmed papain at 37°C for 20 min, gently swirling the tube every 5 min.
3. Wash the brains extensively but carefully with prewarmed plating media 3 × while keeping them intact, leaving 2 mL of plating media on the tissue.
4. Triturate tissue carefully through a 1 mL regular plastic pipette tip several times (homogenize tissue), let the dissociated tissue sit for 60 s, then transfer 1 mL of the supernatant to a new tube and add 1 mL fresh plating media to the remaining tissue. Repeat this 3–5 times.
5. Dilute the cells collected in the supernatant to 10 mL with plating media and run it through a 40 μm mesh-sized cell strainer to filter out clumped tissue or small pieces of the meninges.
6. Centrifuge the cells at 420 × *g* for 4 min.
7. Aspirate the supernatant and resuspend the cells carefully in ~5 mL warm plating media.
8. Check the live/dead cell ratio by using trypan blue (0.4%) with an automated cell counter or hematocytometer. A healthy cell suspension should have >90% live cells.
9. Dilute cells with warm plating media to achieve a final seeding density. 5-Fluoro-2′-deoxyuridine in the plating media prevents glial cells or any other dividing cell types (e.g., cells from blood vessels) from overgrowing the neuronal culture. FBS is needed for the survival of ~10% astroglia that will support neuronal cell survival at plating and attachment, and it is not needed thereafter. B27 supports neuronal survival, especially during the first 4 days of culture. Initial seeding numbers are influenced by neuronal cell type, the type of plates used, and the coating

employed for attachment and survival. Each pup brain yields ~4–6 × $10^6$ cells.

10. Add AVV9 containing the iCRE recombinase to the diluted cells at a final concentration of 1.5 × $10^8$ viral particles/mL, mix gently and incubate for a few minutes, and then distribute the homogenous cell culture to the desired plates in the desired volume/wells. Using a robot for plating such as the Biomek $FX^P$ will increase consistency of plating across the plates. We add 80 μL/15,000 cells/well (0.1875 × $10^6$ cells/mL) on poly-D-lysine, precoated Aurora 384-well plates or 14 μL/7000 cells/well on precoated 1536-well Aurora plates. These give a nice confluency of monolayer cells which is needed for a healthy neuronal cell culture.
11. After 8 days, perform a 50% media change with feeding media and repeat it every 4 days thereafter.

### 2.2.4 Notes

- Isolation of the forebrains (step 1) should not exceed 20–25 min since it will affect the viability of the cells.
- Make sure to wash out papain well.
- Trituration should be gentle, but the brain tissue should be totally homogenized at the end of the trituration cycles.
- Make sure to use warm plating media throughout the whole process.
- Cell number/well and the type of coating used for plating will substantially influence the growth and health of the neurons and should be optimized as an initial step for screens.
- The number of viral particles added per well determines the number of neurons expressing the fluorophores and should be optimized for each batch of packaged virus and each screen.

## 2.3 Compound Transfer

Automated and standardized compound transfer is essential for a reproducible high-throughput screen. Reproducibility not only depends on the method of delivery of compounds to the assay plates, but also on the type of library source plates, and how the compounds were prepared and how they were stored.

Compounds are usually dissolved at 10 m*M* concentration in DMSO: $H_2O$ = 9:1, where the goal is to maximize compound dissolution and stability, and minimize the amount of solvent delivered onto the assay plates containing the cells. In case of primary neuronal cultures, cells will not tolerate more than 0.5% DMSO in the media, especially for 24–48-h long treatments

(Zhang et al., 2017). Compound transfer is also influenced by the material of the source plate that is used and its storage (room temperature, 4°C, −20°C, or −80°C). Polypropylene and cycloolefin plates are preferable over polystyrene plates because polar molecules bind them less, they are more resistant to solvents and have high thermostability. Cycloolefin plates seem to perform even better than polypropylene plates when compounds were frozen previously (Smith et al., 2013). Compound potency seems to be the highest when plates have never been frozen, which is usually not feasible since source plates are usually kept frozen for the long term. If the compound plate is used frequently, it is advisable to store it well-sealed at room temperature then to subject it to multiple freeze–thaw cycles. Compounds can be stored at room temperature in DMSO:$H_2O$ = 9:1 up to 6 months without significant loss of potency.

The delivery of compounds can be accomplished by pintooling them from a source plate into a destination assay plate containing the cell culture (Fig. 4). In this contact-based method, each pin containing a slot or groove submerges into the liquid of a well and delivers a preset volume (few nanoliters to couple of hundreds of nanoliters) via capillary action and surface tension. The pins themselves are in a floating head, preventing pin damage and allowing the soaking of all the tips into the solution. Transfer depends very much on the properties of the solvent used for the source plates

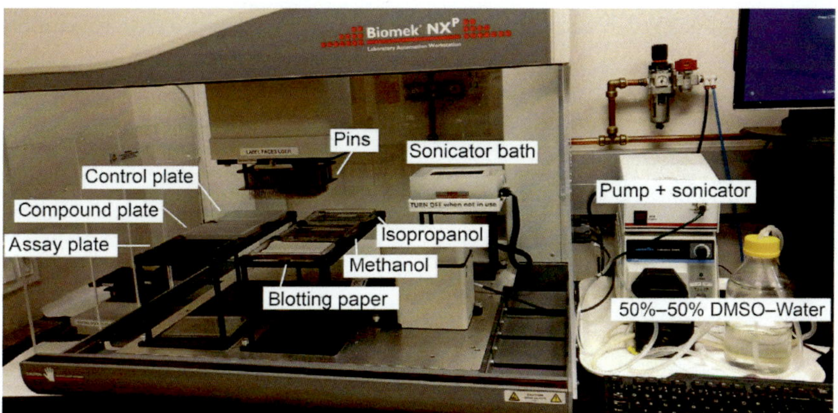


**Fig. 4** Deck layout of compound transfer via pintooling. The pump is continuously circulating the DMSO:water mixture in the sonicator bath, where the pins are washed between each transfer, then soaked in isopropanol and methanol and dried on blotting paper. Pins pick up compounds first from either the control or compound plates and transfer them into the wells of the assay plate containing neurons. Usually pintooling is performed under nonsterile conditions, so minimizing the time the cells spend without a cover is important.

(Dunn & Feygin, 2000), the surface properties of the pins and the source plate, the amount of liquid it submerges into and the method of the cleaning of the tips (Cleveland & Koutz, 2005). As long as these variables are kept constant, pintooling is considered to be highly accurate and reproducible (Smith et al., 2013). Nevertheless, one should check precise transfer volumes using standardized procedures before starting a screen using pintooling (Cleveland & Koutz, 2005). One of its main drawbacks is its transfer of fixed volumes, which means the source concentrations may need to be adjusted to fit desired final concentrations. However, adjusting source concentrations is usually not applicable with preset concentrations and volumes of drug libraries available commercially.

The noncontact acoustic droplet ejection method offers the possibility of creating assay ready plates (ARPs), offering the advantage of custom designed layouts of the compounds from a fixed source plate. It uses acoustic energy to propel droplets (picoliters to nanoliters of volumes) with very high accuracy from a source plate into a destination plate that lies over the source plate in an inverted position. The source plate is stationary, while the transducer and the destination plate move to different locations, allowing the transfer of compounds from any well of the source plate to any other well of the destination plate. Although ARPs can be frozen and kept for a long periods of time, some evaporation of the droplets may occur during storage depending on temperature and humidity. Drying of the compounds onto the assay plates may lead to poor recovery and increased assay variability. Side-by-side comparisons of delivering compounds by pintooling vs using ARPs have produced the conclusion that ARPs provide lower potencies, reduced hit rates and higher false negative hit rates (Smith et al., 2013).

We have used pintooling from predefined library source plates for screens, and an automated 8-channel head liquid handling workstation for subsequent cherry picking compounds for rescreening or setting up new experimental layouts. Compounds are added to neurons on DIV13 at a final concentration of 12.5 μ*M* and then incubated for 24–48 h to measure compounds-induced changes in neuronal mitochondrial dynamics.

### 2.3.1 Equipment and Tools

- Liquid handling workstation with 100 nL pintool head (Biomek $FX^P$ or equivalent)
- 384-well polypropylene compound plates (Greiner Bio-One 784201 or equivalent)
- 384-well assay plates
- Blotting paper

### 2.3.2 Buffers and Reagents

- Methanol, isopropanol, DMSO:$H_2O$ = 1:1

### 2.3.3 Procedure

1. Soak pins in a bath sonicator containing DMSO:$H_2O$ = 1:1 for 2 min.
2. Submerge pins into methanol then isopropanol and dry pins on blotter paper.
3. Submerge pins into the wells of the compound plate containing 8 μL of 10 m*M* compounds dissolved in DMSO, so that pins touch the bottom and are pushed up in their floating head, thereby making sure all pins come in contact with the compound irrespective of the volume in each well.
4. One can use additional compound plates containing control compounds occupying different wells than the experimental compounds, making sure the pins are submerged into the control liquid without the pins touching the bottom of the plate.
5. Wash the pins into the destination assay plates containing DIV13 cultured neurons, making sure pins do not touch the bottom of the wells. Incubate for 24–48 h.

### 2.3.4 Notes

- Ideally, the sonicator bath is attached to a peristaltic pump, which continuously circulates the DMSO:$H_2O$ = 1:1 between the bath and a sealable reservoir bottle. By doing this, the volume of the washing mixture is increased considerably, eliminating the possibility of cross contaminations. If the equipment is not in use, store the DMSO:$H_2O$ mixture in the reservoir bottle.
- DMSO:$H_2O$ can be reused multiple times. Methanol and isopropanol should be changed at least daily, depending on the pintooling cycles
- Pins should be washed before and after pintooling as well.
- The procedure can be performed similarly with 1536-well assay plates using a smaller volume 384-well pintool head (e.g., 10 nL) and pintooling from the four quadrants of the control plate to the assay plate.

## 2.4 Imaging

The resolution requirements for a HCS screen depend on the size of the objects of interest. Resolution in $x$ and $y$ is dependent on the emission wavelength ($\lambda_{em}$) of the light source and the numerical aperture (NA) of the objective (Rayleigh resolution$_{x-y}$ = $(0.61^*\lambda_{em})/\mathrm{NA}$). Resolution in the $z$ direction if collecting stacks of $z$-plane images is also influenced by the refractive index (Ri) of the mounting media (Rayleigh resolution$_z$ = $(2^*\lambda_{em}{}^*\text{Ri})/\text{NA}^2$). The other parameter one must consider is the pixel size within the images, which is usually smaller than the resolution and depends on the pixel size of the camera and the magnification of the objective (pixel size$_{image}$ = pixel size$_{camera}$/magnification). When imaging small objects, like mitochondria (axonal mitochondria ~0.5–2 μm long; Kiryu-Seo, Ohno, Kidd, Komuro, & Trapp, 2010), it is critical to ensure that each object is made up of a numerous of pixels. Otherwise identification and measurement of the object will differ significantly from its real size. Using high magnification objectives to increase resolution decreases the size of the field captured compared to lower magnification objectives (field size = (pixel size$_{image}$$^*$number of pixels$_{image}$)$^2$). The net effect of this is that more fields within each well need to be imaged when using high magnification objectives to capture a sufficient number of objects to detect statistically significant differences between control and experimental wells.

A confocal fluorescent and automated screening microscope equipped with a 60× magnification (NA = 0.95) air objective offers sufficient resolution needed to detect fluorescently labeled objects as small as mitochondria and neurites in living neurons. Environmental control that includes adjustable temperature, $CO_2$, and humidity levels help to keep neurons alive and healthy during imaging. The highest signal-to-noise ratio possible is preferred to aid in the identification of objects of interest, and this can be elevated by increasing exposure time and laser intensity or imaging several planes in the $z$ plane. However, these steps all contribute to phototoxicity, photobleaching, and quenching of the fluorophores. Additionally, increasing exposure times, imaging several $z$-planes, and several fields within each well increases imaging time significantly. These are all parameters that have to be optimized in pilot experiments when developing the assay for a screen. Binning pixels (e.g., 2 × 2) increases sensitivity fourfold, but decreases resolution twofold. Brightness or fluorescence intensity of a given fluorophore should also be considered besides excitation and detection wavelengths when developing an assay, which is proportional to the product of its molar extinction coefficient ($\varepsilon$, efficiency of absorbing excitation light at a given wavelength) and quantum yield ($\phi$, ratio of number of photons emitted to absorbed). Brightness of the tagGFP2 protein we used for targeting mitochondria is 33.9 ((56,500 M$^{-1}$ cm$^{-1}$ × 0.6)/1000), which is not much brighter (5%) than EGFP, but it is much more photostable (Subach et al., 2008).

Using an air objective illuminating the samples from the bottom is preferred over oil objectives. Reliable autofocusing of the imaging system is a must in HCS and also contributes to the speed of imaging. Autofocus performance may vary with objective and plate type and should be validated empirically before starting the screen. Finding the plane of interest automatically across a plate type prone to nonuniformity in the elevation of bottom height or bottom thickness can be a challenge for the hardware and the software of the microscope. Hardware or laser autofocus identifies the change in the refractive index of the air-to-plate bottom interface, and the bottom surface inside each well between the plate floor and medium where the cells are attached. The optimal plane of imaging can be fine-tuned by a given offset value. Software or image-based autofocus on the other hand uses algorithms to find the highest contrast/strongest signal along the $z$-axis. This is not ideal if the objects of interest vary in brightness and reside at multiple $z$-elevations relative to the plate bottom, such as the dim mitochondria in flattened axons relative to bright mitochondria localized in the elevated soma. Software autofocus also adds a considerable time to image acquisition, which depends on the magnitude of the search range it uses to find the sharpest plane, and also on whether it uses the previous in-focus plane to start the search. Additionally, software autofocus repeatedly exposes cells to large doses of energy as it captures fluorescent images to determine the optimal focal plane, causing increased phototoxicity and photobleaching. In contrast, hardware autofocus uses low-energy infrared light for detecting differences in refractive index.

The time for collecting images after compound treatment, which depends on the time-course of the drug-induced phenotype, is another important parameter that needs to be optimized. If one begins a screen with an excellent positive control this can be used to establish the optimal incubation time of compounds in pilot experiments. However, after obtaining a set of positives from a screen, it is often necessary to collect images at multiple times after compound treatment in order to establish the maximum signal for each compound. For mitochondria in primary neurons, we have found that collecting images 24–48 h after compound addition offers reliable signals (Fig. 5).

### 2.4.1 Equipment

- InCell 6000 (automated confocal screening microscope) with environmental control (or equivalent)

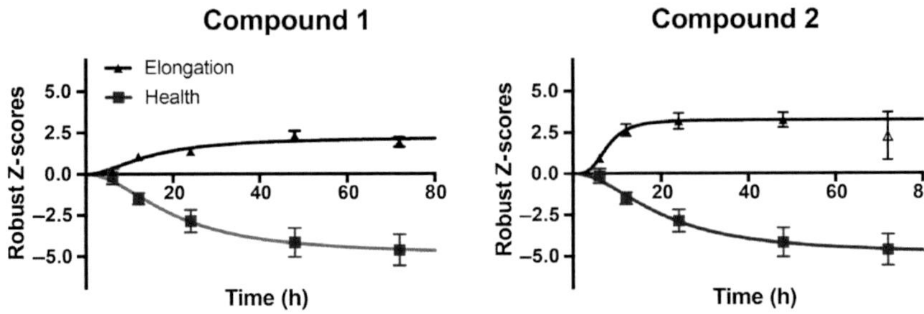


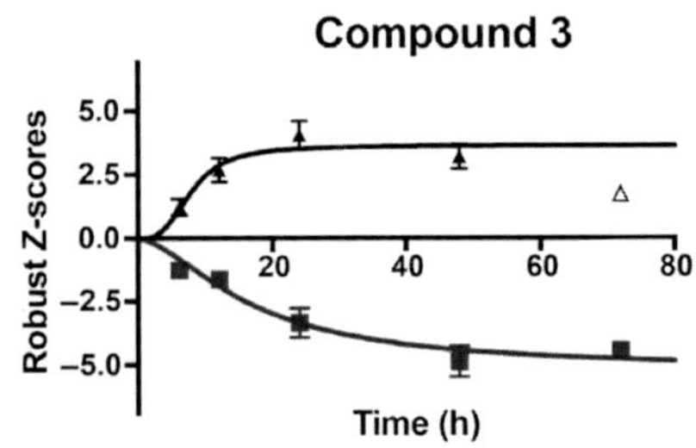


| *Half-times* (*h*) | Elongation | Health |
|---|---|---|
| Compound 1 | 14.8 | 20.1 |
| Compound 2 | 7.8 | 20.1 |
| Compound 3 | 7.8 | 16.7 |

**Fig. 5** Time-courses of compound modulation of mitochondrial elongation and health. Neurons were treated with 12.5 μ*M* of the compounds and imaged 6, 12, 24, 48, and 72 h later; data were normalized to DMSO treated control neurons using a robust Z-scoring method. Data are mean $\pm$ s.e.m. that were fit with a variable slope sigmoidal curve, yielding half-time values for each parameter. Open datapoints were not included in the fit. Half-time values for the two parameters were determined from the fits for each compound and for each parameter. The optimal time for image collection is a function of both the parameter of interest and the compound itself.

### *2.4.2 Specifications and Settings*

- Camera: sCMOS 5.5 Mp, 6.5 μm pixel size
- Objectives: For general use include 4 ×, 10 ×, 20 ×, 60 ×. For objects in the size range of axonal mitochondria: 60 × air objective, 0.95 NA (Rayleigh Resolution$_{x-y}$: 0.33 μm, Rayleigh Resolution$_z$: 1.13 μm, pixel size: 0.11 μm, image size: 2048 × 2048 pixels, field size: 0.22 mm × 0.22 mm)
- Imaging: $\lambda_{exc} = 488$ nm, $\lambda_{em} = 515$–535 nm, exposure time: 0.4 s, power: 50% (12.5 mW), aperture: 1.0 Au, 3D mode: 3 slices, $\Delta z = 0.7$ μm, 4 fields/well, hardware autofocus, offset: 2.5

### *2.4.3 Notes*

- With these settings, a 384-well plate can be imaged in ~40 min.
- The parameters of the plate used in the screen need to be precisely defined for the automated screening microscope precisely.

## 2.5 Automated Image Analysis

To extract biologically meaningful data from the captured images and quantify the various compound induced phenotypes, a robust and automated image analysis pipeline is essential in any HCS project. The main steps of such pipeline usually include: (1) image processing, (2) object segmentation, and (3) feature extraction. In image processing, images are often corrected for noise (e.g., background noise or experimental artifacts) and uneven illumination. This is usually achieved by applying linear or nonlinear filters. Next, regions of interest (ROIs) are defined (e.g., cells or subcellular structures) in the images (fields) through the process of segmentation. This is most commonly performed by pixel binarization or using a thresholding algorithm, such as Otsu's Method (Otsu, 1979), which calculates a threshold value separating the foreground and background pixels by minimizing the variance within the two groups. Once the objects are identified, their features (e.g., shape, intensity, or texture) can be extracted by quantitative measurements.

Image analysis pipelines are constructed by employing one or several image processing algorithms. Different ways of implementing the same algorithms are readily available in image processing libraries such as the MATLAB® Image Processing Toolbox or the open source Python Scikit-image. The use of these packages requires programming expertise. Commercially available HCS imaging systems are usually packaged with their own image analysis software that can also be used for constructing image analysis pipelines. In addition, there are free and open source image analysis software available. Two of the most widely used are ImageJ and CellProfiler. ImageJ can be used for HCS, but its design is more applicable for the analysis of individual images or image stacks rather than the hundreds or thousands of serial images obtained from HCS. Although ImageJ offers a programmable macro mode that supports many scripting languages, many of the macro functions activate GUI elements and cannot run in headless mode (van der Walt et al., 2014). CellProfiler is a free, open-source software for quantitative analysis of biological images designed for biologists without programming expertise in constructing automated image analysis pipelines (Carpenter et al., 2006). It is a collection of advanced image analysis algorithms in forms of modules, which modules can be placed in a sequential order to create an image analysis pipeline. For automated analysis of mitochondria in primary neurons, we constructed an image analysis pipeline using CellProfiler.

Neuronal mitochondria can be divided into three classes that include somatic, axonal, and dendritic mitochondria. The morphology of these classes is different: somatic mitochondria form a highly interconnected, irregularly shaped network in the soma while axonal and dendritic mitochondria are individual tubular structures located in neurites. Axonal mitochondria are short (usually $<1.4$ μm) and dendritic mitochondria are larger (usually $>2.4$ μm) (Kiryu-Seo et al., 2010; Lewis, Kwon, Lee, Shaw, & Polleux, 2018). We separated the mitochondria labeled with the same fluorophore into those three classes based on morphology and focused on the mitochondria found in neurites.

In the first part of our image analysis pipeline, densely packed somatic mitochondria are identified and removed from the original image (Fig. 6). This is achieved by five sequential steps: (1) erosion, (2) Gaussian filtering to erode tubular structures (axonal and dendritic mitochondria), (3) global Otsu thresholding for the segmentation of somatic mitochondria, (4) expansion of the somatic ROI, and (5) removal of the segmented somatic mitochondria from the original image. Then, top-hat filtering is applied for background subtraction and the resulting images are median filtered to reduce noise. Segmentation of GFP-labeled objects is achieved by adaptive Otsu thresholding, while the resulting objects are intensity filtered to remove faint objects. Axonal and dendritic mitochondria are then separated based on their length. Length measurement is achieved by first skeletonizing the objects and then measuring their area. The width of a skeleton is 1 pixel, so the area in this case is equal to the length. Then, the segmented GFP-labeled objects are relabeled using the separated axonal and dendritic skeletons, resulting in the axonal and dendritic mitochondrial binaries. Besides the already performed length measurements of the skeletons, additional measurements like the area occupied by the axonal and the dendritic binaries, which is the sum of the individual mitochondrial areas, as well as the circularity of the axonal binaries are also measured. Then, the individual length and circularity values are aggregated using their median, and the data are exported.

### *2.5.1 Step-By-Step Procedure for CellProfiler*

The names of input and output images are arbitrary. Algorithm and parameter names are consistent with the naming convention used in CellProfiler.

- Erosion: structuring element = disk, size = 11 (input = raw, output = soma_erosion)

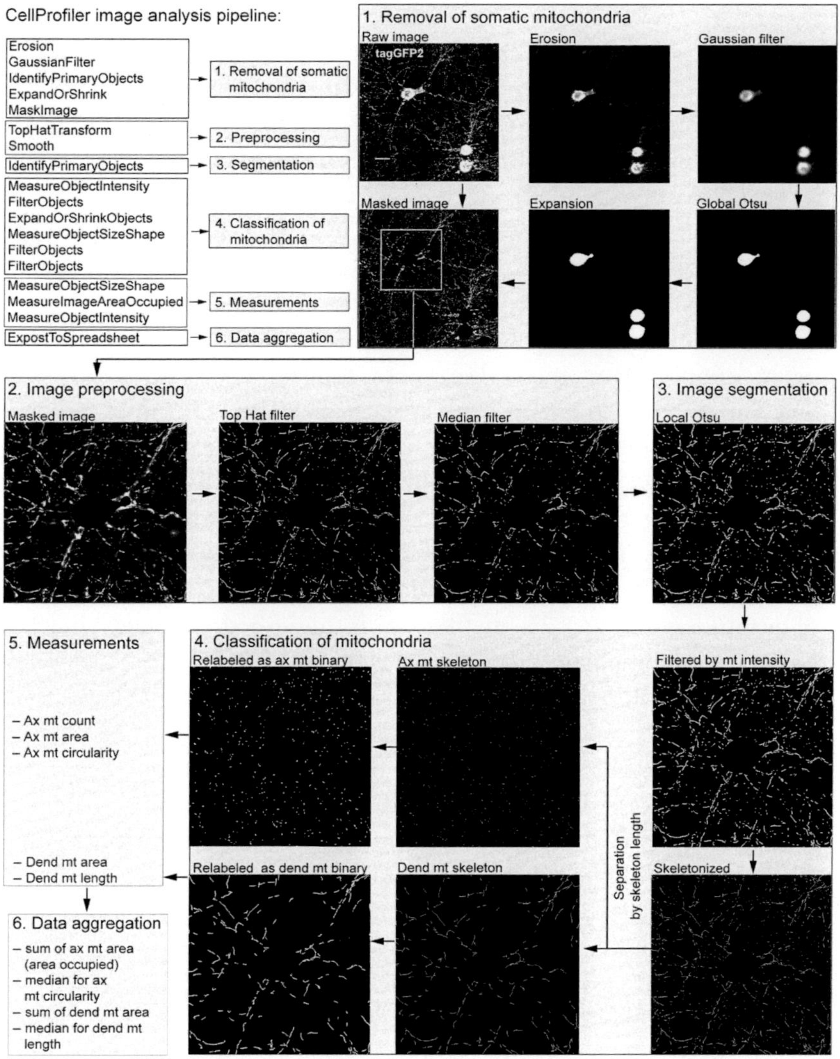


**Fig. 6** Image analysis pipeline. Raw images of mitochondrial tagGFP2 expressed in primary neurons were processed by a CellProfiler image analysis pipeline. *ax* = axonal; *dend* = dendritic; *mt* = mitochondria. Scalebar is 25 μm.

- GaussianFilter: sigma = 20 (input = soma_erosion, output = soma_gaussian)
- IdentifyPrimaryObjects: threshold strategy = global, thresholding method = Otsu, three classes, assign pixels in the middle intensity class to the background, threshold smoothing scale = 1.3488, threshold correction

factor = 1, lower and upper bounds on threshold = 0.0001 and 1, method to distinguished clumped objects = None, maximum number of objects = 20 (input = soma_gaussian, output = soma_binary)

- ExpandOrShrinkObjects: expand objects by a specific number of pixels, number of pixels by which to expand = 5 (input = soma_binary, output =soma_binary_expanded)
- MaskImage: select the input image = raw, name the output image = raw_masked, use objects or an image as objects, select object for mask = soma_binary_expanded, invert the mask = yes
- TopHatTransform: white top-hat transform, structuring element = disk, size = 3 (input = raw_masked, output = tophat)
- Smooth: median filter, typical artifact diameter = 3 (input = tophat, output = median)
- IdentifyPrimaryObjects: threshold strategy = adaptive, thresholding method = Otsu, three classes, assign pixels in the middle intensity class to the background, threshold smoothing scale = 0, threshold correction factor = 0.6, lower and upper bounds on threshold = 0.0003 and 1, size of adaptive window = 25, method to distinguished clumped objects = None (input = median, output = mito_binary)
- MeasureObjectIntensity: select an image = raw, select objects to measure = mito_binary
- FilterObjects: objects to filter = mito_binary, output objects = mito_binary_skeleton, filtering method = limits, category = intensity, measurement = MeanIntensity, minimum value = 0.004 (input = mito_binary, output = mito_binary_int_thresh)
- ExpandOrShrinkObjects: skeletonize each object (input = mito_binary_int_thresh, output = mito_binary_skeleton)
- MeasureObjectSizeShape (input = mito_binary_skeleton)
- FilterObjects: objects to filter = mito_binary_skeleton, output objects = axonal_mito_skeleton, filtering mode = measurements, filtering method = limits, category = AreaShape, measurement = area, minimum value = 4, maximum value = 13, select additional objects to relabel = mito_binary_int_thresh, name the relabeled objects = axonal_mito_binary
- FilterObjects: objects to filter = mito_binary_skeleton, output objects = dend_mito_skeleton, filtering mode = measurements, filtering method = limits, category = AreaShape, measurement = area, minimum value = 22, select additional objects to relabel = mito_binary_int_thresh, name the relabeled objects = dend_mito_binary

- MeasureObjectSizeShape: select objects = mito_binary_skeleton, dend_mito_binary, dend_mito_skeleton)
- MeasureImageAreaOccupied: measure the area occupied in objects, objects to measure = mito_binary_int_thresh, axonal_mito_binary, dend_mito_binary

### 2.5.2 Notes

- Parameters of the different modules should be optimized for each assay.
- If the intensity values of the images vary from field to field and/or well to well, intensity normalization can be applied.

## 2.6 Analyzing the Data

After obtaining raw data generated from the image analysis pipeline, it is processed further in multiple steps. HCS data are somewhat different from HTS data, in that it also provides cellular level data (individual measurements of the segmented objects) besides well-level values (aggregated individual values). So, data analysis is a two-part process, first cellular measurements are converted to well-level aggregates, then well-level data are assessed further similar to other screening technologies (normalization, assay, statistics, etc.).

### 2.6.1 Individual Object Measurements of Cell-Level Data

Cell-level analysis, examining the individual measurements of each object as a population, involves examining hundreds or thousands of data points for each image. Thus, the number of objects can be extremely high even when comparing two single images, making statistical comparisons strong and valid. In addition, even if the mean values of a population fail to change with compound treatment, there may be changes in the distribution pattern of the population. So, the first task with the large data of individual object measurements is to aggregate it to a few concise, informative summary statistics without losing the effect a compound might have on the population. If the data generated are influenced by independent random factors with additive effects on variability, a symmetric, bell-shaped normal distribution of population will emerge, where the mean, median, and mode values are the same. However, biological processes often deviate from normal distributions and produce skewed population distributions.

The most popular summary statistics are the mean and standard deviation (SD) of the population and the coefficient of variation (CV). Since CV is normalized to the sample mean, it is unitless and allows comparisons of the dispersion of the data. The mean, SD, and CV are all sensitive to outliers and are appropriate aggregate values for describing normal distributions. In contrast, the median, median absolute deviation (MAD), and interquartile range (IQR) are derived from rank ordering the elements of the population. These measures are less influenced by outliers and thus appropriate as aggregated values for skewed distributions. The shape of the distribution of the population can be described by skewness and kurtosis. Skewness measures the degree of asymmetry in a distribution, positive values refer to right-skewed distributions while negative values indicate left-skewed data. A symmetrical distribution has a skewness of 0. Kurtosis measures how heavy the tails are in a distribution. A normal distribution has a kurtosis of 3; values greater than this indicate heavy tails and a narrow peak, while lower values indicate a wide peak and thinner tails.

Transformation of nonnormally distributed data provides an option for generating normal distribution (log: for positively skewed data, square root: for Poisson distribution, power: negatively skewed data, reciprocal: extremely right skewed data) if the transformation makes biological sense. It would enable the use of parametric tests which are more sensitive, more powerful, and easier to interpret than nonparametric statistics. On the other hand, the distribution of the population may change due to experimental artifacts (autofocus or image analysis errors, edge effects) that increase the number of outliers or due to compound treatment. Therefore, careful consideration needs to be taken of whether data transformation can be applied across the entire dataset. Changes in population distribution will also change the variance of the samples, prohibiting the use of many statistical tests for comparisons that assume a constant variance across samples.

Our individual measures of elongation (dendritic mitochondrial length) and health (axonal mitochondrial circularity) generate nonnormally distributed data, whose transformation would not make biological sense. Thus, median values are used to generate summary statistics for well-level data. The sum of individual axonal and dendritic mitochondrial area (area occupied by mitochondria) is used as well-level data to assess mitochondrial content in axons and dendrites (Figs. 7 and 8).

### *2.6.2 Well-Level Data*

Once the aggregated values from individual cell-level measurements are generated, they can be used to calculate summary statistics. Trends and artifacts can be seen when plotting data for each well (numbers of outliers, positional effects across the plate, etc.). Individual object measures are not always

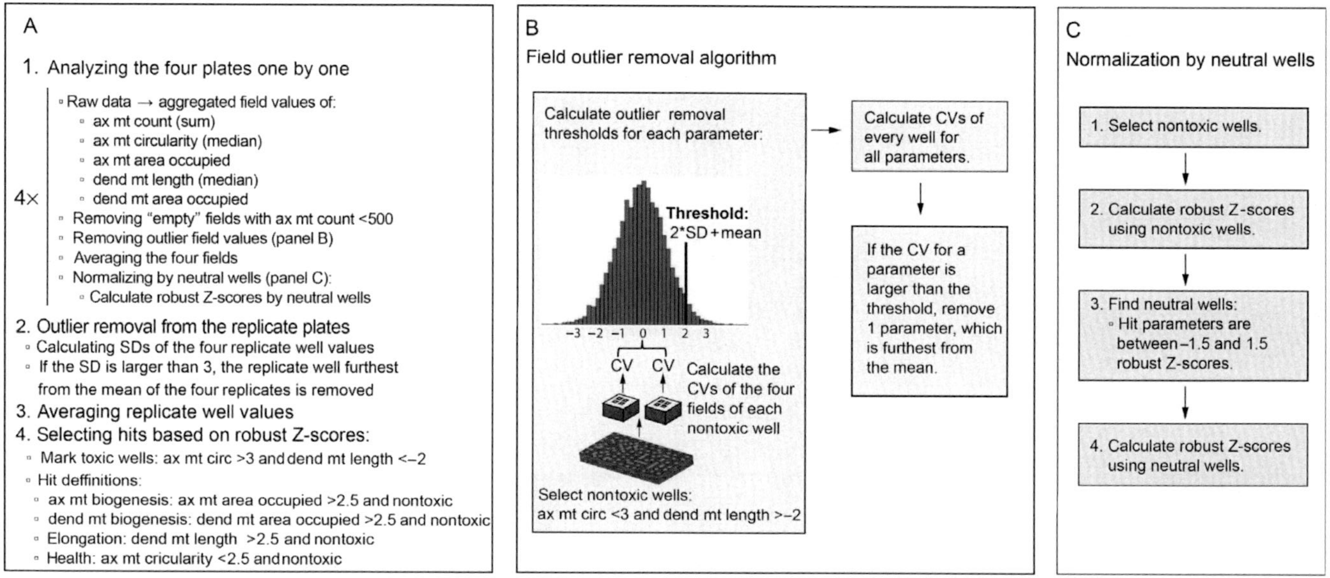


**Fig. 7** Data analysis pipeline. Summary of the data analysis (A), and detailed description and illustration of the field outlier removal (B) and normalization (C) performed within step 1 of the data analysis pipeline. *ax* = axonal; *dend* = dendritic; *mt* = mitochondria; *circ* = circularity.

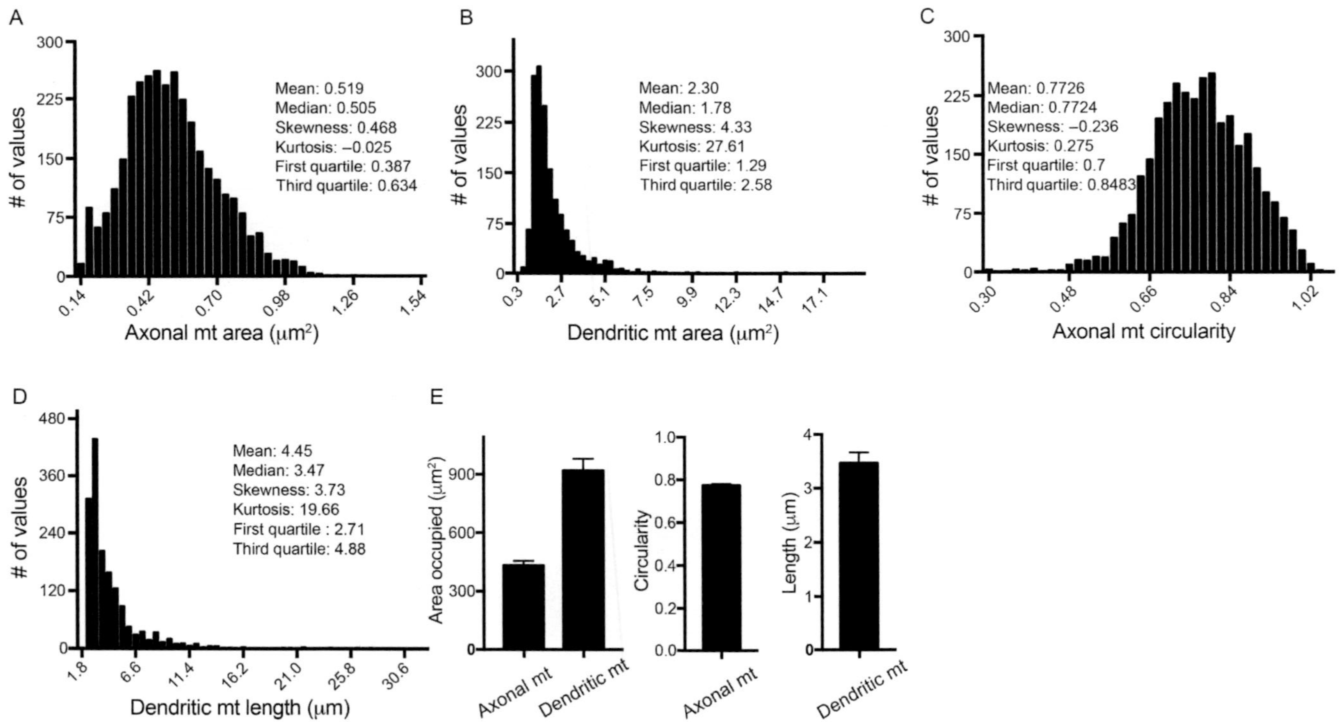


**Fig. 8** Cell-level and well-level data of mitochondrial features in primary neurons. (A–D) Cell-level measurements of individual mitochondria of vehicle-treated (DMSO) DIV15 neurons from four imaged fields show that distributions of axonal mitochondrial area and circularity are the closest to a Gaussian distribution ($R^2_{\text{Gaussian, ax mt area}} = 0.97$, $R^2_{\text{Gaussian, ax mt circ}} = 0.98$), while the two dendritic features are highly right-skewed. (E) Well-level data were generated from the cell-level data as the average of the sum of the areas of individual mitochondria across each of the four imaged fields and the average of the field medians of axonal mitochondrial circularity and dendritic mitochondrial length. Well-level data are presented as the mean ± SD of the four aggregated field values. *mt*, mitochondria.

provided automatically by HCS software, only summary statistics. Examining how far the means and medians are from each other can be an implication of distribution and skewness of the cell-level data. By examining the first and third quartiles in the summary statistics one can assess whether the majority of the signal comes from a minority of the population. In case it comes from the minority, images and the image analysis process should be rechecked for possible artifacts and cell-level data should be examined.

Normalization of the raw data is necessary to compare screening data across plates and across multiple experiments. Normalization can be performed to within-plate controls, to all wells on a plate (provided the majority of compounds are inactive) or to a subset of within-plate compounds ("neutral" compounds). The use of neutral compounds in a plate offers a convenient approach, since they are scattered across the plate which will attenuate edge or column/row effects. Often, the layout of the libraries for a primary screen does not allow the experimenter to distribute controls scattered across the plate, relegating them primarily to positions along the outer columns. Even alternating positive, vehicle-treated, and negative controls along these columns can produce serious issues when used to normalize experimental wells (Bray & Carpenter, 2012; Malo, Hanley, Cerquozzi, Pelletier, & Nadon, 2006). When normalizing values to within-plate compounds, the use of robust descriptive statistics is preferred because of its relative insensitivity to outliers.

Several methods exist for normalizing screening data (Goktug, Chai, & Chen, 2013), among the most used are percent of control, *Z*-score, robust *Z*-score, and *B*-score (Table 1). Percent of control is simple, but it does not take into consideration the variability of controls. *Z*-score (used for normally distributed data) accounts for the variability of controls, as it tells how many

**Table 1** Methods for Normalizing Screening Data

| | |
|---|---|
| Percent of Control | $(S_i/C_{mean}) \times 100$ |
| *Z*-score | $(S_i - C_{mean})/C_{SD}$ |
| Robust *Z*-score | $(S_i - C_{median})/C_{MAD}$; $C_{MAD} = 1.4826 \times \text{median}(\lvert C_i - C_{median} \rvert)$ |
| *B*-score | $r_{ijp}/\text{MAD}_p$; $r_{ijp} = S_{ijp} - \mu_p - \text{row}_i - \text{col}_j$ (median polish); $\text{MAD}_p = 1.4826 \times \text{median}(\lvert (r_{ijp})_{all} - \text{median}((r_{ijp})_{all}) \rvert)$ |

$S_i$ = activity of the i^th^ sample; $C$ = control; $C_i$ = i^th^ control; SD = standard deviation; MAD = median absolute deviation; $r_{ijp}$ = residual in $i^{th}$ row, $j^{th}$ column, and $p^{th}$ plate; $S_{ijp}$ = sample value at $i^{th}$ row, $j^{th}$ column, and $p^{th}$ plate; $\mu_p$ = plate average; $\text{row}_i$ = median of $i^{th}$ row; $\text{col}_j$ = median of $j^{th}$ column.

standard deviations a compound are away from the mean of the controls. In robust *Z*-score calculations (for nonnormally distributed data), mean and SD are replaced by median and MAD, respectively; it is more resistant to outliers yielding more reliable results. All of these methods correct for plate-to-plate variability but assume a random distribution of error within each plate. Besides normalization, the *B*-score method uses a two-way median polish algorithm to minimize within-plate column and row effects (Brideau, Gunter, Pikounis, & Liaw, 2003). Other noise reduction methods based on Tukey median polish algorithm or using polynomial least squares fit have also been developed to correct for systematic within plate effects (Makarenkov et al., 2007; Mpindi et al., 2015). The Loess local polynomial fit is based on the least squares polynomial approximation and provides more reliable data and less false positives on plates with higher hit rates than 20% than the *B*-score method, which could be attributed to *B*-score's dependency on the median polish algorithm (Mpindi et al., 2015). Combining a polynomial least squares fit method with normalization of data to scattered controls may provide the best option for screens with high hit rates and also for generating accurate dose–response curves.

We have normalized our screening data by using robust *Z*-scores and within plate neutral compounds scattered randomly across each plate (~100 compounds/384-well plate).

### 2.6.3 Assay Performance and Hit Selection

The robustness of a screen highly depends on the nature of selected positive and negative controls. Positive controls with huge effect sizes but without biological relevance may provide false expectations and few or no hits at all. Therefore, strong positive controls with more modest effect sizes are preferred in a HCS (Davies & Shamu, 2014).

Assay performance may be quantified based on signal-to-background (S/B) or signal-to-noise ratio (S/N), where S/B is the ratio of the mean of the effect of the positive controls to the negative controls and S/N is the difference between the mean of the positive and negative controls divided by the variance of the negative controls. The most widely used HTS assay performance measure is the *Z*-factor (Zhang, Chung, & Oldenburg, 1999). It involves the means (μ) and standard deviations (SD) of both positive (p) and negative (n) controls $(1-(3\,(SD_p - SD_n)/|\mu_p - \mu_n|))$. Generally, assays with Z-factor values above 0.5 are considered reliable and robust, most screening facilities use this value as an acceptance criterion to run an assay as a HTS. This parameter plays a less important role for HCS data, since assays performed on living cells are usually noisier than data generated from biochemical or other assays employed in the more classical HTS.

Several hit selection methods have been described in the literature (Bhinder & Djaballah, 2012; Birmingham et al., 2009; Malo et al., 2006; Zhang, 2011), many having been developed originally for RNAi screens. The simplest method to select hits is to identify the upper 1% of the normalized and rank ordered data for further rescreening. Other methods use a predefined threshold to select hits, based on *Z*-score, robust *Z*-score, or quartiles (Zhang et al., 2006). *Z*-Score or robust *Z*-score methods define the threshold for hits usually to be at least 3 SD or 3 MAD away from the population mean or median, respectively, although one can be more or less stringent depending on the assay. The quartile-based method usually used in RNAi screens seems to perform better with extremely skewed data and selects more true hits than *Z*-score-based methods. Calculating strictly standardized mean difference (SSMD) values for each well, also developed originally for RNAi screens, takes into account the variability of both a hit compound and the control given that both groups have several replicates. This can lead to fewer false negatives or false positives (Zhang et al., 2007).

Hit detection also depends on compound concentration, which influences its potency and variability. Variables that include solvent evaporation, compound solubility, and transfer efficiency will all affect the measured potency of a compound. Replicate plate screening provides a way of normalizing for this variability. For our screens for mitochondrial dynamics, we utilized four replicate plates with each compound plate represented in one well per plate. Once compounds are confirmed in a rescreen, dose–response curves are generated to yield $EC_{50}$ values (half maximal effect concentration). A dose–response relationship assumes that there exists a low dose where the compound has no effect, as well as concentration producing a maximum response beyond which increasing compound concentration is not accompanied by further increase of the effect.

Here we present our data analysis pipeline with outlier removal optimized for four replicate fields/well and four replicate plates with one well/plate (Fig. 7).

### 2.6.4 Step-By-Step Data Analysis Procedure Using Four Fields/Well and Four Replicate Plates

- Remove fields which do not contain a sufficient number of objects (e.g., >500 axonal mitochondrial/field for the mitochondrial dynamics assay).
- Remove outlier field values (Fig. 7B):

  - Select nontoxic wells:
    - In our example, a well is considered toxic if there exists dendritic mitochondrial fragmentation and increased axonal mitochondrial circularity. To identify such wells, field values are averaged (without outlier removal), and robust *Z*-scores are calculated using the whole plate population. A well is classified as nontoxic if axonal mitochondrial circularity $<3$ and dendritic mitochondrial length $>-2$ robust *Z*-scores.
  - Calculate outlier field removal thresholds for each feature:
    - CVs of the field values are calculated for the nontoxic wells.
    - Mean and standard deviations of the CVs are calculated.
    - Outlier removal thresholds are calculated by taking the mean $+2$SD of the CVs for the given parameter.
  - Remove outliers:
    - Recalculate the CVs for every well.
    - If the CV for a given feature is larger for a well than the outlier removal threshold for that feature, one field value is removed from the four, which is furthest from the mean.
- Average field values into well values.
- Normalize data by neutral wells (Fig. 7C):
  - Select nontoxic wells as described earlier.
  - Select neutral wells from nontoxic wells:
    - A well is considered to be neutral if its robust *Z*-scores are between $-1.5$ and 1.5 for the hit parameters.
  - Calculate robust *Z*-scores using neutral wells.
- Repeat the same data processing for all the four plates.
- Remove outlier well values for each feature across the replicate plates:
  - Calculate *Z*-scores for the replicate well values using the mean and standard deviation of the four replicate wells.
  - Identify those replicates, whose *Z*-score $>3$ for a given feature.
  - Remove one value from the four, which is furthest from the mean.
- Average the replicate wells for each feature.
- Select the hits:
  - Health: axonal mitochondrial circularity $<-2.5$ robust *Z*-score
  - Elongation: dendritic mitochondrial length $>2.5$ robust *Z*-score
  - Axonal mitochondrial content: axonal mitochondrial area occupied $>2.5$ robust *Z*-score
  - Dendritic mitochondrial content: dendritic mitochondrial area occupied $>$ robust *Z*-score

### 2.6.5 Notes

- The Pandas library of Python is a great tool to implement data analysis pipelines.

## 3. SUMMARY AND CONCLUSIONS

High content screens offer a powerful platform for probing biological processes at a "cell systems" level of analysis. This can be accomplished using many types of cells including primary neurons. The approach does require a high level of sophisticated image analysis and data processing knowledge, along with hardware for liquid handling robotics and a confocal screening microscope for the collection of images from microtiter plates. However, the approach offers a reward in a wealth of cellular data useful as basic science knowledge on the effects of small molecules, libraries of RNAi vectors, etc., on processes of interest to the investigator.

## REFERENCES


Aschauer, D. F., Kreuz, S., & Rumpel, S. (2013). Analysis of transduction efficiency, tropism and axonal transport of AAV serotypes 1, 2, 5, 6, 8 and 9 in the mouse brain. *PLoS One*, *8*(9), e76310. https://doi.org/10.1371/journal.pone.0076310.

Bereiter-Hahn, J. (1976). Dimethylaminostyrylmethylpyridiniumiodine (DASPMI) as a fluorescent probe for mitochondria in situ. *Biochimica et Biophysica Acta*, *423*(1), 1–14.

Bhinder, B., & Djaballah, H. (2012). A simple method for analyzing actives in random RNAi screens: Introducing the "H Score" for hit nomination & gene prioritization. *Combinatorial Chemistry & High Throughput Screening*, *15*(9), 686–704.

Birmingham, A., Selfors, L. M., Forster, T., Wrobel, D., Kennedy, C. J., Shanks, E., et al. (2009). Statistical methods for analysis of high-throughput RNA interference screens. *Nature Methods*, *6*(8), 569.

Bray, M. A., & Carpenter, A. (2012). Advanced assay development guidelines for image-based high content screening and analysis. In G. S. Sittampalam, N. P. Coussens, K. Brimacombe, A. Grossman, M. Arkin, D. Auld, C. Austin, J. Baell, B. Bejcek, T. D. Y. Chung, J. L. Dahlin, V. Devanaryan, T. L. Foley, M. Glicksman, M. D. Hall, J. V. Hass, J. Inglese, P. W. Iversen, S. D. Kahl, S. C. Kales, M. Lal-Nag, Z. Li, J. McGee, O. McManus, T. Riss, O. J. Trask, Jr., J. R. Weidner, M. Xia, & X. Xu (Eds.), *Assay guidance manual.* Bethesda (MD): Eli Lilly & Company and the National Center for Advancing Translational Sciences.

Brideau, C., Gunter, B., Pikounis, B., & Liaw, A. (2003). Improved statistical methods for hit selection in high-throughput screening. *Journal of Biomolecular Screening*, *8*(6), 634–647. https://doi.org/10.1177/1087057103258285.

Carpenter, A. E., Jones, T. R., Lamprecht, M. R., Clarke, C., Kang, I. H., Friman, O., et al. (2006). CellProfiler: Image analysis software for identifying and quantifying cell phenotypes. *Genome Biology*, 7(10), R100. https://doi.org/10.1186/gb-2006-7-10-r100.

Chazotte, B. (2011). Labeling mitochondria with MitoTracker dyes. *Cold Spring Harbor Protocols*, *2011*(8), 990–992. https://doi.org/10.1101/pdb.prot5648.

Chen, H., & Chan, D. C. (2009). Mitochondrial dynamics—Fusion, fission, movement, and mitophagy—In neurodegenerative diseases. *Human Molecular Genetics, 18*(R2), R169–R176. https://doi.org/10.1093/hmg/ddp326.

Cleveland, P. H., & Koutz, P. J. (2005). Nanoliter dispensing for uHTS using pin tools. *Assay and Drug Development Technologies, 3*(2), 213–225. https://doi.org/10.1089/adt.2005.3.213.

Cummings, J. (2018). Lessons learned from Alzheimer disease: Clinical trials with negative outcomes. *Clinical and Translational Science, 11*(2), 147–152. https://doi.org/10.1111/cts.12491.

Daub, A., Sharma, P., & Finkbeiner, S. (2009). High-content screening of primary neurons: Ready for prime time. *Current Opinion in Neurobiology, 19*(5), 537–543. https://doi.org/10.1016/j.conb.2009.10.002.

Davies, A., & Shamu, C. (2014). *An introduction to high content screening: Imaging technology, assay development, and data analysis in biology and drug discovery*. John Wiley & Sons.

Dunn, D. A., & Feygin, I. (2000). Challenges and solutions to ultra-high-throughput screening assay miniaturization: Submicroliter fluid handling. *Drug Discovery Today, 5*(12 Suppl. 1), 84–91.

Ehrenberg, B., Montana, V., Wei, M. D., Wuskell, J. P., & Loew, L. M. (1988). Membrane potential can be determined in individual cells from the nernstian distribution of cationic dyes. *Biophysical Journal, 53*(5), 785–794. https://doi.org/10.1016/S0006-3495(88)83158-8.

Emaus, R. K., Grunwald, R., & Lemasters, J. J. (1986). Rhodamine 123 as a probe of transmembrane potential in isolated rat-liver mitochondria: Spectral and metabolic properties. *Biochimica et Biophysica Acta, 850*(3), 436–448.

Goktug, A. N., Chai, S. C., & Chen, T. (2013). Data analysis approaches in high throughput screening. In H. El-Shemy (Ed.), *Drug discovery*. IntechOpen.

Hammond, S. L., Leek, A. N., Richman, E. H., & Tjalkens, R. B. (2017). Cellular selectivity of AAV serotypes for gene delivery in neurons and astrocytes by neonatal intracerebroventricular injection. *PLoS One, 12*(12), e0188830. https://doi.org/10.1371/journal.pone.0188830.

Hawkes, N. (2018). Pfizer abandons research into Alzheimer's and Parkinson's diseases. *BMJ, 360*, k122. https://doi.org/10.1136/bmj.k122.

Johri, A., & Beal, M. F. (2012). Mitochondrial dysfunction in neurodegenerative diseases. *The Journal of Pharmacology and Experimental Therapeutics, 342*(3), 619–630. https://doi.org/10.1124/jpet.112.192138.

Kiryu-Seo, S., Ohno, N., Kidd, G. J., Komuro, H., & Trapp, B. D. (2010). Demyelination increases axonal stationary mitochondrial size and the speed of axonal mitochondrial transport. *The Journal of Neuroscience, 30*(19), 6658–6666. https://doi.org/10.1523/JNEUROSCI.5265-09.2010.

Korchak, H. M., Rich, A. M., Wilkenfeld, C., Rutherford, L. E., & Weissmann, G. (1982). A carbocyanine dye, DiOC6(3), acts as a mitochondrial probe in human neutrophils. *Biochemical and Biophysical Research Communications, 108*(4), 1495–1501.

Lesuisse, C., & Martin, L. J. (2002). Long-term culture of mouse cortical neurons as a model for neuronal development, aging, and death. *Journal of Neurobiology, 51*(1), 9–23.

Lewis, T. L., Kwon, S.-K., Lee, A., Shaw, R., & Polleux, F. (2018). MFF-dependent mitochondrial fission regulates presynaptic release and axon branching by limiting axonal mitochondria size. *bioRxiv*, 276691.

Li, Y., Yang, Y., & Wang, S. (2005). Neuronal gene transfer by baculovirus-derived vectors accommodating a neurone-specific promoter. *Experimental Physiology, 90*(1), 39–44. https://doi.org/10.1113/expphysiol.2004.028217.

Makarenkov, V., Zentilli, P., Kevorkov, D., Gagarin, A., Malo, N., & Nadon, R. (2007). An efficient method for the detection and elimination of systematic error in high-throughput screening. *Bioinformatics, 23*(13), 1648–1657. https://doi.org/10.1093/bioinformatics/btm145.

Malo, N., Hanley, J. A., Cerquozzi, S., Pelletier, J., & Nadon, R. (2006). Statistical practice in high-throughput screening data analysis. *Nature Biotechnology*, *24*(2), 167–175. https://doi.org/10.1038/nbt1186.

Mansouri, M., Bellon-Echeverria, I., Rizk, A., Ehsaei, Z., Cianciolo Cosentino, C., Silva, C. S., et al. (2016). Highly efficient baculovirus-mediated multigene delivery in primary cells. *Nature Communications* 7, 11529. https://doi.org/10.1038/ncomms11529.

Michelfelder, S., & Trepel, M. (2009). Adeno-associated viral vectors and their redirection to cell-type specific receptors. *Advances in Genetics*, *67*, 29–60. https://doi.org/10.1016/S0065-2660(09)67002-4.

Misgeld, T., & Schwarz, T. L. (2017). Mitostasis in neurons: Maintaining mitochondria in an extended cellular architecture. *Neuron*, *96*(3), 651–666. https://doi.org/10.1016/j.neuron.2017.09.055.

Mpindi, J. P., Swapnil, P., Dmitrii, B., Jani, S., Saeed, K., Wennerberg, K., et al. (2015). Impact of normalization methods on high-throughput screening data with high hit rates and drug testing with dose-response data. *Bioinformatics*, *31*(23), 3815–3821. https://doi.org/10.1093/bioinformatics/btv455.

Murlidharan, G., Samulski, R. J., & Asokan, A. (2014). Biology of adeno-associated viral vectors in the central nervous system. *Frontiers in Molecular Neuroscience* 7, 76. https://doi.org/10.3389/fnmol.2014.00076.

Niles, W. D., & Coassin, P. J. (2008). Cyclic olefin polymers: Innovative materials for high-density multiwell plates. *Assay and Drug Development Technologies*, *6*(4), 577–590. https://doi.org/10.1089/adt.2008.134.

Nunnari, J., & Suomalainen, A. (2012). Mitochondria: In sickness and in health. *Cell*, *148*(6), 1145–1159. https://doi.org/10.1016/j.cell.2012.02.035.

Otsu, N. (1979). A threshold selection method from gray-level histograms. *IEEE Transactions on Systems, Man, and Cybernetics*, *9*(1), 62–66.

Rafael, J. (1980). 2-(Dimethylaminostyryl)-1-ethylpyridinium iodide: A fluorescent probe of energetic conditions in brown-adipose-tissue mitochondria. *Hoppe-Seyler's Zeitschrift für Physiologische Chemie*, *361*(3), 437–444.

Rizzuto, R., Brini, M., Pizzo, P., Murgia, M., & Pozzan, T. (1995). Chimeric green fluorescent protein as a tool for visualizing subcellular organelles in living cells. *Current Biology*, *5*(6), 635–642.

Sheng, B., Wang, X., Su, B., Lee, H. G., Casadesus, G., Perry, G., et al. (2012). Impaired mitochondrial biogenesis contributes to mitochondrial dysfunction in Alzheimer's disease. *Journal of Neurochemistry*, *120*(3), 419–429. https://doi.org/10.1111/j.1471-4159.2011.07581.x.

Singh, S., Carpenter, A. E., & Genovesio, A. (2014). Increasing the content of high-content screening: An overview. *Journal of Biomolecular Screening*, *19*(5), 640–650. https://doi.org/10.1177/1087057114528537.

Smiley, S. T., Reers, M., Mottola-Hartshorn, C., Lin, M., Chen, A., Smith, T. W., et al. (1991). Intracellular heterogeneity in mitochondrial membrane potentials revealed by a J-aggregate-forming lipophilic cation JC-1. *Proceedings of the National Academy of Sciences of the United States of America*, *88*(9), 3671–3675.

Smith, T., Ho, P. I., Yue, K., Itkin, Z., MacDougall, D., Paolucci, M., et al. (2013). Comparison of compound administration methods in biochemical assays: Effects on apparent compound potency using either assay-ready compound plates or pin tool-delivered compounds. *Journal of Biomolecular Screening*, *18*(1), 14–25. https://doi.org/10.1177/1087057112455434.

Subach, O. M., Gundorov, I. S., Yoshimura, M., Subach, F. V., Zhang, J., Gruenwald, D., et al. (2008). Conversion of red fluorescent protein into a bright blue probe. *Chemistry & Biology*, *15*(10), 1116–1124. https://doi.org/10.1016/j.chembiol.2008.08.006.

Swinney, D. C., & Anthony, J. (2011). How were new medicines discovered? *Nature Reviews. Drug Discovery*, *10*(7), 507–519. https://doi.org/10.1038/nrd3480.

Taylor, H., Haskins, J., & Giuliano, K. (2007). *High content screening: A powerful approach to systems cell biology and drug discovery totowa.* New Jersery: Humana Press Inc.
van der Walt, S., Schonberger, J. L., Nunez-Iglesias, J., Boulogne, F., Warner, J. D., Yager, N., et al. (2014). scikit-image: Image processing in Python. *PeerJ 2*, e453. https://doi.org/10.7717/peerj.453.
Wang, D., Wang, J., Bonamy, G. M., Meeusen, S., Brusch, R. G., Turk, C., et al. (2012). A small molecule promotes mitochondrial fusion in mammalian cells. *Angewandte Chemie (International Ed. in English)*, *51*(37), 9302–9305. https://doi.org/10.1002/anie.201204589.
Westermann, B. (2012). Bioenergetic role of mitochondrial fusion and fission. *Biochimica et Biophysica Acta*, *1817*(10), 1833–1838. https://doi.org/10.1016/j.bbabio.2012.02.033.
Wu, Z., Asokan, A., & Samulski, R. J. (2006). Adeno-associated virus serotypes: Vector toolkit for human gene therapy. *Molecular Therapy*, *14*(3), 316–327. https://doi.org/10.1016/j.ymthe.2006.05.009.
Zanella, F., Lorens, J. B., & Link, W. (2010). High content screening: Seeing is believing. *Trends in Biotechnology*, *28*(5), 237–245. https://doi.org/10.1016/j.tibtech.2010.02.005.
Zhang, X. D. (2011). Illustration of SSMD, Z score, SSMD*, Z* score, and t statistic for hit selection in RNAi high-throughput screens. *Journal of Biomolecular Screening*, *16*(7), 775–785. https://doi.org/10.1177/1087057111405851.
Zhang, J. H., Chung, T. D., & Oldenburg, K. R. (1999). A simple statistical parameter for use in evaluation and validation of high throughput screening assays. *Journal of Biomolecular Screening*, *4*(2), 67–73. https://doi.org/10.1177/108705719900400206.
Zhang, C., Deng, Y., Dai, H., Zhou, W., Tian, J., Bing, G., et al. (2017). Effects of dimethyl sulfoxide on the morphology and viability of primary cultured neurons and astrocytes. *Brain Research Bulletin*, *128*, 34–39. https://doi.org/10.1016/j.brainresbull.2016.11.004.
Zhang, X. D., Ferrer, M., Espeseth, A. S., Marine, S. D., Stec, E. M., Crackower, M. A., et al. (2007). The use of strictly standardized mean difference for hit selection in primary RNA interference high-throughput screening experiments. *Journal of Biomolecular Screening*, *12*(4), 497–509. https://doi.org/10.1177/1087057107300646.
Zhang, X. D., Yang, X. C., Chung, N., Gates, A., Stec, E., Kunapuli, P., et al. (2006). Robust statistical methods for hit selection in RNA interference high-throughput screening experiments. *Pharmacogenomics*, 7(3), 299–309. https://doi.org/10.2217/14622416.7.3.299.

## FURTHER READING

Cagalinec, M., Safiulina, D., Liiv, M., Liiv, J., Choubey, V., Wareski, P., et al. (2013). Principles of the mitochondrial fusion and fission cycle in neurons. *Journal of Cell Science*, *126*(Pt. 10), 2187–2197. https://doi.org/10.1242/jcs.118844.
Ebrahimi-Fakhari, D., Saffari, A., Wahlster, L., Di Nardo, A., Turner, D., Lewis, T. L., Jr., et al. (2016). Impaired mitochondrial dynamics and mitophagy in neuronal models of tuberous sclerosis complex. *Cell Reports*, *17*(4), 1053–1070. https://doi.org/10.1016/j.celrep.2016.09.054.

CHAPTER ELEVEN

# Achieving a Good Crystal System for Crystallographic X-Ray Fragment Screening

**Patrick M. Collins*, Alice Douangamath*, Romain Talon*, Alexandre Dias*, Jose Brandao-Neto*, Tobias Krojer†, Frank von Delft*,†,‡,1**

*Diamond Light Source, Harwell Science and Innovation Campus, Didcot, United Kingdom
†Structural Genomics Consortium, Nuffield Department of Medicine, University of Oxford, Oxford, United Kingdom
‡Department of Biochemistry, University of Johannesburg, Aukland Park, South Africa
[1]Corresponding author: e-mail address: frank.von-delft@diamond.ac.uk

## Contents



## Abstract

The XChem facility at Diamond Light Source offers fragment screening by X-ray crystallography as a general access user program. The main advantage of X-ray crystallography as a primary fragment screen is that it yields directly the location and pose of the fragment hits, whether within pockets of interest or merely on surface sites: this is the key information for structure-based design and for enabling synthesis of follow-up molecules. Extensive streamlining of the screening experiment at XChem

*Methods in Enzymology*, Volume 610
ISSN 0076-6879
https://doi.org/10.1016/bs.mie.2018.09.027

has engendered a very active user program that is generating large amounts of data: in 2017, 36 academic and industry groups generated 35,000 datasets of uniquely soaked crystals. It has also generated a large number of learnings concerning the main remaining bottleneck, namely, obtaining a suitable crystal system that will support a successful fragment screen. Here we discuss the practicalities of generating screen-ready crystals that have useful electron density maps, and how to ensure they will be successfully reproduced and usable at a facility outside the home lab.

## 1. INTRODUCTION

Many biophysical methods have been described for the primary fragment screen, including both ligand- and protein-observed NMR spectroscopy (Harner, Frank, & Fesik, 2013), surface plasmon resonance (Navratilova & Hopkins, 2010), X-ray crystallography (Hartshorn et al., 2005), to name only a few. Invariably, however, knowledge of the fragment-binding mode is crucial, and cocrystal structures of bound hits are generally sought, in order to initiate iterative structure-based design and synthesis cycles. X-ray crystallography is, along with NMR, the oldest primary screen, and, in principle, represents the gold standard, since it provides direct structural readout for follow-up design, and is unmatched in sensitivity (Patel, Bauman, & Arnold, 2014). However, the logistical overheads of collecting X-ray diffraction data from 100s to 1000s of crystals have historically made it prohibitive for most groups. More recently, though, advances in synchrotron science and automation have reduced these overheads, enabling the development of high-throughput X-ray diffraction data collection. Taking advantage of this potential throughput, however, including for experiments such as fragment screening, places much higher demands on crystal production.

There are two basic approaches to obtaining protein-fragment complexes, namely, cocrystallization or soaking (Hassell et al., 2007). Cocrystallization involves mixing of the free protein in solution with the small molecule prior to crystallization, which allows the ligand to bind to the protein without any crystal lattice constraints, and in principle allowing for more conformational freedom of the protein. This is the method of choice when a protein complex with a specific ligand is required. High-throughput methods for cocrystallization with small molecules have been described previously (Gelin et al., 2015). The drawback is that the addition of various small molecules in cocrystallization tends to lower crystallization success rates and induce larger variations in resolution and crystal form, placing a greater burden on downstream analysis in the high-throughput setting. In this approach, false negatives arise from the failure of the protein to crystallize, or the crystals selecting for protein molecules with no compound bound.

Crystal soaking is simpler: compound is added, dissolved or pure, directly to the crystallization drop, and it subsequently diffuses through the crystal solvent channels to reach the binding pockets (Danley, 2006). This approach yields diffracting crystals more reliably, since the many crystals required can be grown from the same condition; X-ray diffraction also tends to be more consistent. Instead, false negatives arise from failure of the compound to reach the binding pocket, or the protein conformation being incompatible with binding, the latter leading either to empty pockets, or crystals dissolving upon compound binding due to lattice disruption. Low compound solubility undermines both cocrystallization and soaking.

The methods described here are tailored to screening by soaking, as this is the approach implemented in the XChem facility at Diamond Light Source, which is a platform for high-throughput fragment screening by crystal structure. The soaking protocol itself has been described previously (Collins et al., 2017) and comprises fast crystal soaking using acoustic droplet ejection techniques. Additionally, the platform relies on machine-assisted crystal harvesting, automated X-ray diffraction data collection, and streamlined data processing (Krojer et al., 2017) with enhanced fragment hit detection (Pearce et al., 2017). As the first ever facility supporting such an experiment, it has enabled the generation of 1000s of protein-fragment complexes for large numbers of industrial and academic users (Delbart et al., 2018; Foley et al., 2017; Harding et al., 2017; Keedy et al., 2018; McIntyre et al., 2017).

The high-throughput nature of crystal-based fragment screening places specific demands on protein crystals that tend not to apply to traditional structure solution projects. The hands-on support required by the nature of the XChem facility has led to the accumulation of a large amount of experience of getting problematic crystal systems to work, and these lessons and common problems are the focus of the rest of the chapter.

## 2. PROCESS OVERVIEW AND CONSTRAINTS ON CRYSTALS

To perform the screen, the 100s of crystals need to be on site; unsurprisingly, therefore, how to ensure this is the case dominates the early conversation between users and XChem staff for every experiment. Users close to the synchrotron may find it easiest to grow crystals at home lab and bring them to Diamond for the experiment. In contrast, for users further afield or without crystallization robotics in-house, crystallization is performed on-site, either by the attending user or by XChem staff. In this case, the crystallization protocol must be smoothly transferable from lab to facility, requiring a level of robustness of protocol not usually addressed, and further discussed below.

The experiment performed at the facility itself proceeds as follows:

1. Crystal plates are imaged in standard crystal plate imagers (Formulatrix). Crystal drop images are analyzed using TeXRank (Ng, Dekker, Kroemer, Osborne, & von Delft, 2014), and suitable crystal wells are selected and marked with target coordinates for acoustic dispensing of fragments.
2. Fragments from a source library in dimethyl sulfoxide (DMSO) solvent are acoustically transferred (Labcyte) to the targeted crystal drop locations, and the crystals allowed to soak (Collins et al., 2017).
3. Crystals are manually harvested into cryo-loops and frozen in liquid nitrogen, the process robotically assisted with a motorized microscope stage (Shifter; Oxford Lab Technologies).
4. X-ray diffraction data are collected in automated, unattended mode at Diamond beamlines (I04-1), with crystals aligned by centering with fast X-ray rastering.
5. Data are automatically integrated, scaled, and electron density maps calculated by molecular substitution.
6. Electron density maps are analyzed using PanDDA (Pearce et al., 2017) for detection and modeling of bound fragments.
7. Models are refined and prepared for deposition to the PDB, with all calculations managed and tracked using XChemExplorer (Krojer et al., 2017).

This process is deliberately highly flexible, since most crystal systems have peculiarities that must be accommodated. Nevertheless, the likelihood of overall success is greatly enhanced if as many as possible of the common constraints can be dealt with—and this generally is feasible, with some focused effort.

- *Reproducibility of crystallization*: many 100s of crystals must grow in a specific crystallization format (for XChem: the SBS-format 3-lens SwissCi plates), with ideal drop volumes of 100–600 nL. Drops with volumes smaller than 100 nL tend to be more difficult to mount crystals from due to rapid evaporation of the drop, and drops over 600 nL require so much compound that the speed of acoustic dispensing (500 nL/s) becomes limiting.

- *Crystal size and robustness*: if crystals are large enough (>30 μm), they become reliable to manipulate during harvesting, and when robust, they can tolerate both fast manual harvesting and the high solvent concentrations in the drop after dissolved compound is added.
- *Consistency of diffraction:* to improve robustness of data analysis, crystals should have consistent X-ray diffraction, ideally with resolution to better than 2.5 Å. Any problems encountered during data analysis slow down the overall experiment enormously and are worth avoiding.
- *Suitable molecular packing*: while accessibility of the binding site is a basic physical constraint, high quality of electron density in the regions of interest and low conformational heterogeneity greatly simply analysis of the data.

# 3. WAYS TO ESTABLISH A SUITABLE CRYSTAL SYSTEM

It is a common beginner's mistake to consider one's crystal system robust, merely because the first structure was achieved quite straightforwardly: in reality, the requirements are far more stringent. The description below therefore covers how to approach the challenge of generating large numbers of suitable crystals, as well as how to ensure that these crystals will also in fact be available at the remote facility where the fragment screen is to be performed.

## 3.1 Crystallization

Quality of crystals is always the dominant consideration when selecting between crystallization conditions should a choice of conditions be available. Beyond that, the nature of the precipitant is important: polyethylene glycol (PEG) conditions are preferable over high salt conditions, since the latter tend to reduce fragment solubility and result in heavy compound precipitation/crystallization on addition to the drop, thereby reducing the effective compound concentration (see also Section 4.6). Volatile solvents such as isopropanol are inconvenient, as they lead to rapid evaporation of the crystal drop and cause difficulties during crystal mounting (see Section 4.4). Finally, the pH should be as close to physiological as possible, to maximize the likelihood of the binding sites having their relevant protonation state.

To provide the best chances of consistent X-ray diffraction after crystal soaking, it is important that all crystallization drops across a plate have identical components. Variations in equilibrated drop volume (for example, from variable reservoir:protein ratios) will result in variable final compound and solvent concentrations after soaking, which can result in inconsistent

X-ray diffraction for sensitive systems. Graduation of components across a plate (such as pH or precipitant concentration gradients) should also be avoided for optimal consistency.

Subtle changes to the crystallization protocol can profoundly change the crystal system's behavior. Therefore, if the crystallization protocol changes in any detail (construct, crystallization components, drop volumes, purification batch, etc.), it is vital that control experiments are run to confirm that behavior remains as expected, including solvent tolerance, before embarking on the large fragment screen. This includes crystal age, with older crystals frequently having dramatic deterioration of diffraction consistency.

Robotic crystallization is essential to achieve consistent small drop volumes and provide the throughput necessary for the many crystallization plates usually required. The XChem facility relies on the resources of the crystallization facility in the neighboring Research Complex at Harwell, with Mosquito liquid handlers (TTP Labtech) available for nanodrop crystallization. The standard experiment is performed in the 96-well, 3-lens crystallization plates (SwissCi), containing 20 μL of reservoir solution, and 150 nL crystal drops in a 1:1 ratio of protein to reservoir (75 nL each).

## 3.2 Seeding

Seeding is a supremely effective tool in the quest for consistent crystallization; it is also the trick most studiously ignored by newcomers to screening, until they have first failed to achieve reproducibility for several rounds of seedless optimization. It is now the single strongest recommendation tendered by XChem staff, being the closest thing to a magic bullet available.

In particular, it offers the necessary close control of the nucleation process for ensuring sufficient numbers of crystals can be grown for a large screen. Our seeding protocol follows that described in detail by D'Arcy, Bergfors, Cowan-Jacob, and Marsh (2014), and once a seed stock is prepared, aliquots are frozen at −80°C. By preparing and freezing a particular seed stock in aliquots, a specific seed dilution can be established that tends to reproduce the initial seeding trial significantly better than preparing fresh seed stocks prior to each round of crystallization.

Seeding is also a route for establishing alternative crystal forms: matrix seeding (Shaw Stewart, Kolek, Briggs, Chayen, & Baldock, 2011) has yielded some remarkable successes, though close attention to the details appears to be crucial, which may be why there are also many anecdotes of it "not working."

## 3.3 Protein Engineering

It is rare that extensive exploration and optimization of crystallization conditions will convert poor, sporadic crystals into a strong, reliable system. Protein engineering provides a far more reliable if labor intensive route, a wisdom long since widely adopted by experienced practitioners.

Structural genomics initiatives in particular have established useful ground rules for discovering well-behaved constructs. At the SGC, the standard approach is to generate many (>10) truncated or mutated versions of the same protein up front and in parallel, to establish immediately whether any constructs have sufficient levels of protein expression to achieve solubility and purity to test for crystallization (Gileadi et al., 2008; Savitsky et al., 2010).

Similarly, having access to multiple starting points can be pivotal to a successful fragment screen. Not only does crystal quality vary, but availability of different crystal forms permits alternatives to be explored if the leading crystal system proves to be refractory further down the screening process, such as having poor solvent tolerance or an occluded binding site (discussed below). The detailed methods outlined by Gileadi et al. (2008) can be set up in any protein production lab and helps generate more options when embarking on a fragment screening campaign.

Another way to help coax well-behaved protein to crystallize or change crystal form is to explore surface site mutations that introduce so-called crystallization epitopes that correlate with crystallization propensity (Price 2nd et al., 2009). In our hands, this approach has yielded many new crystal forms (manuscript in preparation).

## 3.4 Test for Transferability

It is critical to establish that the crystal system established in the home lab can be reproduced at another facility. The main guideline is to minimize the number of processes that must be performed at the remote facility: thus ideally, all components, including the protein, seed stocks, crystallization cocktail, all at the correct concentrations, can be frozen for transport, to be thawed again at the destination for use. The exact dispensing protocol will also have been established at the home lab, ideally using the same robotics, and very thoroughly documented. This way, the only processes that need occur at the facility are the thawing of relevant components, and robotic preparation of the crystallization plate.

Importantly, the exact freeze/thaw protocol should be tested and verified exhaustively at the home lab, to verify that reproducibility is maintained. And finally, the packaging should contain copious amounts of dry ice, at least 20 kg for a *trans*-Atlantic shipment, to provide for the inevitable delays at customs or mistakes by the courier company.

## 4. COMMON PROBLEMS

Here we summarize common problems and how they can be addressed.

### 4.1 Poor Reproducibility

Often, only a small fraction of drops will contain any crystals, or crystals of sufficient size/morphology. The following options can drastically improve reproducibility and productivity of each plate:

1. Seeding, as discussed in Section 3.2.
2. Exploring a different area of the phase diagram, such as a two–fourfold increase in protein concentration (if possible), coupled with an equal fold (or more) decrease in precipitant concentration.
3. A broad optimization screen around the crystallization condition, exploring a wide pH and salt concentration range.
4. Ensuring that the purification protocol is yielding consistent protein quality, with gel filtration as the final step in purification.

Slow appearance of protein crystals (weeks or months) indicates a process that is not driven by vapor diffusion, but instead a drift in pH, oxidation, or evaporation through the plastic: in these cases, looking for very different crystallization conditions will help. Protein degradation is also a very common reason, and often the clearest evidence is a marked difference between what was purified and what is visible in the structure. Then it is highly advisable to adjust the expression construct to match what was observed in the structure: this frequently transforms not only the rate of crystallization but also the quality of the resulting crystals.

Excess nucleation or overly rapid crystal formation may be reduced by small amounts of cryoprotectants (such as glycerol) added to the reservoir, or by lower (or higher) temperatures. Showers of crystals can sometimes be rescued by placing crystal plates containing small crystals through a high temperature cycle, for example, placing them at 37°C for a few hours/overnight before returning to the lower crystallization temperature.

Alternatively, if protein is plentiful and a few quality crystals do appear in each crystallization plate, a valid strategy is simply to prepare large numbers of crystal plates until the desired number of crystals is obtained. In the XChem lab, this slows down the overall soaking and harvesting process only marginally, while enabling a successful fragment screen.

## 4.2 Poor X-Ray Diffraction Consistency

Confirmation of whether X-ray diffraction inconsistency is inherent to the crystal system rather than the crystals' sensitivity to soaking, mounting, and freezing can be confirmed by in situ or room temperature X-ray diffraction experiments. For inherently inconsistent crystals, the same points raised above for crystal reproducibility tend to be relevant, especially seeding. Likewise, it becomes important to use only freshly grown crystals, and assess carefully their decay time in the drop. Sometimes, crystal quality is related to crystal morphology, which requires an observant experimenter to notice; in such cases, more careful crystal selection would help, though often requiring larger numbers of crystallization plates.

If soaking and mounting are the cause of inconsistent diffraction, common factors to check are:

1. Ensure that crystal drops are of consistent volume from the crystallization setup. If drop volumes are variable, but the same volume of solvent (e.g., DMSO) is added, the final solvent concentrations will vary and this could undermine consistency.
2. Ensure that drop ratios are the same. When setting up crystallization plates, it is common to vary the protein:reservoir ratios (e.g., 1:2, 1:1, 2:1 through the three subwells). The drops will achieve different final volumes after vapor diffusion, again resulting in varying solvent concentrations, as for #1 above.
3. Ensure that the crystallization condition is uniform across the plate. A precipitant or pH gradient may result in some crystals being more sensitive to the screening treatments or requiring greater cryoprotection.
4. If soaking using acoustic methods, ensure that targeting and solvent dispensing are occurring at the edges of the crystal drop, and away from crystals (Collins et al., 2017).
5. Check the crystal harvesting technique and ensure a fast and efficient plunge into liquid nitrogen with minimal time passing through the cold gas layer (keep mounting dewar topped up with fresh liquid nitrogen) (Warkentin, Berejnov, Husseini, & Thorne, 2006).
6. Dehydration could be a factor (see Section 4.4).
7. Reduce the amount of solvent, and thus lower the fragment concentration.

Ultimately, though, some natural and stochastic variation in data quality is to be expected across many hundreds of crystals, and therefore the real question to ask is whether the data obtained are of sufficient quality to clearly answer the scientific question. For fragment screening, the main requirement is identification of sufficient hits, in sufficiently clear poses, to progress with hit-to-lead chemistry. This is frequently possible even when attrition of diffraction through the screen was unusually high.

### 4.3 Difficulties With Crystallization Format

Sometimes converting from one crystallization format, such as hanging drop, to one compatible with the XChem facility (sitting drop) may result in large changes to the crystallization outcome, such as clear drops or heavy precipitate. Since the target protein has been shown to crystallize, it is often a matter of optimization (see Section 4.1), with exploration of protein/ precipitant concentrations and seeding often proving successful after 2–3 weeks of focused effort in advance of embarking on the visit to the screening facility.

### 4.4 Drop Issues

Minor problems with the crystallization drop that can interfere with crystal harvesting, such as a skin over the drop, are greatly amplified if the protocol requires many hundreds of crystals to be manually mounted. Skins that cover a drop can be removed manually with a loop or acupuncture needle, but can also sometimes be avoided by ensuring that fresh crystallization plates are used, trying larger or smaller drop volumes, or by adding a 1:1 equivalent volume of reservoir solution prior to soaking and harvesting.

Other problems that can affect crystal mounting are drops that rapidly dry out or result in showers of additional crystals appearing rapidly after unsealing of wells. Similar solutions to those for the skin can resolve such issues, especially the addition of additional reservoir solution, or setting up larger initial crystallization drops. Addition of reservoir prior to soaking can also help to prevent X-ray diffraction inconstancies due to crystals that are sensitive to dehydration. Additionally, when mounting crystals, minimizing the number of wells that are unsealed during mounting, even when using devices for assisted crystal harvesting (the Shifter in the XChem lab), can reduce the effects of dehydration for sensitive systems.

## 4.5 Poor Soaking Tolerance

Some crystals have very low tolerance to being soaked in the presence of fragment carrying solvents, such as DMSO. Such solvent sensitivity can be difficult to overcome in soaking experiments. Addition of reservoir solution (1:1 volume ratio) to crystal drop just prior to solvent addition for crystal soaking can help to stabilize sensitive crystals. Alternatively, a fast soaking scheme, like flash soaking (<5 s), may be required to minimize soaking times and enable crystal survival.

While DMSO is the most common solvent in use in fragment work, alternative solvents such as ethylene glycol are considerably gentler and may help for sensitive crystal systems. This does mean repreparing the fragment library, which may at first appear prohibitively complicated, but this tends to be worth the trouble, especially if it converts an intractable system to a tractable one. Exploration of alternative crystal forms or crystallization conditions is sometimes required to discover a solvent tolerant soaking system.

## 4.6 High Salt Conditions

As discussed in Section 3.1, high concentrations of salt in the crystallization condition may reduce small molecule solubility and consequently reduce the effective soaking concentration. There are a number of examples of successful fragment screens in high salt conditions, and with high hit rates. However, if a low hit rate is observed and a high salt condition is suspected as the cause, optimization of any existing PEG-based crystal hits may afford a low salt crystal form. Alternatively, matrix seeding using crystal seeds from the salt condition into PEG-based crystal screens may provide new hits (D'Arcy et al., 2014). Another alternative is to transfer crystals from the salt-based condition into osmotically matched PEG containing drops. An appropriate PEG condition may be discovered by adding a significant excess (2:1 v/v or larger) of PEG solutions of increasing concentration to a series of drops: one of these additions will maintain the osmotic pressure and thus allow the crystal to survive the change, while achieving the desired reduction in salt concentration. In an appropriately equipped facility, the series of crystals can be quickly tested for diffraction.

# 5. MAP AND MODEL CONSIDERATIONS

Certain aspects of the actual structure and corresponding electron density maps can also have a large influence on the success of the fragment screen. Diffraction resolution is the most obvious, and at resolutions poorer than 2.5 Å, the loss of detail in the maps makes it very hard to confidently assign details of binding for molecules as small as fragments. In addition, the soaking process itself invariably causes a loss of resolution, and therefore crystals really ought to diffract natively at least to 2.3 Å to ensure a painless screening experiment.

Any other pathologies that affect map quality will also affect the ability to assign the weakly binding fragments unambiguously. Thus, features like a large twin fraction, strong anisotropy, or translational noncrystallographic symmetry can cause complications. On the other hand, achieving consistent map quality and resolution greatly helps with downstream processing and map analysis using PanDDA (Pearce et al., 2017).

Crystal conditions that result in multiple space groups or crystal forms from the same condition are less problematic than they might appear, as they can be handled reasonably effectively by downstream processing using XChemExplorer (Krojer et al., 2017), as long as good appropriate input models are available for each crystal form. Nevertheless, this does constitute a complication that many XChem users work hard to avoid.

Finally, the binding site itself needs to be open to the solvent channels, and as far as possible from a crystal contact, as these often involve conformational adjustments that might propagate to the site of interest in subtle ways. It will also ideally not have side chains or stretches of backbone that occupy multiple conformations or are completely disordered: energetically, this increases the entropic penalty of binding; and practically, such variations are very hard to model and therefore produce artifacts. Sometimes a small trial screen of ~100 fragments may reveal that fragments can after all access the binding site; more likely though it will show more clearly that a new crystal form will be needed.

## 6. CONCLUSION

Thanks to modern synchrotron technologies, crystal-based fragment screens are now not only accessible through user programs, but the work involved is now far less than that required in the home lab to achieve a suitable crystal system. Close attention to detail and rigorous experimental practice remains crucial, and certainly XChem users gain a very intensive understanding of the crystallogenic and biophysical behavior of their system, quite apart from the chemistry readout of the screening itself.

Overall, the best approach is to recruit the help of an experienced crystallographer that has herself dealt with the relevant problems before, particularly those working in industry. Serendipitously, the scale of the XChem operation means that the number of crystallographers with the requisite experience will continue to rise, allowing for a self-reinforcing dynamic in the community. It is also to be expected that this increasing competence will prove useful to challenges of crystallogenesis at the other end of the scale, namely, the need for reproducible showers for serial crystallography approaches such as needed for X-ray free electron laser experiments.

## REFERENCES

Collins, P. M., Ng, J. T., Talon, R., Nekrosiute, K., Krojer, T., Douangamath, A., et al. (2017). Gentle, fast and effective crystal soaking by acoustic dispensing. *Acta Crystallographica. Section D, Structural Biology*, *73*(Pt. 3), 246–255. https://doi.org/10.1107/S205979831700331X.

Danley, D. E. (2006). Crystallization to obtain protein-ligand complexes for structure-aided drug design. Acta Crystallographica. Section D. *Biological Crystallography*, *62*(Pt 6), 569–575. https://doi.org/10.1107/S0907444906012601.

D'Arcy, A., Bergfors, T., Cowan-Jacob, S. W., & Marsh, M. (2014). Microseed matrix screening for optimization in protein crystallization: what have we learned? *Acta Crystallographica. Section F, Structural Biology Communications*, *70*(Pt 9), 1117–1126. https://doi.org/10.1107/S2053230X14015507.

Delbart, F., Brams, M., Gruss, F., Noppen, S., Peigneur, S., Boland, S., et al. (2018). An allosteric binding site of the alpha7 nicotinic acetylcholine receptor revealed in a humanized acetylcholine-binding protein. *The Journal of Biological Chemistry*, *293*(7), 2534–2545. https://doi.org/10.1074/jbc.M117.815316.

Foley, D. J., Craven, P. G. E., Collins, P. M., Doveston, R. G., Aimon, A., Talon, R., et al. (2017). Synthesis and demonstration of the biological relevance of sp(3)-rich scaffolds distantly related to natural product frameworks. *Chemistry*, *23*(60), 15227–15232. https://doi.org/10.1002/chem.201704169.

Gelin, M., Delfosse, V., Allemand, F., Hoh, F., Sallaz-Damaz, Y., Pirocchi, M., et al. (2015). Combining 'dry' co-crystallization and in situ diffraction to facilitate ligand screening by X-ray crystallography. *Acta Crystallographica. Section D*, *71*(8), 1777–1787. https://doi.org/10.1107/S1399004715010342.

Gileadi, O., Burgess-Brown, N. A., Colebrook, S. M., Berridge, G., Savitsky, P., Smee, C. E., et al. (2008). High throughput production of recombinant human proteins for crystallography. *Methods in Molecular Biology*, *426*, 221–246. https://doi.org/10.1007/978-1-60327-058-8_14.

Harding, R. J., Ferreira de Freitas, R., Collins, P., Franzoni, I., Ravichandran, M., Ouyang, H., et al. (2017). Small molecule antagonists of the interaction between the histone deacetylase 6 zinc-finger domain and ubiquitin. *Journal of Medicinal Chemistry*, *60*(21), 9090–9096. https://doi.org/10.1021/acs.jmedchem.7b00933.

Harner, M. J., Frank, A. O., & Fesik, S. W. (2013). Fragment-based drug discovery using NMR spectroscopy. *Journal of Biomolecular NMR*, *56*(2), 65–75. https://doi.org/10.1007/s10858-013-9740-z.

Hartshorn, M. J., Murray, C. W., Cleasby, A., Frederickson, M., Tickle, I. J., & Jhoti, H. (2005). Fragment-based lead discovery using X-ray crystallography. *Journal of Medicinal Chemistry*, *48*(2), 403–413. https://doi.org/10.1021/jm0495778.

Hassell, A. M., An, G., Bledsoe, R. K., Bynum, J. M., Carter, H. L., 3rd, Deng, S. J., et al. (2007). Crystallization of protein-ligand complexes. *Acta Crystallographica. Section D, Biological Crystallography*, *63*(Pt. 1), 72–79. https://doi.org/10.1107/S0907444906047020.

Keedy, D. A., Hill, Z. B., Biel, J. T., Kang, E., Rettenmaier, T. J., Brandao-Neto, J., et al. (2018). An expanded allosteric network in PTP1B by multitemperature crystallography, fragment screening, and covalent tethering. *eLife*, 7. https://doi.org/10.7554/eLife.36307.

Krojer, T., Talon, R., Pearce, N., Collins, P., Douangamath, A., Brandao-Neto, J., et al. (2017). The XChemExplorer graphical workflow tool for routine or large-scale protein-ligand structure determination. *Acta Crystallographica. Section D, Structural Biology*, *73*(Pt. 3), 267–278. https://doi.org/10.1107/S2059798316020234.

McIntyre, P. J., Collins, P. M., Vrzal, L., Birchall, K., Arnold, L. H., Mpamhanga, C., et al. (2017). Characterization of three druggable hot-spots in the aurora-A/TPX2 interaction using biochemical, biophysical, and fragment-based approaches. *ACS Chemical Biology*, *12*(11), 2906–2914. https://doi.org/10.1021/acschembio.7b00537.

Navratilova, I., & Hopkins, A. L. (2010). Fragment screening by surface plasmon resonance. *ACS Medicinal Chemistry Letters*, *1*(1), 44–48. https://doi.org/10.1021/ml900002k.

Ng, J. T., Dekker, C., Kroemer, M., Osborne, M., & von Delft, F. (2014). Using textons to rank crystallization droplets by the likely presence of crystals. *Acta Crystallographica. Section D, Biological Crystallography*, *70*(Pt. 10), 2702–2718. https://doi.org/10.1107/S1399004714017581.

Patel, D., Bauman, J. D., & Arnold, E. (2014). Advantages of crystallographic fragment screening: Functional and mechanistic insights from a powerful platform for efficient drug discovery. *Progress in Biophysics and Molecular Biology*, *116*(2–3), 92–100. https://doi.org/10.1016/j.pbiomolbio.2014.08.004.

Pearce, N. M., Krojer, T., Bradley, A. R., Collins, P., Nowak, R. P., Talon, R., et al. (2017). A multi-crystal method for extracting obscured crystallographic states from conventionally uninterpretable electron density. *Nature Communications*, *8*, 15123. https://doi.org/10.1038/ncomms15123.

Price, W. N., 2nd, Chen, Y., Handelman, S. K., Neely, H., Manor, P., Karlin, R., et al. (2009). Understanding the physical properties that control protein crystallization by analysis of large-scale experimental data. *Nature Biotechnology*, *27*(1), 51–57. https://doi.org/10.1038/nbt.1514.

Savitsky, P., Bray, J., Cooper, C. D., Marsden, B. D., Mahajan, P., Burgess-Brown, N. A., et al. (2010). High-throughput production of human proteins for crystallization: The SGC experience. *Journal of Structural Biology*, *172*(1), 3–13. https://doi.org/10.1016/j.jsb.2010.06.008.

Shaw Stewart, P. D., Kolek, S. A., Briggs, R. A., Chayen, N. E., & Baldock, P. F. M. (2011). Random microseeding: A theoretical and practical exploration of seed stability and seeding techniques for successful protein crystallization. *Crystal Growth & Design*, *11*(8), 3432–3441. https://doi.org/10.1021/cg2001442.

Warkentin, M., Berejnov, V., Husseini, N. S., & Thorne, R. E. (2006). Hyperquenching for protein cryocrystallography. *Journal of Applied Crystallography*, *39*(6), 805–811. https://doi.org/10.1107/S0021889806037484.

CHAPTER TWELVE

# Hit-to-Lead: Hit Validation and Assessment

**Kirk E. Hevener*, Russell Pesavento[†,‡], JinHong Ren[†], Hyun Lee[†,§,¶], Kiira Ratia[§,¶], Michael E. Johnson[†,¶,1]**

*Department of Pharmaceutical Sciences, University of Tennessee Health Science Center, Memphis, TN, United States
[†]Center for Biomolecular Sciences, University of Illinois at Chicago, Chicago, IL, United States
[‡]College of Dentistry, University of Illinois at Chicago, Chicago, IL, United States
[§]Research Resources Center, University of Illinois at Chicago, Chicago, IL, United States
[¶]Department of Medicinal Chemistry & Pharmacognosy, University of Illinois at Chicago, Chicago, IL, United States
[1]Corresponding author: e-mail address: mjohnson@uic.edu

## Contents



## Abstract

High-throughput screening assays have become nearly ubiquitous in the search for small compounds or peptides that can modulate biological processes for therapeutic purposes. While many assays have become quite robust, with well-established

*Methods in Enzymology*, Volume 610
ISSN 0076-6879
https://doi.org/10.1016/bs.mie.2018.09.022

protocols, the subsequent steps of validating the hits and choosing the best ones to take forward into leads for further chemical development are less established. In this chapter, we describe a variety of approaches, including chemical assessment, the use of various computational approaches, a variety of counter-screens, and "orthogonal" biophysical assays using nuclear magnetic resonance, surface plasmon resonance, isothermal titration calorimetry or thermal shift assays as methods for validating and assessing the quality of hits.

# 1. INTRODUCTION

A pubmed search for "HTS or high-throughput screen" yields more than 5000 papers utilizing some variant of this approach. The use of HTS assays to identify small molecules as modulators of various biological processes has expanded enormously over the last decade or two, going from being the sole province of large pharmaceutical firms with the resources to develop chemical libraries to the commercial availability of relatively low cost and high quality chemical libraries that are now accessible to academic groups and small biotech companies. A number of primary assays are now routine, and many targets are now accessible by novel assays that are often limited only by the ingenuity of the investigator, and there is at least one specialty journal devoted to the publication of novel assays. Some of these HTS strategies are outlined in earlier Chapters in this volume. The expanding identification of chemical compounds or hits from HTS assays has increased the corresponding need for validation and triaging to provide the best leads for going forward in therapeutic development. In this chapter, we set out some of the chemical criteria for triaging HTS hits, a variety of strategies for validating hits, and some discussion of computational and other methods for the hit-to-lead process through structure-based design.

# 2. HIT ASSESSMENT AND PRIORITIZATION

## 2.1 Hit Quality Assessment

The assessment of hit quality following a high-throughput chemical library screen involves a multifactorial analysis of the chemical and physical properties of all compounds meeting the predefined hit identification criteria. Factors to be assessed that may contribute to the quality of screening hits, and hit advancement decisions, include synthetic or chemical tractability; assessment of potential reactivity, toxicity, and assay interference; an assessment of "drug-likeness"; and calculation of various ligand efficiency indices. Lastly, though not strictly an assessment of hit quality, an analysis of primary screening results to identify potential "latent hits" using a method, such as Compound Set Enrichment, is also good practice (Varin et al., 2010). It should be noted that often some or all these factors are considered in the design of modern chemical screening libraries; however, it is good practice to perform a rigorous hit quality assessment, regardless of the perceived quality of the screening library used to generate the hits.

### 2.1.1 Assessment of Synthetic Accessibility

An assessment of the synthetic accessibility, or tractability, of identified hits is important as this factor influences the ability of the medicinal chemists on the drug discovery team to perform a full structure–activity relationship (SAR) expansion and necessary lead optimization work (Baber & Feher, 2004). A key factor that can influence synthetic accessibility is the presence or absence of key features or functional groups that enable the use of specific chemical reactions to resynthesize the hit or synthesize a series of related analogs for SAR development. For example, the presence of a β-hydroxy ketone in the screening hit indicates that an aldol condensation reaction may be employed in the compound's (or compound analog's) synthesis. Fig. 1 shows several representative reactions that are commonly used by synthetic medicinal chemists and shows enabling chemical features to look for in screening hit compounds during hit quality assessment (Roughley & Jordan, 2011). Retrosynthetic methods, which use known reaction databases and commercially available reagent lists, can be employed to estimate the synthetic accessibility of hit compounds based on an analysis of the hit compound's chemical structure. Several methods based on retrosynthetic analysis have been published, including the WODCA system (Gasteiger & Ihlenfeldt, 1996), the SPROUT/CAESA program (Gillet, Myatt, Zsoldos, & Johnson, 1995), and the RASA method (Huang, Li, & Yang, 2011). Disadvantages of retrosynthetic approaches to estimating synthetic accessibility include the reliance on databases of reactions and reagents that must be updated, the amount of time required if a full synthetic path is being generated, and the variability of confounders such as synthetic yield at each step and reagent costs. Another method that has gained in popularity is the estimation of synthetic accessibility based on the chemical complexity of

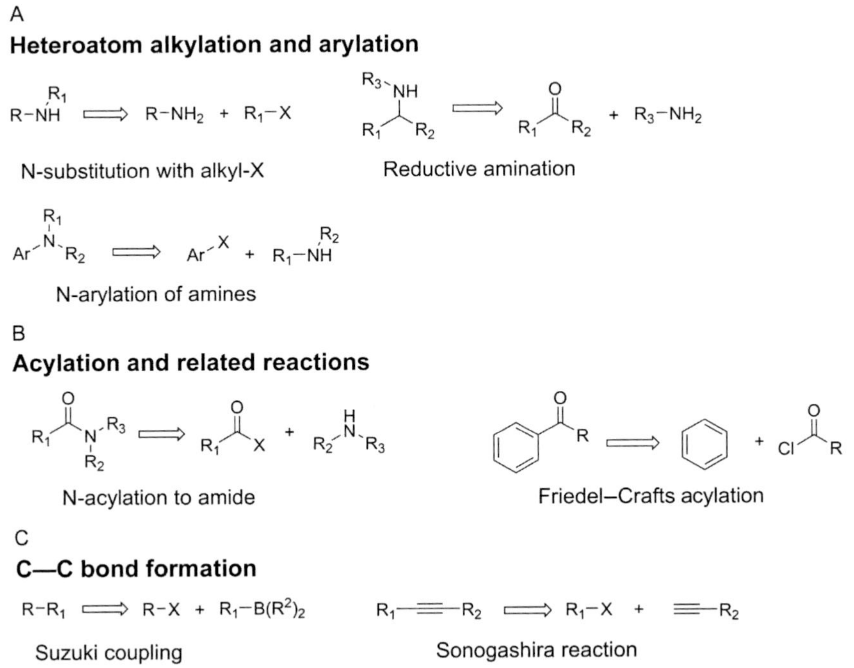


**Fig. 1** Common synthetic reactions used in medicinal chemistry. Shown are representative examples of the most common organic synthetic reactions used in the synthesis of drug candidates, as reported by Roughley and Jordan (2011). Other common reactions not shown include O-substitutions and N-containing heterocycle formations. Retrosynthetic reactions are shown (product on *left*); for full reaction details, readers are referred to the organic chemistry literature.

the hit compound. In this approach, a score or index is calculated based on the complexity of the hit compound, which includes such factors as number of stereocenters, compound size, presence of spiro-rings and nonstandard ring fusions, and number of rotatable bonds. SA-score, a score-based method for estimating the synthetic accessibility of drug-like compounds, is an example of a molecular complexity method (Ertl & Schuffenhauer, 2009). A fairly new approach is the use of machine learning methods trained using known compounds to predict synthetic accessibility using either a retrosynthetic- or complexity-based approach (Podolyan, Walters, & Karypis, 2010). This approach requires a training set of compounds with known synthetic accessibility. Overall, there is a broad variety of software and methods available to assess the ease of synthesis of screening hit compounds and their analogs which can be used to complement (but ideally not replace) the trained eye of an experienced medicinal chemist.

### 2.1.2 Assessment of Reactivity, Toxicity, and Promiscuity

A second important step in assessing the quality of screening hits is careful inspection of the hit compounds for features that may impart reactivity, toxicity, or potential assay interference. The role of reactive functional groups in biological promiscuity, adverse effects, and human toxicity has been described in several reviews of the subject (Bruns & Watson, 2012; Oprea, Bologa, Edwards, Prossnitz, & Sklar, 2005; Rishton, 1997; Williams & Naisbitt, 2002). Examples of chemically reactive functional groups include alkylating and acylating agents, aldehydes and ketones, reactive heterocycles, Michael acceptors, and many others (Rishton, 1997). Examples of known toxicophores include quinones, anilines, nitroaromatic groups, polycyclic aromatic systems, and thiourea compounds (Hakimelahi & Khodarahmi, 2005; Williams & Naisbitt, 2002) (Fig. 2). Early identification of such features is crucial so that compounds possessing them can be removed from consideration in drug discovery campaigns. Several groups have published filtering

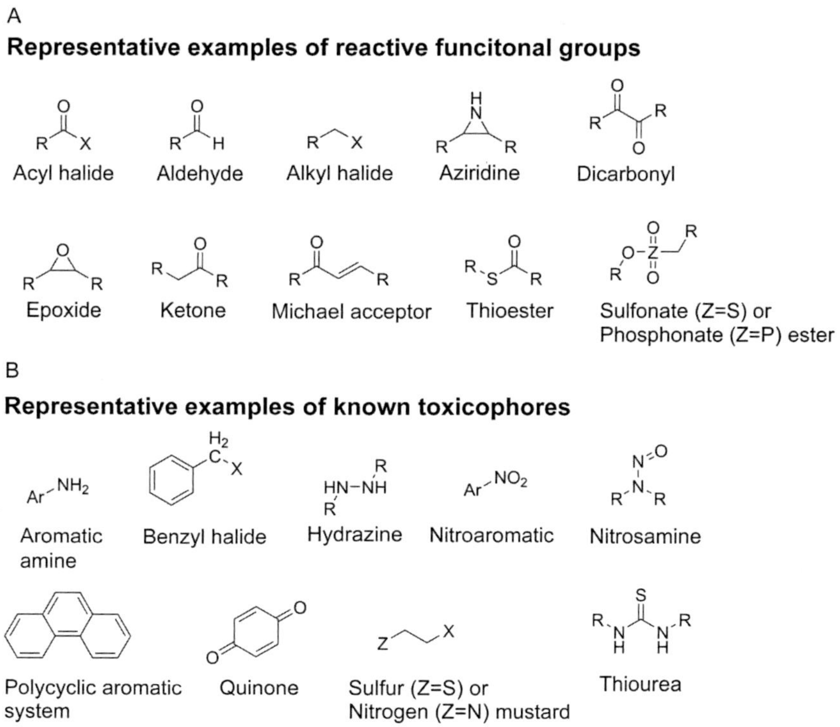


**Fig. 2** Representative examples of known reactive and toxic functional groups. Note that there is some degree of overlap between the two categories as reactivity often leads to toxicity.

rules that can be easily applied computationally to flag compounds with known reactive or toxic functional groups (Bruns & Watson, 2012; Cumming, Davis, Muresan, Haeberlein, & Chen, 2013; Pearce, Sofia, Good, Drexler, & Stock, 2006; Walters, Murcko, & Murcko, 1999). It should be noted that, as above, most modern screening libraries are prefiltered to remove compounds possessing reactive or toxic functional groups. This is generally considered to be the best practice due to the low possibility of successfully advancing a compound possessing these features to the clinic and the high costs associated with pursuing poor quality hits. However, as the application of a computational functional group assessment using predefined SMARTS patterns for reactive groups (or a similar method) is relatively rapid, it is our recommendation that this operation be performed postscreening as well. Lastly, inspection of hit compounds for potential promiscuity or assay interference is important to flag potentially problematic compounds. Screening compounds can interfere with biochemical assays in a variety of ways to produce misleading results, including covalent modification of the target biomolecule, signal interference (e.g., UV absorption or intrinsic fluorescence), formation of colloidal aggregates, and redox activity (Baell & Walters, 2014). The well-known PAINS (pan-assay interference compounds) filters designed by Baell and coworkers can be employed postscreening as a hit quality assessment tool to flag compounds that may prove to be false positives (Baell & Holloway, 2010; Baell & Nissink, 2018). These chemical filters, based on SMARTS patterns of compounds showing activity in multiple orthogonal biochemical assays, have proven utility and are now recommended to be applied and discussed in manuscripts submitted to several journals published by the American Chemical Society (Aldrich et al., 2017). However, as it has been noted recently, flagging a compound as a "PAINS" molecule does not guarantee failure of the compound in the lead optimization process and several marketed drugs possess known PAINS features (Kenny, 2017). For this reason, we recommend the application of PAINS (or similar assay interference) filters after experimental library screening to flag compounds for more careful evaluation and more cautious treatment if chosen to advance to the optimization stage.

### *2.1.3 Assessment of Drug-Likeness*

Another key component of screening hit quality assessment is the assessment of the compound or compounds for "drug-likeness" (Bickerton, Paolini, Besnard, Muresan, & Hopkins, 2012). The concept of drug-likeness generally applies to the physicochemical properties of a compound that lead toward favorable in vivo pharmacokinetic and pharmacodynamic effects, which includes compound absorption, distribution, metabolism, and excretion. A related concept of "fragment-likeness" applies to smaller, less complex molecules typically used in so-called fragment-based drug discovery (Congreve, Carr, Murray, & Jhoti, 2003). These latter concepts are outside the scope of this discussion. Physicochemical properties that can influence PK/PD factors include compound solubility and lipophilicity, the presence and number of polar or ionizable functional groups, compound flexibility, compound size and shape, the presence and type of ring structures, and the number and ratio of heteroatoms to carbon atoms. There are many useful programs and online tools that can be used to assess the drug-likeness of screening hit compounds; some representative examples are provided in Table 1. Proprietary software in several drug discovery modeling packages can be used to predict ADME properties and assess drug-likeness, including

**Table 1** Selected Software and Web-Based Tools for Assessment of Drug-Likeness and Prediction of ADMET Properties

| Software/Tool | Classification | Website or Reference |
|---|---|---|
| ACD/Percepta | Software | http://www.acdlabs.com/products/percepta/predictors.php |
| ADMET Predictor™ | Software | http://www.simulations-plus.com/software/admet-property-prediction-qsar |
| BIOVIA Discovery Studio | Software | http://accelrys.com/products/collaborative-science/biovia-discovery-studio |
| Drug Likeness Tool (DruLiTo) | Software | http://www.niper.gov.in/pi_dev_tools/DruLiToWeb/DruLiTo_index.html |
| Schrödinger/QikProp | Software | https://www.schrodinger.com/qikprop |
| Molecular Discovery/VolSurf+ | Software | http://www.moldiscovery.com/software/vsplus |
| MolSoft—Molecular Property Prediction | Server | http://molsoft.com/mprop |
| Open Molecules—Property Explorer | Server | http://www.openmolecules.org/propertyexplorer/applet.html |
| OSIRIS Property Explorer | Server | http://www.organic-chemistry.org/prog/peo |
| SwissADME | Server | http://www.swissadme.ch |

Schrödinger's QikProp program, the VolSurf+ program from Molecular Discovery, and the ADMET prediction tool in Biovia's Discovery Studio. Web-based tools are also available, such as the SwissADME tool (Daina, Michielin, & Zoete, 2017). In addition to predictive software, several groups have published useful drug-like indices that can be rapidly applied to screening libraries during their preparation or to hit compounds during the assessment stage. The so-called Lipinski Rules (or Lipinski's Rule of 5) was one of the very first published methods to evaluate drug-likeness (Lipinski, 2000). The rules, developed by evaluating a set of known drugs, stated that most orally active agents possessed no more than 5 hydrogen bond donors, 10 hydrogen bond acceptors, molecular weights less than 500 Da, and an octanol–water partition coefficient (log*P*) no greater than 5. Oprea and coworkers published a similar set of drug-like indices which also included rotatable bond counts and ring counts (Oprea, Gottfries, Sherbukhin, Svensson, & Kuhler, 2000). The Veber Rules included indices for rotatable bonds, total polar surface area, and the sum of hydrogen bond donors and acceptors (Veber et al., 2002), while the indices published by Walters included several of the above as well as a term for formal charge (Walters & Murcko, 2002). Some more recent publications took into account the molecular unsaturation and heteroatom proportion (Zheng et al., 2005), and carbon and hydrogen to nonhalogen heavy atom ratios (Chen et al., 2005). Overall, the assessment of a hit compound's drug-likeness is a critical step in the determination of which compounds to advance to the costly stages of lead optimization as this factor can be very predictive of poor pharmacokinetic properties that may contribute to the failure of a drug candidate.

### 2.1.4 Assessment of Ligand Efficiency

Ligand efficiency is a value that expresses the binding energy of a compound normalized by the compound's size, typically expressed as the number of heavy (nonhydrogen) atoms (Hopkins, Groom, & Alex, 2004). This is an important property to consider when assessing the quality of screening hits, as larger compounds tend to show greater binding energy due to the larger number of interactions they make but may not necessarily be the most efficient binders. This influences the likelihood that a lead compound can be further optimized into a drug candidate, a process that often adds further bulk to the compound. Hence, smaller but more efficient binders are often desirable as they have a greater likelihood of successfully advancing through the lead optimization stage. The concept of ligand efficiency led, in part, to the fragment-based drug discovery methods mentioned earlier

(Carr, Congreve, Murray, & Rees, 2005). A common method of calculating ligand efficiency (LE) is to divide the Gibbs free energy of binding ($\Delta G$) by the number of heavy atoms ($N$) as: LE = $(\Delta G)/N$. Using the thermodynamic equation for Gibbs free energy, $\Delta G = -RT\ln K_i$, and substituting $IC_{50}$ for $K_i$ (a less rigorous approximation), ligand efficiency can be calculated as follows: $LE = 1.4(pIC_{50})/N$ (Shultz, 2014). Several variations of the ligand efficiency concept have been reported and are now routinely used in hit quality assessment (Shultz, 2013). The ligand-lipophilicity efficiency (LLE), or lipophilic efficiency (LipE), assesses the binding energy of a compound relative to the compound's lipophilicity as: $LipE = pIC_{50} - LogP$ (Leeson & Springthorpe, 2007). Other ligand efficiency metrics include the percentage efficiency index, size-independent ligand efficiency, the binding efficiency index, and the surface binding efficiency index (Abad-Zapatero & Metz, 2005; Nissink, 2009). While there has been considerable debate in the drug discovery community regarding the strengths and weaknesses of the different ligand efficiency indices, there is a strong consensus that the evaluation of ligand efficiency in some manner is an important component of screening hit quality assessment and a useful contribution to the hit advancement decision-making process (Lowe, 2013).

## 2.2 Ligand-Based Hit Analysis

The hit validation process involves the confirmation of hit activity using secondary assays and confirmation of on-target activity and selectivity using counter-screens and biophysical assays, while hit quality assessment involves the determination of synthetic accessibility, potential reactivity or toxicity, drug-likeness, and ligand efficiency. Complementing the hit validation and quality assessment steps are several best practice techniques and methods that can be used to aid the drug discovery team in the selection of hits to advance to the lead optimization stage. These include diversity analysis and clustering techniques, literature mining and scaffold analysis, various SAR expansion techniques, and early ADMET studies. Collectively, these methods may be considered ligand-based hit analysis. These methods are generally applicable to both experimental and virtual screening hits following activity testing and hit validation (Zhu et al., 2013).

### *2.2.1 Diversity Analysis*

An important step in the identification of hits for advancement is the analysis of the diversity of the validated hit compounds. Depending on the chemical diversity of the experimental screening library, and the number of validated hit compounds, there may be many structurally related hits possessing the same or similar chemical scaffolds. As the number of hits advanced to the lead optimization stage is often dependent on the resources and infrastructure available to the drug discovery team, there is typically a selection process whereby only the most promising hits are advanced to optimization. In this case, it is desirable to advance not only the highest quality hits (see Section 2.1) but also as diverse a selection as possible to ensure a high likelihood of successfully identifying a clinical candidate. This is because when a lead compound fails in preclinical or clinical stages of drug development, there is a higher possibility that structurally related compounds will also fail, an ancillary of the neighborhood behavior concept (Patterson, Cramer, Ferguson, Clark, & Weinberger, 1996). A ligand-based diversity analysis applied to the validated hit compounds is a useful method to aid the team in selecting hit compounds for advancement. There are a large variety of diversity analysis methods that can be applied to hit compounds using one of several open-source or proprietary cheminformatics software packages. A common method of similarity/dissimilarity analysis is the calculation of "2D fingerprints," binary vectors where bits (0 or 1) represent the presence or absence of specific chemical features within the compounds, followed by the calculation of similarity or distance coefficients using these fingerprints, such as the Tanimoto (Jaccard) or Dice (Hodgkin) Coefficients (Willett, 2011). There are many types of 2D fingerprint calculations and a large number of similarity/distance coefficient calculations that can be selected (Sastry, Lowrie, Dixon, & Sherman, 2010). In addition to 2D fingerprints, other types of chemical descriptors can be used to calculate similarity/distance coefficients, including atom-pairs, topological indices, shape indices, continuous descriptors (e.g., cLogP & molar refractivity), and subgraph similarity (Raymond & Willett, 2002). Another useful method is to employ a compound "clustering" approach, where the hit compounds are clustered into bins based upon their chemical similarity. This allows for the selection of a representative compound (or compounds) from each cluster, usually based on the best activity or hit quality. Clusters are typically built based upon similarity indices calculated for each compound using the generated descriptors mentioned earlier. There are a number of different clustering algorithms that can be used, including hierarchical methods, where one can choose the number of clusters to be generated, and nonhierarchical methods, such as Jarvis-Patrick and K-means clustering (Bacha, Brunck, & Delany, 2007; Raymond, Blankley, & Willett, 2003). It should be noted that the performance of a selected clustering method depends, in part, on the choice of descriptors and similarity/distance coefficients used. A clustering approach to diversity analysis and hit selection is a good approach when there is a large number of validated hits resulting from the experimental screen.

### 2.2.2 Literature Mining

Another important component of a rigorous hit analysis protocol is a thorough evaluation of the chemical literature. Validated hit compounds should be routinely checked against the available literature and chemical databases to investigate the novelty of the compound and the scaffold. Checking for any previous reports of biological activity, patents, and adverse or toxic effects is one way of evaluating the intellectual property landscape and informing the process of hit selection for advancement. Many compounds found in screening libraries will be indexed in a variety of chemical databases which often include useful information such as literature associated with the compound, patent reports, physicochemical properties, commercial vendors, and synthetic routes. Chemical databases such as CAS SciFinder, ChEMBL, PubChem, and Reaxys (Beilstein) can be very useful sources of information on identified hit compounds and checking them can potentially save the research team a great deal of time and resources spent pursuing compounds that have already been previously reported or have known liabilities. Table 2 details some common chemical databases and lists some useful features of each. Many of these chemical databases include tools for structure-based scaffold and similarity searching, which should be employed along with the exact structure search to check the literature for reports on related compounds as well as for the SAR expansion studies described next.

### 2.2.3 SAR Expansion Methods

Exploring the SAR of a lead series is a key part of the drug discovery process, often considered to be part of the lead optimization stage. Ultimately, all SAR studies will require the skills of an experienced synthetic medicinal chemist. However, SAR expansion can often be initiated during the hit analysis stage using commercially available compounds and the chemoinformatic resources available in several of the chemical databases mentioned earlier. These early SAR studies can be less resource intensive and may be used as a means of hit compound or hit series analysis. This is demonstrated in our recent studies exploring a novel benzimidazole scaffold with inhibitory activity against the bacterial FabI enzyme (Hevener et al., 2012; Mehboob et al., 2012). A “clean” SAR, with clearly interpretable,

**Table 2** Selection of Useful Chemical Databases and Data Repositories

| Database | Website | Open Access | Chemical Structure Search | Patent Literature Search | Scientific Literature Search | Biological Activity Search | Reaction Search | Commercial Vendor Search |
|---|---|---|---|---|---|---|---|---|
| CAS SciFinder | https://scifinder.cas.org | | X | X | X | X | X | X |
| ChEMBL[a] | https://www.ebi.ac.uk/chembl | X | X | | X | X | | |
| ChemSpider | http://www.chemspider.com/ | X | X | | X | | | X |
| eMolecules | https://www.emolecules.com | X | X | | | | | X |
| PubChem | https://pubchem.ncbi.nlm.nih.gov | X | X | X | X | X | | X |
| Reaxys (Beilstein) | https://www.reaxys.com | | X | X | X | X | X | X |

[a]Bento et al. (2014).

continuous activity changes associated with compound structural modification, is indicative of well-behaved, on-target activity, while erratic SAR or the identification of so-called activity cliffs may be indicative of mechanism hopping (Iyer, Stumpfe, & Bajorath, 2011) or indirect activity (Stumpfe & Bajorath, 2012). While this latter SAR behavior may not necessarily invalidate a hit series, erratic SAR is often predictive of lead optimization difficulties at future stages. The scaffold search tools of chemical databases such as CAS SciFinder and ChEMBL should be used to identify commercially available compounds for ordering and testing as part of early SAR studies. The first step is to perform a "scaffold pruning" study, by iteratively removing functional group substitutions from the core scaffold of the initial hit compound as a means of identifying the basic structural requirements for activity, the so-called pharmacophore. This should be followed by an SAR expansion study to identify commercially available compounds possessing the hit scaffold with varying functional group substitutions that can be ordered and tested for activity. Note that it is critical that all compounds ordered from commercial vendors for in-house testing be rigorously assessed for purity and identity using established methods (e.g., LC-MS & NMR). As much as permitted by the commercial availability of the chemical series, a rational and ordered approach to analog selection, such as the Topliss scheme, should be followed along with the input of an experienced medicinal chemist (Topliss, 1972). If enough compounds within the chemical hit series are available, preliminary quantitative structure–activity relationship (QSAR) models can also be developed which can further inform the hit advancement process (Guha, 2013). The same chemical databases typically possess a similarity search feature as well, which can be utilized to identify structurally similar compounds that may not possess the same scaffold. This feature can be employed as a "scaffold hopping" strategy to identify more attractive or alterative hit series (e.g., improved intellectual property landscape). These SAR expansion strategies should be thoroughly employed for each validated hit compound or series until the pool of commercially available analogs has been exhausted.

### 2.2.4 ADMET Predictions

Drug candidate attrition in the later stages of the development process can be very costly. Because poor pharmacokinetic properties (absorption, distribution, metabolism, and excretion) are a major reason for drug candidate failure in clinical trials, there has been a large focus recently on performing predictive ADME studies early in the drug discovery process

(Ban et al., 2017; Wishart, 2007). The theoretical (in silico) calculation of several key ADME predictors can greatly inform the selection process for hit advancement, is relatively inexpensive to perform, and is an important component of ligand-based hit analysis. Table 1 lists several useful online servers and open-source or proprietary software that can be used to calculate a variety of ADME predictors. Examples include software such as Schrödinger's QikProp program, Biovia's Discovery Studio, and Molecular Discovery's VolSurf+ which can rapidly calculate ADME predictors that can be used to prioritize hit selection. Many computational ADME tools calculate molecular descriptors initially, such as cLogP, H-bond donor & acceptor counts, molecular weight, molar refractivity, and polar surface area, which are then used to predict ADME properties using pretrained models built with training sets of compounds possessing known ADME properties (Cheng et al., 2012; Daina et al., 2017; Gleeson, Hersey, & Hannongbua, 2011; Khan, 2010; Shen, Cheng, Xu, Li, & Tang, 2010; Talevi et al., 2011). The ADME models are often created using regression techniques such as multiple linear regression and partial-least squares like QSAR models and can be considered quantitative structure–property relationship (QSPR) models (Cruciani, Pastor, & Guba, 2000). Some modern applications include models created using machine learning methodology such as neural networks and support vector systems (Maltarollo, Gertrudes, Oliveira, & Honorio, 2015; Tao et al., 2015). There is a large variety of ADME models available in the different software packages. Some useful models include Caco-2 and MDCK cell penetration as useful predictors of drug absorption, plasma protein binding, CYP enzyme susceptibility and metabolic lability, blood–brain barrier penetration, and propensity for binding off-target macromolecules such as P-glycoprotein (drug efflux) and the hERG channel (cardiotoxicity). While informative for the hit selection process, particularly when there is a large number of hits to prioritize, it should be emphasized that these ADME predictions are theoretical and should not supplant more rigorous experimental ADME studies in the later stages of the drug discovery process.

# 3. STRUCTURE-BASED STRATEGIES

## 3.1 Virtual Screening/Molecular Docking as a Complementary Strategy to HTS

Where the target structure is known, virtual screening (VS) is a basic tool often used to supplement and/or complement an HTS effort that can make it more effective in time, resource usage, and cost. With a known target structure, VS can readily identify compounds that have shape complementarity appropriate for binding, usually in the range of 5%–20% of a general chemical library, depending on the relative constraints of the binding site, potentially reducing the number of compounds to be initially screened or to be followed up after the initial screen. Specifically, VS can rapidly predict the ligand-binding pose and estimate the binding affinity by force field based, empirical or knowledge-based methods (Lionta, Spyrou, Vassilatis, & Cournia, 2014). However, VS and experimental HTS each have limited accuracy that can be compensated in part by the number of compounds screened, often resulting in the identification of large numbers of hits. Hits identified by either VS or HTS alone therefore need to be followed up with comprehensive experimental validation and/or more extensive calculations before pursuing the hit-to-lead process. Recently, several approaches have shown that the combination of VS and HTS, pursued either sequentially or in parallel, improves lead generation. Generally, three integration strategies have been developed, including parallel, focused, and sequential screening (Polgar & Keseru, 2011).

### 3.1.1 Parallel Screening

The parallel approach simultaneously applies both VS and HTS to the same screening library. An example of this approach is a parallel VS and HTS of ~200,000 compounds against cruzain by Shoichet and colleagues to identify competitive inhibitors (Ferreira et al., 2010); five inhibitor chemotypes were identified in their study: two chemotypes were prioritized by scoring among the top 0.1% of the VS, another two chemotypes were identified from experimental HTS, and a fifth chemotype was identified by both approaches, as shown schematically in Fig. 3. Moreover, the hits predicted by VS that were also HTS-active yielded well-behaved molecules, relatively unobscured by the false positives to which both techniques are individually prone. Thus, VS is a useful tool that is complementary to HTS and provides orthogonal structural support for the prioritization of interesting screening hits for subsequent follow-up.

### 3.1.2 Focused Screening

VS can filter out compounds that have steric clashes with active site residues or that exhibit polarity properties that clash with the active site electrostatic characteristics, allowing the identification of subsets of a database or library that are of a more manageable size for the subsequent experimental HTS,

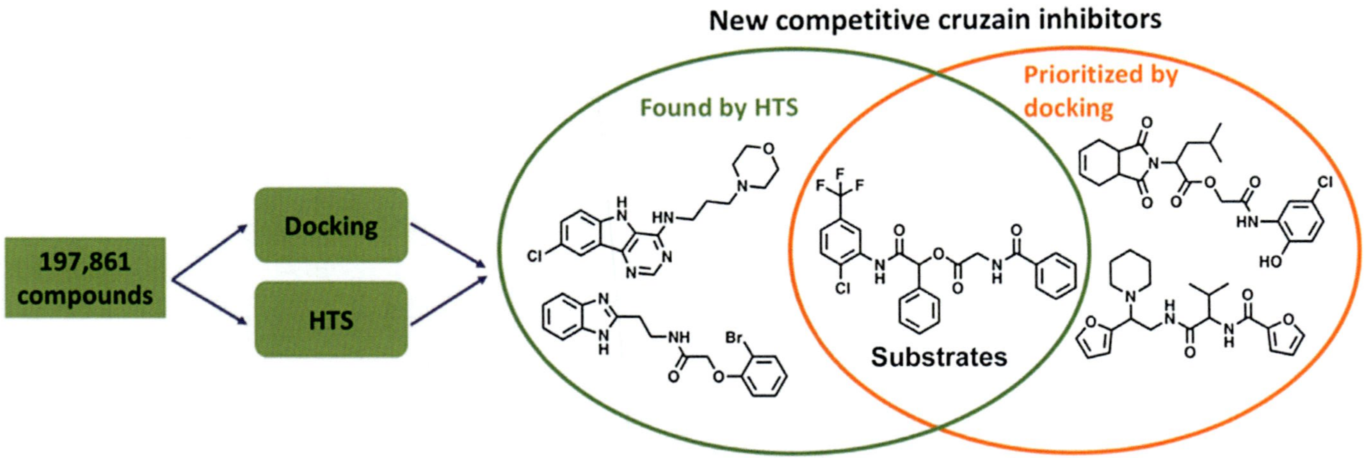


**Fig. 3** Scheme to find new competitive cruzain inhibitors by a parallel strategy of VS and HTS. *This figure is adapted with permission from Ferreira, R. S., Simeonov, A., Jadhav, A., Eidam, O., Mott, B. T., Keiser, M. J., et al. (2010). Complementarity between a docking and a high-throughput screen in discovering new cruzain inhibitors.* Journal of Medicinal Chemistry, *53(13), 4891–4905. https://doi.org/10.1021/jm100488w. © 2010, American Chemical Society.*

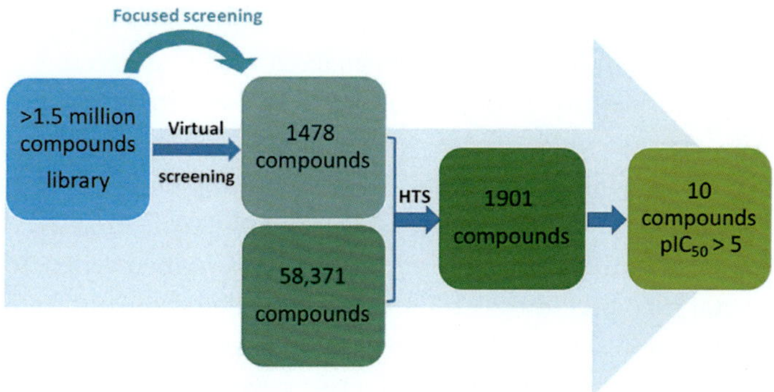


**Fig. 4** Focused screening using VS along with traditional HTS to produce $PI3K_{\gamma}$ inhibitors. *Based on data in Damm-Ganamet, K. L., Bembenek, S. D., Venable, J. W., Castro, G. G., Mangelschots, L., Peeters, D. C., et al. (2016). A prospective virtual screening study: Enriching hit rates and designing focus libraries to find inhibitors of PI3Kdelta and PI3Kgamma.* Journal of Medicinal Chemistry, *59(9), 4302–4313, https://doi.org/10.1021/acs.jmedchem.5b01974.*

resulting in dramatic cost and time savings. Damm-Ganamet et al. report a study using $PI3K_{\gamma}$ and $PI3K_{\delta}$ to discuss the benefits of using VS along with traditional HTS, where they conducted a VS to create a focused library for experimental testing with the goal of identifying a lead compound for the $PI3K_{\gamma}$ campaign shown schematically in Fig. 4 (Damm-Ganamet et al., 2016). Over 1.5 million compounds were screened VS against $PI3K_{\gamma}$, and after various filtering steps, from which 1478 molecules were deemed to have the appropriate size and chemical functionality to test experimentally. Those were then combined with 58,371 compounds from an external kinase focused library and subjected to an HTS assay. Subsequently, 1901 compounds that showed above 50% inhibition in the primary HTS assay at a concentration of $10\,\mu M$ were taken forward for confirmation, resulting in 10 compounds that showed activity at a $pIC_{50}$ of greater than 5. Their studies showed that, in general, the use of VS can lead to small focused libraries exhibiting hit rates that are substantially better than those from an HTS on an entire corporate library potentially containing millions of compounds.

### *3.1.3 Sequential Screening*

Sequential screening or the iterative combination of VS and HTS is an integrated screening strategy. When active molecules are known, similarity search methods can be applied to enrich screening library subsets with molecules having the desired biological activity. If resulting subsets contains hits with acceptable biological activity, the subsequent steps will provide compounds with a better biological activity at a higher probability (Riniker, Wang, Jenkins, & Landrum, 2014; Yao et al., 2017). This iterative approach has the ability to identify analogues at the very early steps and provide an initial SAR. An example of this approach was a campaign to discover small-molecule inhibitors to block ATX–lysophosphatidic acid signaling by modulating autotaxin (ATX), the lysophospholipase D enzyme that produces lysophosphatidic acid (Fells et al., 2013), described schematically in Fig. 5. Several hits, including 918013, that included aromatic sulfonamide derivatives interacting with the hydrophobic pocket were known from a prior HTS for ATX inhibitors. To identify new ATX inhibitors, VS was performed through application of a series of independent two-dimensional similarity searches using three highly potent aromatic sulfonamide inhibitors from prior HTS: 918013, 934313, and 897131 against a library including ~340,000 compounds. The similarity search resulted in 5997 hits based on 918013, 2360 hits based on 934313, and 1754 hits from 897131. Cross-matching of the database results with the three templates yielded 230 overlapping virtual hits. After a series of filters and experimental tests, 16 compounds were identified with $IC_{50}$ values of $<3\ \mu M$ (Fells et al., 2014). Using an SAR examination of the aromatic sulfonamide scaffold, a new and spatially more constrained pharmacophore was developed, and the unique interactions between the aromatic sulfonamide group and residues around the hydrophobic pocket of ATX were identified. Based on subsequent structure-based design, a new ATX inhibitor, 403070, with an

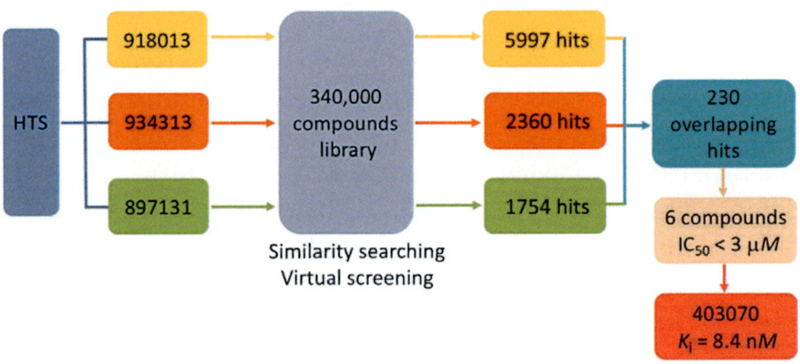


**Fig. 5** Sequential screening using VS along with traditional HTS to find ATX inhibitors. *Based on data in Fells, J. I., Lee, S. C., Fujiwara, Y., Norman, D. D., Lim, K. G., Tsukahara, R., et al. (2013). Hits of a high-throughput screen identify the hydrophobic pocket of autotaxin/lysophospholipase D as an inhibitory surface.* Molecular Pharmacology, 84*(3), 415–424, https://doi.org/10.1124/mol.113.087080.*

inhibition constant ($K_i$) of 8.4 n$M$ and drug-like properties was identified. Docking of 403070 into ATX results in an orientation similar to that of the lead compound, 918013, with the results providing insights for development of further derivatives.

VS is often used in conjunction with experimental HTS, particularly where the initial HTS campaign results in a large number of hits that would be difficult to screen individually in follow-up assays. For example, Fung and colleagues used molecular docking (GOLD) to triage potential PurE inhibitors by docking 525 hits from a high-throughput thermal shift assay (TSA) to the PurE crystal structure (Kim et al., 2015). The docking scores for 75 compounds showed comparable or better predicted scores than known inhibitors. The hits were then filtered down to 60 for subsequent enzyme and MIC testing, which resulted in substantial savings of time and experimental resources. Molecular docking has also been used to validate hits during the last step of candidate hunting. Reithmeier and colleagues recently used HTS to screen a collection of 30,315 compounds toward TRAP 5b, resulting in the identification of 90 primary hits (Reithmeier et al., 2018). After removal of promiscuous compounds, unwanted groups, and false positives, the properties of 52 compounds were further investigated to better understand their mechanism of action, resulting in six prioritized compounds. Structure-based approaches were then implemented through molecular docking to predict the bonding modes of the six compounds with TRAP to further improve the hits toward developing more potent lead inhibitors.

## 3.2 Postprocessing of Molecular Docking: Rescoring by MM-PBSA or MM-GBSA

Although molecular docking has improved significantly in recent years, it is becoming commonly accepted that docking results can be postprocessed with more accurate (and computationally intensive) methods to improve scoring quality (Graves et al., 2008). The Molecular Mechanics Poisson–Boltzmann or Generalized Born Surface Area (MM-PBSA or MM-GBSA) methods are very popular approaches to estimate the binding free energies of small molecules to their biological targets and are commonly used for postprocessing of molecular docking (Virtanen, Niinivehmas, & Pentikainen, 2015). They are typically based on molecular dynamics simulations of the protein–ligand complex and therefore incorporate the flexibilities of both ligands and protein. The solvation effects are also incorporated into the system, which are usually omitted in docking studies. Thus, MM-PBSA or

MM-GBSA is often used as rescoring methods to refine the list of potential hits from high-throughput VS (Genheden & Ryde, 2015). Degliesposti et al. reported the discovery of novel and potent inhibitors of *Plasmodium falciparum* plasmepsin by a protocol designated as BEAR (binding estimation after refinement), using molecular dynamics simulation followed by MM-PBSA and MM-GBSA calculations of binding free energy to refine and rescore the structures obtained from VSs (Degliesposti et al., 2009; Rastelli, Degliesposti, Del Rio, & Sgobba, 2009), with the protocol graphically illustrated in Fig. 6 (Degliesposti, Portioli, Parenti, & Rastelli, 2011). Twenty-six hits from BEAR prediction were experimentally validated to be active as plasmepsin inhibitors, showing the usefulness of the protocol in predicting binding free energies and ranking VS hits. Recently, Hou and colleagues developed a conceptually similar protocol, named MIEC-GBDT (molecular interaction energy components-gradient boosting decision tree) and evaluated its performance in discriminating luciferase inhibitors from noninhibitors by combining binding free energy decomposition and an ensemble learning strategy (Chen et al., 2017). Instead of using MM-GBSA directly to rank binding ability, the electrostatic, van der Waals, polar solvation, and nonpolar solvation (calculated by MM-GBSA decomposition) energy components of for the top 30 highest contributing residues were used to construct the model. The resulting MIEC-GBDT model was used as a powerful tool to identify potential interference compounds in luciferase-based HTS experiments.

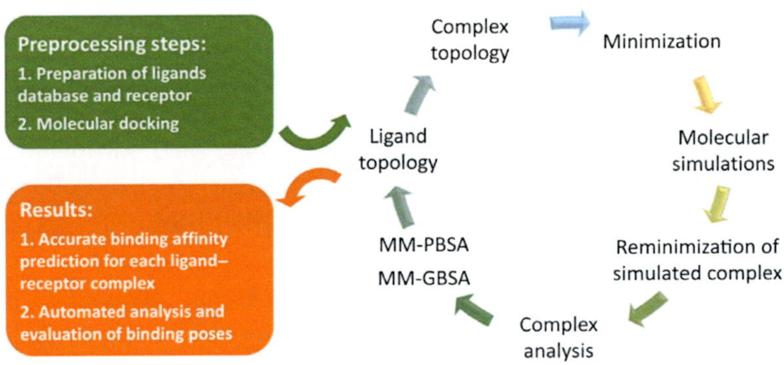


**Fig. 6** BEAR (binding estimation after refinement) work flow. *Based on data in Degliesposti, G., Portioli, C., Parenti, M. D., & Rastelli, G. (2011). BEAR, a novel virtual screening methodology for drug discovery.* Journal of Biomolecular Screening, 16*(1), 129–133, https://doi.org/10.1177/1087057110388276.*

### 3.3 Fragment Molecular Orbital Method to Predict the Binding Affinity

The fragment molecular orbital (FMO) protocol, initially developed by Kitaura and colleagues, enables ab initio quantum mechanical calculations for large biomolecules such as protein–ligand complexes (Kitaura, Ikeo, Asada, Nakano, & Uebayasi, 1999). Traditionally used "visual inspection" and force field-based molecular mechanics calculations for determination of protein–ligand affinity cannot always incorporate the full complexity of molecular interactions, such as CH–π (Ozawa, Okazaki, & Kitaura, 2011), halogen–π (Lu, Zou, Wang, & Yu, 2007), and cation–π (Gallivan & Dougherty, 1999) interactions. Therefore, the FMO method provides a more complete assessment of the interactions formed between the ligand and protein that includes effects that are often omitted by traditional structure-based descriptions mentioned earlier. FMO calculations involve the fragmentation of the system into small manageable fragments, typically amino acid residues and the ligand. The pairwise interaction energies are then calculated to get the total energy of the system. In one example, Heifetz, Ortwine, and colleagues used FMO for HTS hit improvement (Heifetz et al., 2016). They reported that FMO methodology was successfully used as part of a rational structure-based drug design effort to improve the HTS hits of inducible T-cell kinase (ITK) to treat allergic asthma. The FMO approach was used to explore the interactions between ITK and inhibitors to gain a more detailed understanding of the nature and magnitude of the forces governing protein–ligand interactions at specific active-site subsites, and also to proactively apply what was learned to the design of more potent inhibitors. They ultimately obtained ligands with potency in the subnanomolar range by this strategy. They also compared the performance of the FMO with MM-GBSA, with the results indicating that FMO outperformed the MM-GBSA force field-based approach, exhibiting a higher correlation with the experimental results, and demonstrating the value of the more computationally intensive approach.

## 4. NMR SPECTROSCOPY IN HIT SCREENING/VALIDATION

The cost of isolation and purification of the target substrate, combined with extensive instrument time, precludes the use of nuclear magnetic resonance (NMR) as a primary HTS screen for drug development. However, the utility of NMR as a secondary screen of smaller libraries for hit validation is feasible and commonplace. NMR is a sensitive biophysical technique that can provide direct evidence of a target–ligand complex in solution not observed in plate-based spectrophotometric studies. NMR methodology can be used to detect weakly binding ligands ($K_d \cong mM$) while also providing binding location of the substrate on the macromolecular substrate or target. The integration of NMR techniques for screening of smaller libraries and hit validation was first implemented in the late 1990s and has now become widespread (Barile & Pellecchia, 2014; Pellecchia et al., 2008; Sugiki, Furuita, Fujiwara, & Kojima, 2018). The utility of NMR techniques for hit screening/validation typically involves three general methods of analysis and interpretation. Target-based studies focus on chemical perturbations to a known macromolecular substrate (e.g., nucleic acid, protein) identified through mixing of various equivalents of ligands with an isotopically (e.g., $^{15}N$, $^{13}C$) labeled substrate. Ligand-based studies focus on changes to the NMR spectrum of the ligand following introduction of a substoichiometric amount of the target substrate. Third, chemical reaction analysis focuses on perturbations of the NMR spectrum of heteronuclear (or isotopically) labeled substrates following reaction with an enzyme of therapeutic interest (Pellecchia et al., 2008). The utility of NMR methodology to provide screening/hit validation of potential drug candidates is discussed later.

## 4.1 NMR-Assisted FBS With Parallel Techniques/Assays

Fragment-based screening (FBS) and validation was first developed 22 years ago to screen as a method of detecting weakly binding ligands for the design of novel inhibitors (Shuker, Hajduk, Meadows, & Fesik, 1996). FBS utilizes a small library (<15,000 compounds) of low weight molecule ligands (<500 Da) with discrete functional groups that are screened for binding activity to a previously characterized macromolecular substrate (typically either a protein or nucleic acid) (Pellecchia et al., 2008). NMR spectroscopy is sufficiently sensitive to monitor the weak binding affinity that is often found in small molecular fragments. The weak bonding affinity of small molecular fragments often cannot be detected by plate-based spectrophotometric assays most often utilized in HTS. NMR-assisted FBS can be detected via target, ligand, or chemical reaction-based methods. However, ligand-induced chemical shift of the macromolecular substrate NMR signal (referred to as "chemical shift mapping") has emerged as the gold standard in detecting binding affinity as well mapping the binding pocket of a substrate (Pellecchia et al., 2008). FBS is initiated with an initial NMR screen to identify weakly binding fragments to the macromolecule target of interest.

A second screen is carried out utilizing the same or another small molecule library to identify binding affinity to a second, but nearby adjacent binding site. Once small molecule leads/fragments are established for each target site on the substrate of interest, the lead fragments can be tethered or chemically linked to produce novel bidentate ligands with a higher binding affinity than each individual fragment. This method is useful when using small (<30 kDa), water soluble isotopically labeled proteins of therapeutic interest. In most circumstances, the protein–ligand interaction can be detected using isotopically (e.g., $^{15}N$, $^{13}C$) labeled heteronuclear single quantum correlation spectra (HSQC). This technique is known as "chemical shift mapping" and provides a wealth of structural information and binding affinity in solution that cannot be obtained from X-ray crystallography (Pellecchia et al., 2008). FBS utilizing chemical shift mapping to identify novel high affinity, tethered ligands to bind to a macromolecule of therapeutic interest was initially utilized by Fesik and coworkers in 1996. This was the initial "SAR by NMR" approach and is still utilized today to both validate and screen agents for binding activity (Hajduk, Meadows, & Fesik, 1997; Shuker et al., 1996). FBS studies utilizing 2D NMR methods have been extended to a variety of nonprotein targets, including DNA/RNA substrates (Jimenez-Moreno et al., 2016). Target-based NMR analysis is often utilized in parallel with other experimental-based screening techniques (or binding assays) for more confidence. Parallel screening techniques include X-ray crystallography (Upadhyay et al., 2018), isothermal titration calorimetry (ITC) (Upadhyay et al., 2018), as well as surface plasma resonance (SPR) and plate-based spectrophotometric assays (Bresciani et al., 2017). Note that protein–ligand docking studies often accompany NMR-assisted FBS and will be addressed separately in this perspective.

## 4.2 Additional Target-Based NMR Analysis

Since the inception of "SAR by NMR" studies in 1996, several NMR-based methods have been utilized to build upon target-based NMR screening and hit validation (i.e., chemical shift mapping analysis). Pellecchia and coworkers proposed the integration NMR-assisted FBS with combinatorial and positional screening methods utilized in HTS. This method, known as "HTS by NMR," (Wu et al., 2013) has been used to screen/validate a larger library of compounds ($10^5$) that comprise a common scaffold that can be screened as mixtures amenable to target-based NMR studies. Recently, Pellecchia and coworkers have provided a proof of concept study in screening inhibitors via "HTS by NMR" for the development of a series of novel, matrix metalloprotease inhibitors (Baggio, Cerofolini, Fragai, Luchinat, & Pellecchia, 2018). In addition, Shapiro and coworkers developed a method for screening multiple protein targets via heteronuclear 2D NMR against various molecular fragments of a common scaffold. This technique was referred to as "RAMPED UP NMR" (Zartler et al., 2003). The "RAMPED UP NMR" technique adds an additional dimension of chemical selectivity not incorporated into earlier NMR screening/validation studies. To achieve NMR spectral resolution, this technique requires labeling of each protein to be screened with a unique amino acid. A limitation of many screening methods, including FBS, is that it provides no information on potential harmful side reactions of the agents in vivo. An additional NMR-based screening method known as "ALARM" was introduced in 2005 to limit potentially harmful, reactive false positives toward the human La antigen prior to advancing to in vivo studies (Huth et al., 2005).

## 4.3 Ligand-Based NMR Analysis

Ligand-based NMR methodology provides an additional means to screen macromolecular–inhibitor complexes in which the targets are large (>50 kDa) and/or not amenable to isotopic enrichment (or in limited supply) (Schade, 2006). Isotopic enrichment of the ligand is not required and the target substrate is added in a substoichiometric amount reducing the amount of protein/nucleic acid substrate necessary. Ligand-based NMR methodology typically employs 1D ($^1$H) spectra obtained via either saturation transfer difference (STD), transverse relaxation ($T_2$R), or WaterLOGSY techniques (Harner, Mueller, Robbins, & Reily, 2017; Schade, 2006). $^{19}$F NMR studies can also be employed in ligand-based methods of analysis (Jordan et al., 2012; Nagatoishi et al., 2018). Further, the degree of ligand binding can be derived from ligand-based NMR methods. Powers demonstrated ligand-binding affinities may be obtained experimentally via NMR line broadening studies following an initial HTS analysis, and serve as a guide to prioritization (Shortridge, Hage, Harbison, & Powers, 2008). Ligand-based NMR-binding assays are often utilized in parallel with other techniques or binding assays including: SPR (Lei et al., 2016; Nagatoishi et al., 2018; Xiong et al., 2018), spectrophotometric assays (Lei et al., 2016; Xiong et al., 2018), isotopic labeling studies (Xiong et al., 2018), X-ray crystallography (Korepanova et al., 2018), microscale thermophoresis (Coletti et al., 2017), and ITC (Nagatoishi et al., 2018). Ligand-based NMR techniques have been utilized to screen/validate a variety of therapeutically relevant target–inhibitor complexes, including reversible inhibitors of xanthine oxidase (Evenas et al., 2014), beta-lactam potentiation targets in MRSA strains (Hou et al., 2016), and several potential anticancer targets (Korepanova et al., 2018; Tomala et al., 2018; Xiong et al., 2018). As expected, the combination of *ligand- and target-based NMR methodology*, in conjunction with parallel screening techniques or binding assays, is used to yield a more robust hit validation set (Aretz et al., 2017; Chilingaryan et al., 2018; Tomala et al., 2018; Vulpetti et al., 2017).

## 4.4 NMR Validation of Protein–Protein Interaction Modulators

NMR techniques for screening/hit validation of inhibitors or agonists have shown utility in the rapidly developing field of protein–protein (or macromolecular) interactions (PPI). Due to the large size of the target substrates involved, plate-based spectrophotometric methods utilized in HTS may produce several false negatives. Again, the sensitivity of NMR-based methodology has proven to be useful in screening small molecules or peptide-based agents that reduce (or stabilize) protein–protein interactions perturbing the complex for therapeutic benefit (Barile & Pellecchia, 2014; Wu et al., 2015). Screening and validation of PPI inhibitors or agonists typically utilize target-based NMR methods, including both 1D $^{1}$H-*aliph* and 2D [$^{1}$H-$^{15}$N/$^{13}$C] therapeutically relevant macromolecular targets (Wu et al., 2015). It is important to note NMR methods of PPI hit validation are often accompanied by other screening techniques and binding assays, such as SPR (Fry et al., 2013), X-ray crystallography (Dias et al., 2014; Fry et al., 2013), and ITC (Dias et al., 2014) for more definitive hit validation. The scope of NMR utility validating PPIs has widened in recent years. Shekhtman and coworkers published a method for utilizing $^{15}$N-labeled protein in screening a small peptide library for binding affinity to the FKBP-FRB complex *in a yeast cell line* (*Saccharomyces cerevisiae*) (Xie, Thapa, Reverdatto, Burz, & Shekhtman, 2009). This technique utilizes a small molecule interactor library (<300 peptides), known as "SMILI," in a cell-NMR-based screening assay to screen for agents that facilitate heterodimerization of a complex relevant to human immunosuppression. Pellecchia reported utilizing the "HTS by NMR" methodology for the screening/design of a novel peptide sequence for interrupting the docking of a bacterial (*Yersinia*) protein YopH-NT to its cellular target (Bottini et al., 2016). In addition, PPI screening/validation has been employed to screen for agents that inhibit antigen–RNA interactions. Utilizing a combination of both ligand-based NMR methods ($^{1}$H-STD) and chemical shift mapping ($^{1}$H-N$^{15}$-HSQC) (Wang, Bhattacharya, & Ivanov, 2015), Wang provided hit validation for agents that disrupt Hur/RNA complex interactions, which have been hypothesized to have utility as anticancer agents. A similar strategy, employing both ligand- and target-based NMR methodology as screening aids, was utilized by Craik et al. to screen for antagonists of herpes virus dimeric proteases in the potential treatment of Kaposi's sarcoma (Gable et al., 2016), as well as by Jahnke and Ottmann to screen for potential agonists in the treatment of cystic fibrosis (Sijbesma et al., 2017).

### 4.5 Chemical Reactivity-Based NMR Methodology

NMR methodology utilizing substrate–enzyme chemical reactivity as the benchmark parameter can also be utilized to screen and validate hits from HTS (Pellecchia et al., 2008), the principle being that enzymatic activity on a substrate "mimic" will induce unique perturbations observable (and identifiable) by NMR methods. This technique is applicable to several classes of enzymes, including proteases, phosphatases, lipases, and others, provided that suitable substrate mimics exist (Dalvit et al., 2003). Both 1D and 2D ligand-based NMR methodology may be utilized to monitor the enzymatic reaction with the isotopically enriched ligand. Selectively incorporating NMR active nuclei into the ligand substrate facilitates both the ease of NMR detection and spectral interpretation. This typically involves the incorporation of nuclei (or functional groups) that exhibit large isotropic chemical shifts sensitive to small changes in the nearby chemical environment (Dalvit et al., 2003; Manzenrieder, Frank, & Kessler, 2008).

### 4.6 Summary

In summary, NMR methodology is widely utilized in hit validation/screening as it provides a sensitive biophysical probe for monitoring, quantifying, and mapping the target–ligand complex in the *solution state*. NMR screening/validation is particularly important in identifying weakly binding inhibitors and/or agonists not detectable by plate-based spectrophotometric assays. Most often, NMR-assisted screening and validation is accompanied by parallel techniques or binding assays.

## 5. HIT-TO-LEAD: ASSESSMENT, PRIORITIZATION, AND IMPROVEMENT OF DRUG-LIKE PROPERTIES

### 5.1 Hit Validation

The experimental postscreen hit validation process typically consists of a suite of assays designed to eliminate false positives, to confirm activity with the intended target, and to establish an initial ranking of compounds by activity. As each HTS assay and target is unique, so is the follow-up strategy, and a systematic method of hit validation, including well-optimized orthogonal/secondary assays, should ideally be in place before the primary screen is completed.

It should be noted that many of the steps outlined below are not confirmatory, but due to their high-throughput nature, can triage a large number of false positives from progressing to more labor-intensive or costly confirmatory assays. True hit *confirmation* is established by verifying target engagement by the hit in an assay that is "different in biological context and process" from the primary assay (NIH, 2017).

### *5.1.1 Retesting of Hits in the Primary Assay*

Automated screens, though rapid and efficient, are not free from error, and a large percentage of initial hits are not reproducible in a simple retest in the primary assay. This first step, reproducing the HTS assay results with hit compounds, serves to filter out "nonhits" arising from technical failures. A common source of error during HTS is the liquid handling component, which must be optimized and carefully calibrated for each solution, aspiration and dispense volume, labware, and tip type. Systematic pipetting errors are more easily identifiable through statistical analysis than the smaller, more random errors that lead to false positives, but these nonhits are easily removed from consideration upon retest (Shun, Lazo, Sharlow, & Johnston, 2011). Another source of error is compound cross-contamination, which occurs when an active compound contaminates a well containing an inactive compound. For this reason, the primary screen retest is typically performed with compounds that have been cherry-picked from untouched, in-house reserves of the library samples. Depending on the number of initial hits, the compounds can be tested at a single concentration, typically the concentration used in the screen, or they can be assayed at multiple concentrations, generating a concentration response curve.

The number of false positives identified at this stage is dependent on the hit cutoff criteria and the noise associated with the assay data (Zhang, Wu, & Sills, 2005). Assays with low $Z'$-factors will generate more false positives, as will relaxed hit cutoff criteria. Conversely, HTS campaigns performed in duplicate or at multiple compound concentrations, known as quantitative HTS, produce a significantly lower false-positive hit rate (Inglese et al., 2006).

### 5.1.2 Counter-Assays

Once hits have been confirmed in the primary assay, additional false positives can be weeded out with a counter-screen, which tests the effect of a compound on the assay technology or format. In general, counter-screens are performed in the absence of the target protein and thus are sometimes referred to as "target-minus" assays (Strovel et al., 2004; Thorne, Auld, & Inglese, 2010). As an example, let us take the case where HTS is performed with a coupled enzyme assay to detect the ATP generated by the target enzyme pyruvate kinase (PK). When luciferase is coupled to the PK reaction, luciferase produces a bioluminescent signal, in the presence of its substrate luciferin, which is proportional to the ATP produced by the PK reaction (Inglese et al., 2006). Bioluminescence-based assays are highly utilized in HTS in many formats (Fan & Wood, 2007), due to superior signal-to-noise ratios, but they are prone to generating false positives by compounds that inhibit the luciferase enzyme. An appropriate counter-assay for the luciferase-based PK HTS would be to test active compounds against just the luciferase reaction. This counter-assay should mimic the primary assay as closely as possible, where all primary assay components are present except for PK. ATP should be added to a concentration that is consistent with the amount generated by PK in the timeframe of the assay. Compounds showing activity in the counter-screen are nonhits and should be eliminated from further analysis.

Though not necessarily a target-minus assay, a frequent counter-screen for cell-based HTS is to measure compound cytotoxicity, a common source of false positives. This step can be multiplexed into the primary screen, run alongside the screen, or tested postscreen. If selectivity across a family of related proteins is critical, isoforms of the target protein can be counter-screened in analogous assay formats, eliminating nonselective compounds and assay-interfering compounds in one step. For primary screens conducted with commercially available reagents, most kit manufacturers sell assay components designed to identify false positives specific to their assay technology. For example, Perkin Elmer's AlphaScreen can be counter-assayed using a peptide (biotinylated-His) that links donor and acceptor beads, mimicking a protein–protein interaction.

### 5.1.3 Orthogonal Assays

Orthogonal assays test the activity of hits against the target protein in an assay with a different output from the primary assay (Thorne et al., 2010). These assays are useful for eliminating compounds that interfere with the optical properties of an assay's readout. Because a significant number of compounds in screening collections fluoresce or absorb in the wavelength ranges typical of assay sensors, the assay response to a compound can be artificially increased or decreased, leading to false positives (Inglese et al., 2007). A potential orthogonal assay for a fluorescence-based protease assay would be an HPLC-based method to detect protease cleave products. Because biophysical techniques are largely insensitive to the optical properties of a compound, they are commonly employed as orthogonal methods to confirm a direct interaction between a hit compound and target protein (see Section 5.2).

### 5.1.4 Secondary Assays

While counter- and orthogonal assays can eliminate a large number of false positives, they cannot always eliminate compounds that nonspecifically interact with the protein of interest. Work pioneered by the Shoichet lab identified a sizable class of promiscuous compounds that aggregate in solution and lead to target inhibition (McGovern, Caselli, Grigorieff, & Shoichet, 2002; McGovern, Helfand, Feng, & Shoichet, 2003). The group's recommendation to include a mild detergent, such as Triton-X 100, in biochemical HTS buffers is widely practiced. The point here is that compounds can be nonspecific, by many mechanisms (Aldrich et al., 2017), and it is often extremely difficult to differentiate a true hit from an artifact with just one type of assay. Therefore, hits should be validated in a different context than the primary assay. For biochemical primary assays, a relevant cell-based assay is standard, and for cell-based assays with an established target, biochemical assays should be employed to verify target engagement by the hit compound.

An important aspect of hit follow-up includes confirming the molecular identity of the hit compound, especially when commercially available compound collections are screened. At the minimum, compounds should be repurchased from the original source and analyzed for purity and identity, and ideally, compounds should be resynthesized in-house before embarking on costly or time-consuming secondary assays.

### 5.1.5 Hit Validation Example: JHDM1A Post-HTS Workflow

The following example nicely illustrates the steps involved in hit follow-up from a relatively small screen for histone demethylase inhibitors. To carry out HTS against Jumoji C domain-containing histone demethylase 1A (JHDM1A), Wang et al. designed a fluorescence polarization-based primary assay (Wang, Marholz, & Wang, 2015). The group screened 14,400 compounds in duplicate to identify compounds that displaced a fluorescently labeled probe from the active site of JHDM1A. 258 hit compounds were counter-screened in a fluorescence polarization (FP) assay lacking the JHDM1A protein to identify false positives that interfered with the FP signal. Of the 258 hits tested, 100 compounds showed negligible assay interference and were selected for confirmation in the primary assay at four concentrations (20, 10, 5, 2.5 μ*M*). Following this retest/dose–response, 34 compounds advanced to a cell-based immunofluorescence secondary assay which measured the effect of compounds on JHDM1A-mediated histone demethylation. Five compounds were shown to induce H3K36me2 hypermethylation, the expected outcome of JHDM1A inhibition. These five hits were analyzed by high-performance liquid chromatography/mass spectrometry (HPLC/MS) for purity and molecular mass. One compound's molecular mass could not be confirmed by HPLC/MS and was therefore removed from further consideration. The remaining four compounds were repurchased from the supplier and then analyzed for selectivity for JHDM1A over JMJD2A, a related demethylase.

## 5.2 Orthogonal Biophysical Analyses for Hit Confirmation and Validation

In many cases, direct binding analysis using independent methods is necessary to confirm that selected hit compounds really interact with the target protein of interest because certain types of hits can simply be assay specific. Biophysical technologies became useful tools in drug discovery more than 25 years ago. There are several biophysical methods that can be used for HTS hit confirmation and validation, including SPR, ITC, TSA, NMR (discussed in Section 4).

### 5.2.1 *Surface Plasmon Resonance*

SPR uses polarized light to measure a change in angle that is proportional to a change in refractive index of a metal surface and its vicinity upon the binding of any partners (Besenicar, Macek, Lakey, & Anderluh, 2006; Cooper, 2004; Phillips & Cheng, 2007; Piliarik, Vaisocherova, & Homola, 2009). SPR can monitor the real-time biomolecule interactions such as protein–protein, protein–peptide, protein–DNA/RNA, and protein compound. It can provide quantitative information on three parameters, binding affinity ($K_D$) at equilibrium, association rate ($k_a$), and dissociation rate ($k_d$). SPR is a powerful technique that can be used in every stage of the drug discovery process (Navratilova & Hopkins, 2010; Patching, 2014), including medium-throughput automated primary compound screening, secondary hit confirmation and validation, hit characterization via kinetics, and mechanism of action by competition SPR. Detection limits are instrument dependent. The Biacore T200 instrument discussed in this chapter as an example can detect from large protein complexes (>100 kDa) to very small fragment compounds (<100 Da). SPR can accurately determine binding affinity ranging from p*M* to ~2 m*M*. SPR can be used as a secondary screen when the primary screen is a fluorescence-based enzymatic assay (Lee et al., 2015, 2017) for instance. Once the number of hits is narrowed down to 50 or less after primary HTS and target-minus assays, $K_D$ values of the hits can be determined by SPR following the procedures described later. Usually results will be categorized into three patterns, specific binding, nonspecific binding, and no binding (Fig. 7). Nonspecific binders and nonbinders can be eliminated, and hits with $K_D$ values below 20 μ*M* can be usually considered as real candidates for further analysis, although the cutoff values can vary depending on what the goal is and also the number of hits one would like to follow up. The following is a general protocol for SPR measurement.

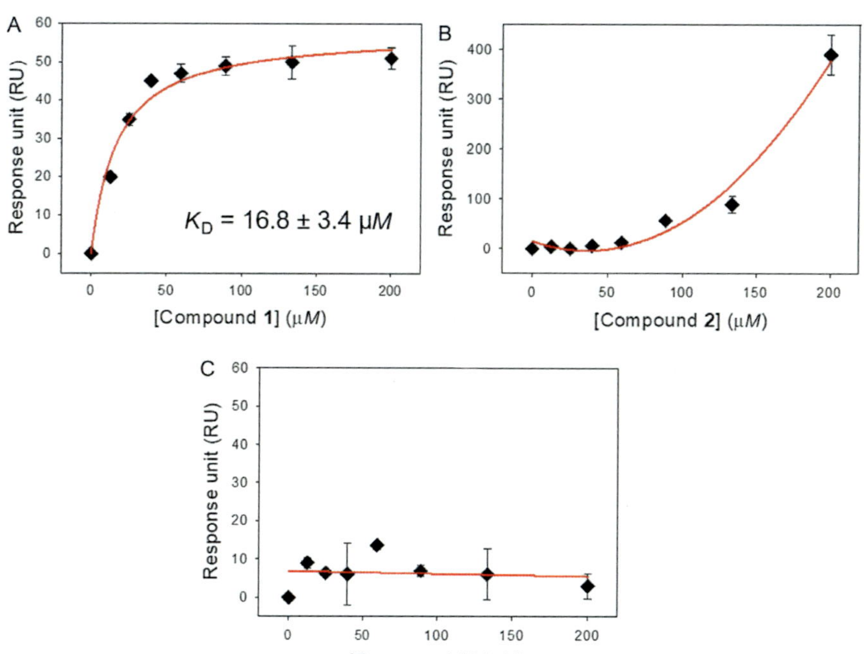


**Fig. 7** Three potential binding patterns from SPR. (A) A steady-state affinity fitting curve of a specific binding compound. (B) Nonspecific binding pattern of a compound. (C) An example of no binding pattern.

#### 5.2.1.1 Equipment

- GE Healthcare Biacore T200

#### 5.2.1.2 Buffers and Reagents

- Purified enzymes or protein samples
- PBS-P (10 m$M$ phosphate, pH 7.4, 2.7 m$M$ KCl, 137 m$M$ NaCl, and 0.05% surfactant P-20), PBS-EP (PBS-P + 3 m$M$ EDTA), HBS-P (10 m$M$ HEPES, pH 7.4, 150 m$M$ NaCl, and 0.05% surfactant P-20), or HBS-EP (HBS-P + 3 m$M$ EDTA)
- Hit compound stocks in 100% DMSO
- CM5 sensor chip (or SA sensor chip if protein is biotinylated)

#### 5.2.1.3 Procedures

1. Prepare proteins at 25–50 μg/mL in 10 m$M$ sodium acetate at a pH lower than the theoretical pI of the protein. Immobilize the target protein on CM5 chip to 10,000–15,000 RU depending on the protein size.
2. Equilibrate and clean the immobilized sensor surface by injecting binding buffer containing 2% DMSO for at least 2 h at 30 μL/min. The response unit difference during the last 1 min should be lower than 1 RU.
3. Prepare 50 × of testing compound serial dilution solutions (0–100 μ$M$ at twofold dilution) with 100% DMSO in a 384-well PCR plate. Dispense 98 μL of binding buffer without DMSO into a 384-well plate. Add 2 μL each of 50 × serial diluted compound solutions to each well and mix them thoroughly by either pipetting or shaking the plate.
4. Inject a series of increasing concentrations of selected hit compounds with 45–60 s of association and 30–45 s of dissociation times at 30 μL/min flow rate at 25°C.
5. Analyze the data using Biacore T200 evaluation software v3. Double subtraction by both a reference channel and zero concentration signals should be done first and fit the data by either steady-state affinity or a kinetic model such as a 1 to 1 Langmuir model.

#### 5.2.1.4 Notes

1. Higher immobilization levels are recommended for larger sized proteins to increase the total number of immobilized protein molecules, which will result in a higher response for detecting compounds.
2. It is important to make sure there is no precipitation in compound solutions. Some inhibitors may have poor solubility in an aqueous buffer, especially at high concentrations. It is not recommended to inject precipitated high concentrations to SPR instruments.

3. DMSO is a solvent with a very high refractive index, and only 1% DMSO shows ~1200 RU. Therefore, it is very critical to have exactly the same concentration of DMSO in all testing samples and binding buffer. The SPR-binding buffer can contain 2%–5% DMSO depending on the tolerance of each protein of interest.
4. Solvent correction is necessary to make sure DMSO signal correction is exact. Follow the manufacture's instruction for solvent correction.

### *5.2.2 Isothermal Titration Calorimetry*

ITC measures the heat changes upon the binding of two or more molecules in solution. ITC can provide the association constant ($K_a$), reaction stoichiometry ($n$), the heat capacity ($\Delta Cp$) of the reaction, binding free energy ($\Delta G$), entropy ($\Delta S$), and enthalpy ($\Delta H$) (Fisher & Singh, 1995; Leavitt & Freire, 2001; Pierce, Raman, & Nall, 1999). Unlike SPR, which requires immobilization of one of the binding partners on a sensor surface, neither immobilization nor modification of the molecules is necessary for ITC. In addition, the ITC technique is not limited by the protein or ligand size and not subject to interference by any potential optical properties of the samples, making it a valuable technique in drug discovery (Holdgate, Meek, & Grimley, 2018). However, ITC requires relatively high concentrations of macromolecule samples. ITC can accurately determine binding affinities ranging from low n*M* to ~5 m*M*. Similar to SPR, ITC can be used not only as a secondary hit confirmation method (Linkuviene, Krainer, Chen, & Matulis, 2016) but also as a tertiary method for detailed hit characterization (Mashalidis, Sledz, Lang, & Abell, 2013).

In ITC experiments, all samples such as proteins and testing compounds must be dissolved in exactly the same buffer composition (buffer, pH, salt concentration, etc.)., The pH, especially, should not differ by more than 0.05 pH units. Degassing all testing samples is necessary before loading them into the ITC instrument. Unstable reducing agents such as DTT should be avoided in the binding buffer, and either TCEP or β-mercaptoethanol is recommended if the reaction requires a reducing agent. The concentration of proteins required for ITC analysis is typically higher than 5 μ*M*, and titrating compound concentrations can vary up to 10–15 times the protein concentration. A standard ITC experiment should have two titrations, one titrating into the binding buffer only (control) and the other titrating into the protein solution.

### *5.2.3 Thermal Shift Assay*

Fluorescence TSA is another method that can be used to investigate specific binding of a compound to a target protein. The TSA technique uses an energetic coupling of protein unfolding reaction and the ligand binding (Lo et al., 2004; Pantoliano et al., 2001). The melting temperature ($T_m$) of a protein can be increased when small molecules bind to the protein and stabilizes them, and hence $T_m$ values can be monitored in the absence and in the presence of a compound. The binding affinity ($K_d$) can be estimated from titration with a series of increasing compound concentrations, similar to SPR and ITC. Due to its high-throughput capability, TSA is being used as a primary HTS screen as well as a secondary confirmation method (Sorrell, Greenwood, Birchall, & Chen, 2010). There are many useful references for TSA, one of which is a Nature protocol containing very detailed information on how to optimize TSA and analyze the data (Niesen, Berglund, & Vedadi, 2007). A $T_m$ shift of a protein in the presence of a known inhibitor is shown in Fig. 8A as an example. Ideally, $T_m$ changes ($\Delta T_m$) as compound concentration increases can be fitted to determine $K_d$ values (Fig. 8B), and cutoff of $K_d$ values can be varied.

#### 5.2.3.1 Equipment

- Applied Biosystem ViiA7 RT-PCR

#### 5.2.3.2 Buffers and Reagents

- 1–5 μ*M* protein
- A series of increasing concentrations of testing compounds
- SYPRO Orange dye (5000 ×)
- White or opaque 96-well or 384-well PCR plates
- Binding buffer (PBS, HEPES, Tris, or any other buffer)

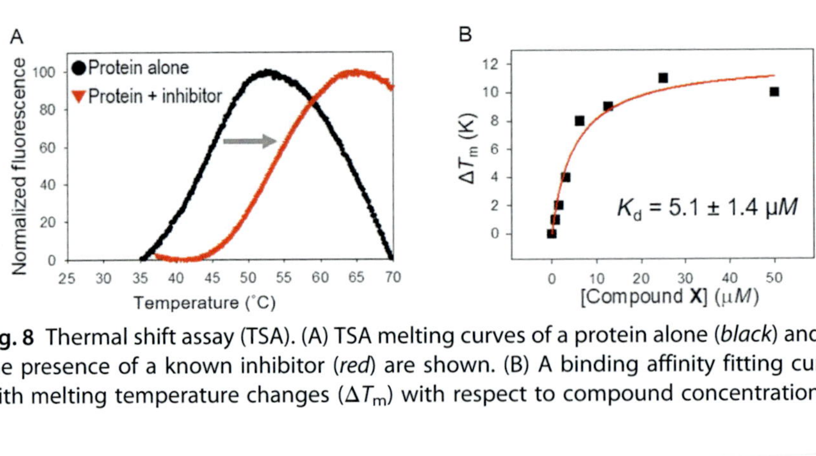


**Fig. 8** Thermal shift assay (TSA). (A) TSA melting curves of a protein alone (*black*) and in the presence of a known inhibitor (*red*) are shown. (B) A binding affinity fitting curve with melting temperature changes ($\Delta T_m$) with respect to compound concentrations.

#### 5.2.3.3 Procedures

1. Prepare proteins at 5 μ*M* concentration in several different buffers to find out which buffer shows the best melting curve. Dispense proteins into 96-well (10 μL each) or 384-well (5 μL each) PCR plates.
2. Add 10 μL each (96-well) or 5 μL each (384-well) of 10 × SYPRO Orange dye and mix the plate thoroughly.
3. Seal the plate with a heat resistant cover and monitor melting curves using an Applied Biosystem ViiA7 RT-PCR instrument at excitation/emission wavelengths of 475/590 nm. The melting curve is typically monitored from 25°C to 95°C with 0.07°C/s temperature increment.
4. Once the appropriate buffer is selected, optimize protein and SYPRO Orange dye concentrations again using the optimized buffer.
5. Prepare an assay plate with a series of increasing concentrations of compounds in a 384-well PCR plate.
6. Monitor melting curves from 25°C to 95°C with 0.07°C/s temperature increment.
7. Analyze the data using Protein Thermal Shift software and calculate $T_m$ values. Calculate $K_d$ values by fitting $T_m$ values and compound concentration.

#### 5.2.3.4 Notes

1. It is important to test a known inhibitor before testing new compounds if possible.
2. Each plate should have enough positive controls and negative controls in case there are many hit compounds.

## 6. CONCLUSIONS

As can be seen from the variety of approaches described earlier, the hit-to-lead process can involve a variety of approaches. The first goal following an HTS campaign is to ensure that the resulting hits are "real," i.e., that they are actually interacting with the target of interest, and that the interaction is specific to the biological process for which modulation is desired. This validation involves a number of steps, including chemical assessment, counter-screens, and orthogonal assays before going on to lead expansion, as described in Sections 2–5. One additional aspect of hit-to-lead development, not described here, is testing compounds in cell-based assays to ensure maximum target specificity, and minimizing off-target effects. Thus, even after a lead is selected, the process of validating the subsequent chemical expansion will continue, as described elsewhere in this volume.

## REFERENCES

Abad-Zapatero, C., & Metz, J. T. (2005). Ligand efficiency indices as guideposts for drug discovery. *Drug Discovery Today*, *10*(7), 464–469. https://doi.org/10.1016/S1359-6446(05)03386-6.

Aldrich, C., Bertozzi, C., Georg, G. I., Kiessling, L., Lindsley, C., Liotta, D., et al. (2017). The ecstasy and agony of assay interference compounds. *Journal of Medicinal Chemistry*, *60*(6), 2165–2168. https://doi.org/10.1021/acs.jmedchem.7b00229.

Aretz, J., Baukmann, H., Shanina, E., Hanske, J., Wawrzinek, R., Zapol'skii, V. A., et al. (2017). Identification of multiple druggable secondary sites by fragment screening against DC-SIGN. *Angewandte Chemie (International Ed. in English)*, *56*(25), 7292–7296. https://doi.org/10.1002/anie.201701943.

Baber, J. C., & Feher, M. (2004). Predicting synthetic accessibility: Application in drug discovery and development. *Mini Reviews in Medicinal Chemistry*, *4*(6), 681–692.

Bacha, P., Brunck, T., & Delany, J. (2007). *Clustering methods and their uses*. Santa Fe, NM. Retrieved from www.daylight.com/cheminformatics/whitepapers/ClusteringWhitePaper.pdf.

Baell, J. B., & Holloway, G. A. (2010). New substructure filters for removal of pan assay interference compounds (PAINS) from screening libraries and for their exclusion in bioassays. *Journal of Medicinal Chemistry*, *53*(7), 2719–2740. https://doi.org/10.1021/jm901137j.

Baell, J. B., & Nissink, J. W. M. (2018). Seven year itch: Pan-assay interference compounds (PAINS) in 2017-utility and limitations. *ACS Chemical Biology*, *13*(1), 36–44. https://doi.org/10.1021/acschembio.7b00903.

Baell, J., & Walters, M. A. (2014). Chemistry: Chemical con artists foil drug discovery. *Nature*, *513*(7519), 481–483. https://doi.org/10.1038/513481a.

Baggio, C., Cerofolini, L., Fragai, M., Luchinat, C., & Pellecchia, M. (2018). HTS by NMR for the identification of potent and selective inhibitors of metalloenzymes. *ACS Medicinal Chemistry Letters*, *9*(2), 137–142. https://doi.org/10.1021/acsmedchemlett.7b00483.

Ban, F., Dalal, K., Li, H., LeBlanc, E., Rennie, P. S., & Cherkasov, A. (2017). Best practices of computer-aided drug discovery: Lessons learned from the development of a preclinical candidate for prostate cancer with a new mechanism of action. *Journal of Chemical Information and Modeling*, *57*(5), 1018–1028. https://doi.org/10.1021/acs.jcim.7b00137.

Barile, E., & Pellecchia, M. (2014). NMR-based approaches for the identification and optimization of inhibitors of protein-protein interactions. *Chemical Reviews*, *114*(9), 4749–4763. https://doi.org/10.1021/cr500043b.

Bento, A. P., Gaulton, A., Hersey, A., Bellis, L. J., Chambers, J., Davies, M., et al. (2014). The ChEMBL bioactivity database: An update. *Nucleic Acids Research*, *42*(Database issue), D1083–D1090. https://doi.org/10.1093/nar/gkt1031.

Besenicar, M., Macek, P., Lakey, J. H., & Anderluh, G. (2006). Surface plasmon resonance in protein-membrane interactions. *Chemistry and Physics of Lipids*, *141*(1–2), 169–178. https://doi.org/10.1016/j.chemphyslip.2006.02.010.

Bickerton, G. R., Paolini, G. V., Besnard, J., Muresan, S., & Hopkins, A. L. (2012). Quantifying the chemical beauty of drugs. *Nature Chemistry*, *4*(2), 90–98. https://doi.org/10.1038/nchem.1243.

Bottini, A., Wu, B., Barile, E., De, S. K., Leone, M., & Pellecchia, M. (2016). High-throughput screening (HTS) by NMR guided identification of novel agents targeting the protein docking domain of YopH. *ChemMedChem*, *11*(8), 919–927. https://doi.org/10.1002/cmdc.201500441.

Bresciani, A., Missineo, A., Gallo, M., Cerretani, M., Fezzardi, P., Tomei, L., et al. (2017). Nuclear factor (erythroid-derived 2)-like 2 (NRF2) drug discovery: Biochemical toolbox to develop NRF2 activators by reversible binding of Kelch-like ECH-associated protein 1 (KEAP1). *Archives of Biochemistry and Biophysics*, *631*, 31–41. https://doi.org/10.1016/j.abb.2017.08.003.

Bruns, R. F., & Watson, I. A. (2012). Rules for identifying potentially reactive or promiscuous compounds. *Journal of Medicinal Chemistry*, *55*(22), 9763–9772. https://doi.org/10.1021/jm301008n.

Carr, R. A., Congreve, M., Murray, C. W., & Rees, D. C. (2005). Fragment-based lead discovery: Leads by design. *Drug Discovery Today*, *10*(14), 987–992. https://doi.org/10.1016/S1359-6446(05)03511-7.

Chen, F., Sun, H., Liu, H., Li, D., Li, Y., & Hou, T. (2017). Prediction of luciferase inhibitors by the high-performance MIEC-GBDT approach based on interaction energetic patterns. *Physical Chemistry Chemical Physics*, *19*(15), 10163–10176. https://doi.org/10.1039/c6cp08232g.

Chen, G., Zheng, S., Luo, X., Shen, J., Zhu, W., Liu, H., et al. (2005). Focused combinatorial library design based on structural diversity, druglikeness and binding affinity score. *Journal of Combinatorial Chemistry*, 7(3), 398–406. https://doi.org/10.1021/cc049866h.

Cheng, F., Li, W., Zhou, Y., Shen, J., Wu, Z., Liu, G., et al. (2012). admetSAR: A comprehensive source and free tool for assessment of chemical ADMET properties. *Journal of Chemical Information and Modeling*, *52*(11), 3099–3105. https://doi.org/10.1021/ci300367a.

Chilingaryan, Z., Headey, S. J., Lo, A. T. Y., Xu, Z. Q., Otting, G., Dixon, N. E., et al. (2018). Fragment-based discovery of inhibitors of the bacterial DnaG-SSB interaction. *Antibiotics (Basel)*, 7(1), pii: E14 (online). https://doi.org/10.3390/antibiotics7010014.

Coletti, A., Camponeschi, F., Albini, E., Greco, F. A., Maione, V., Custodi, C., et al. (2017). Fragment-based approach to identify IDO1 inhibitor building blocks. *European Journal of Medicinal Chemistry*, *141*, 169–177. https://doi.org/10.1016/j.ejmech.2017.09.044.

Congreve, M., Carr, R., Murray, C., & Jhoti, H. (2003). A 'rule of three' for fragment-based lead discovery? *Drug Discovery Today*, *8*(19), 876–877. https://doi.org/10.1016/S1359-6446(03)02831-9.

Cooper, M. A. (2004). Advances in membrane receptor screening and analysis. *Journal of Molecular Recognition*, *17*(4), 286–315. https://doi.org/10.1002/jmr.675.

Cruciani, G., Pastor, M., & Guba, W. (2000). VolSurf: A new tool for the pharmacokinetic optimization of lead compounds. *European Journal of Pharmaceutical Sciences*, *11*(Suppl. 2), S29–S39.

Cumming, J. G., Davis, A. M., Muresan, S., Haeberlein, M., & Chen, H. (2013). Chemical predictive modelling to improve compound quality. *Nature Reviews. Drug Discovery*, *12*(12), 948–962. https://doi.org/10.1038/nrd4128.

Daina, A., Michielin, O., & Zoete, V. (2017). SwissADME: A free web tool to evaluate pharmacokinetics, drug-likeness and medicinal chemistry friendliness of small molecules. *Scientific Reports*, 7, 42717. https://doi.org/10.1038/srep42717.

Dalvit, C., Ardini, E., Flocco, M., Fogliatto, G. P., Mongelli, N., & Veronesi, M. (2003). A general NMR method for rapid, efficient, and reliable biochemical screening. *Journal of the American Chemical Society*, *125*(47), 14620–14625. https://doi.org/10.1021/ja038128e.

Damm-Ganamet, K. L., Bembenek, S. D., Venable, J. W., Castro, G. G., Mangelschots, L., Peeters, D. C., et al. (2016). A prospective virtual screening study: Enriching hit rates and designing focus libraries to find inhibitors of PI3Kdelta and PI3Kgamma. *Journal of Medicinal Chemistry*, *59*(9), 4302–4313. https://doi.org/10.1021/acs.jmedchem.5b01974.

Degliesposti, G., Kasam, V., Da Costa, A., Kang, H. K., Kim, N., Kim, D. W., et al. (2009). Design and discovery of plasmepsin II inhibitors using an automated workflow on large-scale grids. *ChemMedChem*, *4*(7), 1164–1173. https://doi.org/10.1002/cmdc.200900111.

Degliesposti, G., Portioli, C., Parenti, M. D., & Rastelli, G. (2011). BEAR, a novel virtual screening methodology for drug discovery. *Journal of Biomolecular Screening*, *16*(1), 129–133. https://doi.org/10.1177/1087057110388276.

Dias, D. M., Van Molle, I., Baud, M. G., Galdeano, C., Geraldes, C. F., & Ciulli, A. (2014). Is NMR fragment screening fine-tuned to assess druggability of protein–protein interactions? *ACS Medicinal Chemistry Letters*, *5*(1), 23–28. https://doi.org/10.1021/ml400296c.

Ertl, P., & Schuffenhauer, A. (2009). Estimation of synthetic accessibility score of drug-like molecules based on molecular complexity and fragment contributions. *Journal of Cheminformatics*, *1*(1), 8. https://doi.org/10.1186/1758-2946-1-8.

Evenas, J., Edfeldt, F., Lepisto, M., Svitacheva, N., Synnergren, A., Lundquist, B., et al. (2014). HTS followed by NMR based counterscreening. Discovery and optimization of pyrimidones as reversible and competitive inhibitors of xanthine oxidase. *Bioorganic & Medicinal Chemistry Letters*, *24*(5), 1315–1321. https://doi.org/10.1016/j.bmcl.2014.01.050.

Fan, F., & Wood, K. V. (2007). Bioluminescent assays for high-throughput screening. *Assay and Drug Development Technologies*, *5*(1), 127–136. https://doi.org/10.1089/adt.2006.053.

Fells, J. I., Lee, S. C., Fujiwara, Y., Norman, D. D., Lim, K. G., Tsukahara, R., et al. (2013). Hits of a high-throughput screen identify the hydrophobic pocket of autotaxin/lysophospholipase D as an inhibitory surface. *Molecular Pharmacology*, *84*(3), 415–424. https://doi.org/10.1124/mol.113.087080.

Fells, J. I., Lee, S. C., Norman, D. D., Tsukahara, R., Kirby, J. R., Nelson, S., et al. (2014). Targeting the hydrophobic pocket of autotaxin with virtual screening of inhibitors identifies a common aromatic sulfonamide structural motif. *The FEBS Journal*, *281*(4), 1017–1028. https://doi.org/10.1111/febs.12674.

Ferreira, R. S., Simeonov, A., Jadhav, A., Eidam, O., Mott, B. T., Keiser, M. J., et al. (2010). Complementarity between a docking and a high-throughput screen in discovering new cruzain inhibitors. *Journal of Medicinal Chemistry*, *53*(13), 4891–4905. https://doi.org/10.1021/jm100488w.

Fisher, H. F., & Singh, N. (1995). Calorimetric methods for interpreting protein-ligand interactions. *Methods in Enzymology*, *259*, 194–221.

Fry, D. C., Wartchow, C., Graves, B., Janson, C., Lukacs, C., Kammlott, U., et al. (2013). Deconstruction of a nutlin: Dissecting the binding determinants of a potent protein–protein interaction inhibitor. *ACS Medicinal Chemistry Letters*, *4*(7), 660–665. https://doi.org/10.1021/ml400062c.

Gable, J. E., Lee, G. M., Acker, T. M., Hulce, K. R., Gonzalez, E. R., Schweigler, P., et al. (2016). Fragment-based protein–protein interaction antagonists of a viral dimeric protease. *ChemMedChem*, *11*(8), 862–869. https://doi.org/10.1002/cmdc.201500526.

Gallivan, J. P., & Dougherty, D. A. (1999). Cation-pi interactions in structural biology. *Proceedings of the National Academy of Sciences of the United States of America*, *96*(17), 9459–9464.

Gasteiger, J., & Ihlenfeldt, W. D. (1996). Computer-assisted planning of organic syntheses: The second generation of programs. *Angewandte Chemie (International Ed. in English)*, *34*, 2613–2633. https://doi.org/10.1002/anie.199526131.

Genheden, S., & Ryde, U. (2015). The MM/PBSA and MM/GBSA methods to estimate ligand-binding affinities. *Expert Opinion on Drug Discovery*, *10*(5), 449–461. https://doi.org/10.1517/17460441.2015.1032936.

Gillet, V. J., Myatt, G., Zsoldos, Z., & Johnson, A. P. (1995). SPROUT, HIPPO and CAESA: Tools for de novo structure generation and estimation of synthetic accessibility. *Perspectives in Drug Discovery and Design*, *3*, 34–50. https://doi.org/10.1007/Bf02174466.

Gleeson, M. P., Hersey, A., & Hannongbua, S. (2011). In-silico ADME models: A general assessment of their utility in drug discovery applications. *Current Topics in Medicinal Chemistry*, *11*(4), 358–381.

Graves, A. P., Shivakumar, D. M., Boyce, S. E., Jacobson, M. P., Case, D. A., & Shoichet, B. K. (2008). Rescoring docking hit lists for model cavity sites: Predictions and experimental testing. *Journal of Molecular Biology*, *377*(3), 914–934. https://doi.org/10.1016/j.jmb.2008.01.049.

Guha, R. (2013). On exploring structure-activity relationships. *Methods in Molecular Biology*, *993*, 81–94. https://doi.org/10.1007/978-1-62703-342-8_6.

Hajduk, P. J., Meadows, R. P., & Fesik, S. W. (1997). Discovering high-affinity ligands for proteins. *Science*, *278*(5337), 497–499.

Hakimelahi, G. H., & Khodarahmi, G. A. (2005). The identification of toxicophores for the prediction of mutagenicity, hepatotoxicity and cardiotoxicity. *Journal of the Iranian Chemical Society*, *2*(4), 244–267. https://doi.org/10.1007/Bf03245929.

Harner, M. J., Mueller, L., Robbins, K. J., & Reily, M. D. (2017). NMR in drug design. *Archives of Biochemistry and Biophysics*, *628*, 132–147. https://doi.org/10.1016/j.abb.2017.06.005.

Heifetz, A., Trani, G., Aldeghi, M., MacKinnon, C. H., McEwan, P. A., Brookfield, F. A., et al. (2016). Fragment molecular orbital method applied to lead optimization of novel interleukin-2 inducible T-cell kinase (ITK) inhibitors. *Journal of Medicinal Chemistry*, *59*(9), 4352–4363. https://doi.org/10.1021/acs.jmedchem.6b00045.

Hevener, K. E., Mehboob, S., Su, P. C., Truong, K., Boci, T., Deng, J., et al. (2012). Discovery of a novel and potent class of *F. tularensis* enoyl-reductase (FabI) inhibitors by molecular shape and electrostatic matching. *Journal of Medicinal Chemistry*, *55*(1), 268–279. https://doi.org/10.1021/jm201168g.

Holdgate, G. A., Meek, T. D., & Grimley, R. L. (2018). Mechanistic enzymology in drug discovery: A fresh perspective. *Nature Reviews. Drug Discovery*, *17*(1), 78. https://doi.org/10.1038/nrd.2017.257.

Hopkins, A. L., Groom, C. R., & Alex, A. (2004). Ligand efficiency: A useful metric for lead selection. *Drug Discovery Today*, *9*(10), 430–431. https://doi.org/10.1016/S1359-6446(04)03069-7.

Hou, Y., Mayhood, T., Sheth, P., Tan, C. M., Labroli, M., Su, J., et al. (2016). NMR binding and functional assays for detecting inhibitors of *S. aureus* MnaA. *Journal of Biomolecular Screening*, *21*(6), 579–589. https://doi.org/10.1177/1087057116640199.

Huang, Q., Li, L. L., & Yang, S. Y. (2011). RASA: A rapid retrosynthesis-based scoring method for the assessment of synthetic accessibility of drug-like molecules. *Journal of Chemical Information and Modeling*, *51*(10), 2768–2777. https://doi.org/10.1021/ci100216g.

Huth, J. R., Mendoza, R., Olejniczak, E. T., Johnson, R. W., Cothron, D. A., Liu, Y., et al. (2005). ALARM NMR: A rapid and robust experimental method to detect reactive false positives in biochemical screens. *Journal of the American Chemical Society*, *127*(1), 217–224. https://doi.org/10.1021/ja0455547.

Inglese, J., Auld, D. S., Jadhav, A., Johnson, R. L., Simeonov, A., Yasgar, A., et al. (2006). Quantitative high-throughput screening: A titration-based approach that efficiently identifies biological activities in large chemical libraries. *Proceedings of the National Academy of Sciences of the United States of America*, *103*(31), 11473–11478. https://doi.org/10.1073/pnas.0604348103.

Inglese, J., Johnson, R. L., Simeonov, A., Xia, M., Zheng, W., Austin, C. P., et al. (2007). High-throughput screening assays for the identification of chemical probes. *Nature Chemical Biology*, *3*(8), 466–479. https://doi.org/10.1038/nchembio.2007.17.

Iyer, P., Stumpfe, D., & Bajorath, J. (2011). Molecular mechanism-based network-like similarity graphs reveal relationships between different types of receptor ligands and structural changes that determine agonistic, inverse-agonistic, and antagonistic effects. *Journal of Chemical Information and Modeling*, *51*(6), 1281–1286. https://doi.org/10.1021/ci2001378.

Jimenez-Moreno, E., Montalvillo-Jimenez, L., Santana, A. G., Gomez, A. M., Jimenez-Oses, G., Corzana, F., et al. (2016). Finding the right candidate for the right position: A fast NMR-assisted combinatorial method for optimizing nucleic acids binders. *Journal of the American Chemical Society*, *138*(20), 6463–6474. https://doi.org/10.1021/jacs.6b00328.

Jordan, J. B., Poppe, L., Xia, X., Cheng, A. C., Sun, Y., Michelsen, K., et al. (2012). Fragment based drug discovery: Practical implementation based on (1)(9)F NMR spectroscopy. *Journal of Medicinal Chemistry*, *55*(2), 678–687. https://doi.org/10.1021/jm201441k.

Kenny, P. W. (2017). Comment on the ecstasy and agony of assay interference compounds. *Journal of Chemical Information and Modeling*, *57*(11), 2640–2645. https://doi.org/10.1021/acs.jcim.7b00313.

Khan, M. T. (2010). Predictions of the ADMET properties of candidate drug molecules utilizing different QSAR/QSPR modelling approaches. *Current Drug Metabolism*, *11*(4), 285–295.

Kim, A., Wolf, N. M., Zhu, T., Johnson, M. E., Deng, J., Cook, J. L., et al. (2015). Identification of bacillus anthracis PurE inhibitors with antimicrobial activity. *Bioorganic & Medicinal Chemistry*, *23*(7), 1492–1499. https://doi.org/10.1016/j.bmc.2015.02.016.

Kitaura, K., Ikeo, E., Asada, T., Nakano, T., & Uebayasi, M. (1999). Fragment molecular orbital method: An approximate computational method for large molecules. *Chemical Physics Letters*, *313*(3–4), 701–706. https://doi.org/10.1016/s0009-2614(99)00874-x.

Korepanova, A., Longenecker, K. L., Pratt, S. D., Panchal, S. C., Clark, R. F., Lake, M., et al. (2018). Fragment-based discovery of a potent NAMPT inhibitor. *Bioorganic & Medicinal Chemistry Letters*, *28*(3), 437–440. https://doi.org/10.1016/j.bmcl.2017.12.023.

Leavitt, S., & Freire, E. (2001). Direct measurement of protein binding energetics by isothermal titration calorimetry. *Current Opinion in Structural Biology*, *11*(5), 560–566.

Lee, H., Lei, H., Santarsiero, B. D., Gatuz, J. L., Cao, S., Rice, A. J., et al. (2015). Inhibitor recognition specificity of MERS-CoV papain-like protease may differ from that of SARS-CoV. *ACS Chemical Biology*, *10*(6), 1456–1465. https://doi.org/10.1021/cb500917m.

Lee, H., Ren, J., Nocadello, S., Rice, A. J., Ojeda, I., Light, S., et al. (2017). Identification of novel small molecule inhibitors against NS2B/NS3 serine protease from Zika virus. *Antiviral Research*, *139*, 49–58. https://doi.org/10.1016/j.antiviral.2016.12.016.

Leeson, P. D., & Springthorpe, B. (2007). The influence of drug-like concepts on decision-making in medicinal chemistry. *Nature Reviews. Drug Discovery*, *6*(11), 881–890. https://doi.org/10.1038/nrd2445.

Lei, H., Jones, C., Zhu, T., Patel, K., Wolf, N. M., Fung, L. W., et al. (2016). Identification of B. anthracis N(5)-carboxyaminoimidazole ribonucleotide mutase (PurE) active site binding compounds via fragment library screening. *Bioorganic & Medicinal Chemistry*, *24*(4), 596–605. https://doi.org/10.1016/j.bmc.2015.12.029.

Linkuviene, V., Krainer, G., Chen, W. Y., & Matulis, D. (2016). Isothermal titration calorimetry for drug design: Precision of the enthalpy and binding constant measurements and comparison of the instruments. *Analytical Biochemistry*, *515*, 61–64. https://doi.org/10.1016/j.ab.2016.10.005.

Lionta, E., Spyrou, G., Vassilatis, D. K., & Cournia, Z. (2014). Structure-based virtual screening for drug discovery: Principles, applications and recent advances. *Current Topics in Medicinal Chemistry*, *14*(16), 1923–1938. https://doi.org/10.2174/1568026614666140929124445.

Lipinski, C. A. (2000). Drug-like properties and the causes of poor solubility and poor permeability. *Journal of Pharmacological and Toxicological Methods*, *44*(1), 235–249. https://doi.org/10.1016/S1056-8719(00)00107-6.

Lo, M. C., Aulabaugh, A., Jin, G., Cowling, R., Bard, J., Malamas, M., et al. (2004). Evaluation of fluorescence-based thermal shift assays for hit identification in drug discovery. *Analytical Biochemistry*, *332*(1), 153–159. https://doi.org/10.1016/j.ab.2004.04.031.

Lowe, D. (2013). *Too many metrics.* Retrieved from http://blogs.sciencemag.org/pipeline/archives/2013/08/22/too_many_metrics.

Lu, Y.-X., Zou, J.-W., Wang, Y.-H., & Yu, Q.-S. (2007). Substituent effects on noncovalent halogen/π interactions: Theoretical study. *International Journal of Quantum Chemistry*, *107*(6), 1479–1486. https://doi.org/10.1002/qua.21279.

Maltarollo, V. G., Gertrudes, J. C., Oliveira, P. R., & Honorio, K. M. (2015). Applying machine learning techniques for ADME-Tox prediction: A review. *Expert Opinion on Drug Metabolism & Toxicology*, *11*(2), 259–271. https://doi.org/10.1517/17425255.2015.980814.

Manzenrieder, F., Frank, A. O., & Kessler, H. (2008). Phosphorus NMR spectroscopy as a versatile tool for compound library screening. *Angewandte Chemie (International Ed. in English)*, *47*(14), 2608–2611. https://doi.org/10.1002/anie.200705256.

Mashalidis, E. H., Sledz, P., Lang, S., & Abell, C. (2013). A three-stage biophysical screening cascade for fragment-based drug discovery. *Nature Protocols*, *8*(11), 2309–2324. https://doi.org/10.1038/nprot.2013.130.

McGovern, S. L., Caselli, E., Grigorieff, N., & Shoichet, B. K. (2002). A common mechanism underlying promiscuous inhibitors from virtual and high-throughput screening. *Journal of Medicinal Chemistry*, *45*(8), 1712–1722.

McGovern, S. L., Helfand, B. T., Feng, B., & Shoichet, B. K. (2003). A specific mechanism of nonspecific inhibition. *Journal of Medicinal Chemistry*, *46*(20), 4265–4272. https://doi.org/10.1021/jm030266r.

Mehboob, S., Hevener, K. E., Truong, K., Boci, T., Santarsiero, B. D., & Johnson, M. E. (2012). Structural and enzymatic analyses reveal the binding mode of a novel series of *Francisella tularensis* enoyl reductase (FabI) inhibitors. *Journal of Medicinal Chemistry*, *55*(12), 5933–5941. https://doi.org/10.1021/jm300489v.

Nagatoishi, S., Yamaguchi, S., Katoh, E., Kajita, K., Yokotagawa, T., Kanai, S., et al. (2018). A combination of (19)F NMR and surface plasmon resonance for site-specific hit selection and validation of fragment molecules that bind to the ATP-binding site of a kinase. *Bioorganic & Medicinal Chemistry*, *26*(8), 1929–1938. https://doi.org/10.1016/j.bmc.2018.02.041.

Navratilova, I., & Hopkins, A. L. (2010). Fragment screening by surface plasmon resonance. *ACS Medicinal Chemistry Letters*, *1*(1), 44–48. https://doi.org/10.1021/ml900002k.

Niesen, F. H., Berglund, H., & Vedadi, M. (2007). The use of differential scanning fluorimetry to detect ligand interactions that promote protein stability. *Nature Protocols*, *2*(9), 2212–2221. https://doi.org/10.1038/nprot.2007.321.

NIH. (2017). *Assay development and screening for discovery of chemical probes or therapeutic agents (R01).* Retrieved from https://grants.nih.gov/grants/guide/pa-files/PAR-17-438.html.

Nissink, J. W. (2009). Simple size-independent measure of ligand efficiency. *Journal of Chemical Information and Modeling*, *49*(6), 1617–1622. https://doi.org/10.1021/ci900094m.

Oprea, T. I., Bologa, C. G., Edwards, B. S., Prossnitz, E. R., & Sklar, L. A. (2005). Post-high-throughput screening analysis: An empirical compound prioritization scheme. *Journal of Biomolecular Screening*, *10*(5), 419–426. https://doi.org/10.1177/1087057104272660.

Oprea, T. I., Gottfries, J., Sherbukhin, V., Svensson, P., & Kuhler, T. C. (2000). Chemical information management in drug discovery: Optimizing the computational and combinatorial chemistry interfaces. *Journal of Molecular Graphics & Modelling*, *18*(4–5), 512–524. 541. https://doi.org/10.1016/S1093-3263(00)00066-8.

Ozawa, T., Okazaki, K., & Kitaura, K. (2011). CH/pi hydrogen bonds play a role in ligand recognition and equilibrium between active and inactive states of the beta2 adrenergic receptor: An ab initio fragment molecular orbital (FMO) study. *Bioorganic & Medicinal Chemistry*, *19*(17), 5231–5237. https://doi.org/10.1016/j.bmc.2011.07.004.

Pantoliano, M. W., Petrella, E. C., Kwasnoski, J. D., Lobanov, V. S., Myslik, J., Graf, E., et al. (2001). High-density miniaturized thermal shift assays as a general strategy for drug discovery. *Journal of Biomolecular Screening*, *6*(6), 429–440. https://doi.org/10.1177/108705710100600609.

Patching, S. G. (2014). Surface plasmon resonance spectroscopy for characterisation of membrane protein-ligand interactions and its potential for drug discovery. *Biochimica et Biophysica Acta*, *1838*(1 Pt. A), 43–55. https://doi.org/10.1016/j.bbamem.2013.04.028.

Patterson, D. E., Cramer, R. D., Ferguson, A. M., Clark, R. D., & Weinberger, L. E. (1996). Neighborhood behavior: A useful concept for validation of "molecular diversity" descriptors. *Journal of Medicinal Chemistry*, *39*(16), 3049–3059. https://doi.org/10.1021/jm960290n.

Pearce, B. C., Sofia, M. J., Good, A. C., Drexler, D. M., & Stock, D. A. (2006). An empirical process for the design of high-throughput screening deck filters. *Journal of Chemical Information and Modeling*, *46*(3), 1060–1068. https://doi.org/10.1021/ci050504m.

Pellecchia, M., Bertini, I., Cowburn, D., Dalvit, C., Giralt, E., Jahnke, W., et al. (2008). Perspectives on NMR in drug discovery: A technique comes of age. *Nature Reviews. Drug Discovery*, 7(9), 738–745. https://doi.org/10.1038/nrd2606.

Phillips, K. S., & Cheng, Q. (2007). Recent advances in surface plasmon resonance based techniques for bioanalysis. *Analytical and Bioanalytical Chemistry*, *387*(5), 1831–1840. https://doi.org/10.1007/s00216-006-1052-7.

Pierce, M. M., Raman, C. S., & Nall, B. T. (1999). Isothermal titration calorimetry of protein–protein interactions. *Methods*, *19*(2), 213–221. https://doi.org/10.1006/meth.1999.0852.

Piliarik, M., Vaisocherova, H., & Homola, J. (2009). Surface plasmon resonance biosensing. *Methods in Molecular Biology*, *503*, 65–88. https://doi.org/10.1007/978-1-60327-567-5_5.

Podolyan, Y., Walters, M. A., & Karypis, G. (2010). Assessing synthetic accessibility of chemical compounds using machine learning methods. *Journal of Chemical Information and Modeling*, *50*(6), 979–991. https://doi.org/10.1021/ci900301v.

Polgar, T., & Keseru, G. M. (2011). Integration of virtual and high throughput screening in lead discovery settings. *Combinatorial Chemistry & High Throughput Screening*, *14*(10), 889–897. https://doi.org/10.2174/138620711797537148.

Rastelli, G., Degliesposti, G., Del Rio, A., & Sgobba, M. (2009). Binding estimation after refinement, a new automated procedure for the refinement and rescoring of docked ligands in virtual screening. *Chemical Biology & Drug Design*, *73*(3), 283–286. https://doi.org/10.1111/j.1747-0285.2009.00780.x.

Raymond, J. W., Blankley, C. J., & Willett, P. (2003). Comparison of chemical clustering methods using graph- and fingerprint-based similarity measures. *Journal of Molecular Graphics & Modelling*, *21*(5), 421–433.

Raymond, J. W., & Willett, P. (2002). Effectiveness of graph-based and fingerprint-based similarity measures for virtual screening of 2D chemical structure databases. *Journal of Computer-Aided Molecular Design*, *16*(1), 59–71.

Reithmeier, A., Lundback, T., Haraldsson, M., Frank, M., Ek-Rylander, B., Nyholm, P. G., et al. (2018). Identification of inhibitors of Tartrate-resistant acid phosphatase (TRAP/ACP5) activity by small-molecule screening. *Chemical Biology & Drug Design*, *92*(1), 1255–1271. https://doi.org/10.1111/cbdd.13187.

Riniker, S., Wang, Y., Jenkins, J. L., & Landrum, G. A. (2014). Using information from historical high-throughput screens to predict active compounds. *Journal of Chemical Information and Modeling*, *54*(7), 1880–1891. https://doi.org/10.1021/ci500190p.

Rishton, G. M. (1997). Reactive compounds and in vitro false positives in HTS. *Drug Discovery Today*, *2*(9), 382–384. https://doi.org/10.1016/S1359-6446(97)01083-0.

Roughley, S. D., & Jordan, A. M. (2011). The medicinal chemist's toolbox: An analysis of reactions used in the pursuit of drug candidates. *Journal of Medicinal Chemistry*, *54*(10), 3451–3479. https://doi.org/10.1021/jm200187y.

Sastry, M., Lowrie, J. F., Dixon, S. L., & Sherman, W. (2010). Large-scale systematic analysis of 2D fingerprint methods and parameters to improve virtual screening enrichments. *Journal of Chemical Information and Modeling*, *50*(5), 771–784. https://doi.org/10.1021/ci100062n.

Schade, M. (2006). NMR fragment screening: Advantages and applications. *IDrugs*, *9*(2), 110–113.

Shen, J., Cheng, F., Xu, Y., Li, W., & Tang, Y. (2010). Estimation of ADME properties with substructure pattern recognition. *Journal of Chemical Information and Modeling*, *50*(6), 1034–1041. https://doi.org/10.1021/ci100104j.

Shortridge, M. D., Hage, D. S., Harbison, G. S., & Powers, R. (2008). Estimating protein-ligand binding affinity using high-throughput screening by NMR. *Journal of Combinatorial Chemistry*, *10*(6), 948–958. https://doi.org/10.1021/cc800122m.

Shuker, S. B., Hajduk, P. J., Meadows, R. P., & Fesik, S. W. (1996). Discovering high-affinity ligands for proteins: SAR by NMR. *Science*, *274*(5292), 1531–1534.

Shultz, M. D. (2013). Setting expectations in molecular optimizations: Strengths and limitations of commonly used composite parameters. *Bioorganic & Medicinal Chemistry Letters*, *23*(21), 5980–5991. https://doi.org/10.1016/j.bmcl.2013.08.029.

Shultz, M. D. (2014). Improving the plausibility of success with inefficient metrics. *ACS Medicinal Chemistry Letters*, *5*(1), 2–5. https://doi.org/10.1021/ml4004638.

Shun, T. Y., Lazo, J. S., Sharlow, E. R., & Johnston, P. A. (2011). Identifying actives from HTS data sets: Practical approaches for the selection of an appropriate HTS data-processing method and quality control review. *Journal of Biomolecular Screening*, *16*(1), 1–14. https://doi.org/10.1177/1087057110389039.

Sijbesma, E., Skora, L., Leysen, S., Brunsveld, L., Koch, U., Nussbaumer, P., et al. (2017). Identification of two secondary ligand binding sites in 14-3-3 proteins using fragment screening. *Biochemistry*, *56*(30), 3972–3982. https://doi.org/10.1021/acs.biochem.7b00153.

Sorrell, F. J., Greenwood, G. K., Birchall, K., & Chen, B. (2010). Development of a differential scanning fluorimetry based high throughput screening assay for the discovery of affinity binders against an anthrax protein. *Journal of Pharmaceutical and Biomedical Analysis*, *52*(5), 802–808.

Strovel, J., Sittampalam, S., Coussens, N. P., Hughes, M., Inglese, J., Kurtz, A., et al. (2004). Early drug discovery and development guidelines: For academic researchers, collaborators, and start-up companies. In G. S. Sittampalam, N. P. Coussens, K. Brimacombe, A. Grossman, M. Arkin, D. Auld, C. Austin, J. Baell, B. Bejcek, T. D. Y. Chung, J. L. Dahlin, V. Devanaryan, T. L. Foley, M. Glicksman, M. D. Hall, J. V. Hass, J. Inglese, P. W. Iversen, S. D. Kahl, S. C. Kales, M. Lal-Nag, Z. Li, J. McGee, O. McManus, T. Riss, O. J. Trask, Jr., J. R. Weidner, M. Xia, & X. Xu (Eds.), *Assay guidance manual*. Bethesda (MD): Eli Lilly & Company and the National Center for Advancing Translational Sciences, (pp. 1–43).

Stumpfe, D., & Bajorath, J. (2012). Exploring activity cliffs in medicinal chemistry. *Journal of Medicinal Chemistry*, *55*(7), 2932–2942. https://doi.org/10.1021/jm201706b.

Sugiki, T., Furuita, K., Fujiwara, T., & Kojima, C. (2018). Current NMR techniques for structure-based drug discovery. *Molecules*, *23*(1), 148. https://doi.org/10.3390/molecules23010148.

Talevi, A., Goodarzi, M., Ortiz, E. V., Duchowicz, P. R., Bellera, C. L., Pesce, G., et al. (2011). Prediction of drug intestinal absorption by new linear and non-linear QSPR. *European Journal of Medicinal Chemistry*, *46*(1), 218–228. https://doi.org/10.1016/j.ejmech.2010.11.005.

Tao, L., Zhang, P., Qin, C., Chen, S. Y., Zhang, C., Chen, Z., et al. (2015). Recent progresses in the exploration of machine learning methods as in-silico ADME prediction tools. *Advanced Drug Delivery Reviews*, *86*, 83–100. https://doi.org/10.1016/j.addr.2015.03.014.

Thorne, N., Auld, D. S., & Inglese, J. (2010). Apparent activity in high-throughput screening: Origins of compound-dependent assay interference. *Current Opinion in Chemical Biology*, *14*(3), 315–324. https://doi.org/10.1016/j.cbpa.2010.03.020.

Tomala, M. D., Magiera-Mularz, K., Kubica, K., Krzanik, S., Zieba, B., Musielak, B., et al. (2018). Identification of small-molecule inhibitors of USP2a. *European Journal of Medicinal Chemistry*, *150*, 261–267. https://doi.org/10.1016/j.ejmech.2018.03.009.

Topliss, J. G. (1972). Utilization of operational schemes for analog synthesis in drug design. *Journal of Medicinal Chemistry*, *15*(10), 1006–1011. https://doi.org/10.1021/jm00280a002.

Upadhyay, A. K., Judge, R. A., Li, L., Pithawalla, R., Simanis, J., Bodelle, P. M., et al. (2018). Targeting lysine specific demethylase 4A (KDM4A) tandem TUDOR domain—A fragment based approach. *Bioorganic & Medicinal Chemistry Letters*, *28*, 1708–1713. https://doi.org/10.1016/j.bmcl.2018.04.050.

Varin, T., Gubler, H., Parker, C. N., Zhang, J. H., Raman, P., Ertl, P., et al. (2010). Compound set enrichment: A novel approach to analysis of primary HTS data. *Journal of Chemical Information and Modeling*, *50*(12), 2067–2078. https://doi.org/10.1021/ci100203e.

Veber, D. F., Johnson, S. R., Cheng, H. Y., Smith, B. R., Ward, K. W., & Kopple, K. D. (2002). Molecular properties that influence the oral bioavailability of drug candidates. *Journal of Medicinal Chemistry*, *45*(12), 2615–2623. https://doi.org/10.1021/jm020017n.

Virtanen, S. I., Niinivehmas, S. P., & Pentikainen, O. T. (2015). Case-specific performance of MM-PBSA, MM-GBSA, and SIE in virtual screening. *Journal of Molecular Graphics & Modelling*, *62*, 303–318. https://doi.org/10.1016/j.jmgm.2015.10.012.

Vulpetti, A., Randl, S., Rudisser, S., Ostermann, N., Erbel, P., Mac Sweeney, A., et al. (2017). Structure-based library design and fragment screening for the identification of reversible complement factor D protease inhibitors. *Journal of Medicinal Chemistry*, *60*(5), 1946–1958. https://doi.org/10.1021/acs.jmedchem.6b01684.

Walters, W. P., & Murcko, M. A. (2002). Prediction of 'drug-likeness'. *Advanced Drug Delivery Reviews*, *54*(3), 255–271.

Walters, W. P., Ajay, & Murcko, M. A. (1999). Recognizing molecules with drug-like properties. *Current Opinion in Chemical Biology*, *3*(4), 384–387. https://doi.org/10.1016/S1367-5931(99)80058-1.

Wang, Z., Bhattacharya, A., & Ivanov, D. N. (2015b). Identification of small-molecule inhibitors of the HuR/RNA interaction using a fluorescence polarization screening assay followed by NMR validation. *PLoS One*, *10*(9), e0138780. https://doi.org/10.1371/journal.pone.0138780.

Wang, W., Marholz, L. J., & Wang, X. (2015a). Novel scaffolds of cell-active histone demethylase inhibitors identified from high-throughput screening. *Journal of Biomolecular Screening*, *20*(6), 821–827. https://doi.org/10.1177/1087057115579637.

Willett, P. (2011). Similarity searching using 2D structural fingerprints. *Methods in Molecular Biology*, *672*, 133–158. https://doi.org/10.1007/978-1-60761-839-3_5.

Williams, D. P., & Naisbitt, D. J. (2002). Toxicophores: Groups and metabolic routes associated with increased safety risk. *Current Opinion in Drug Discovery & Development*, *5*(1), 104–115.
Wishart, D. S. (2007). Improving early drug discovery through ADME modelling: An overview. *Drugs in R&D*, *8*(6), 349–362.
Wu, B., Barile, E., De, S. K., Wei, J., Purves, A., & Pellecchia, M. (2015). High-throughput screening by nuclear magnetic resonance (HTS by NMR) for the identification of PPIs antagonists. *Current Topics in Medicinal Chemistry*, *15*(20), 2032–2042.
Wu, B., Zhang, Z., Noberini, R., Barile, E., Giulianotti, M., Pinilla, C., et al. (2013). HTS by NMR of combinatorial libraries: A fragment-based approach to ligand discovery. *Chemistry & Biology*, *20*(1), 19–33. https://doi.org/10.1016/j.chembiol.2012.10.015.
Xie, J., Thapa, R., Reverdatto, S., Burz, D. S., & Shekhtman, A. (2009). Screening of small molecule interactor library by using in-cell NMR spectroscopy (SMILI-NMR). *Journal of Medicinal Chemistry*, *52*(11), 3516–3522. https://doi.org/10.1021/jm9000743.
Xiong, H., Han, J., Wang, J., Lu, W., Wang, C., Chen, Y., et al. (2018). Discovery of 1,8-acridinedione derivatives as novel GCN5 inhibitors via high throughput screening. *European Journal of Medicinal Chemistry*, *151*, 740–751. https://doi.org/10.1016/j.ejmech.2018.02.005.
Yao, T.-T., Xie, J.-F., Liu, X.-G., Cheng, J.-L., Zhu, C.-Y., Zhao, J.-H., et al. (2017). Integration of pharmacophore mapping and molecular docking in sequential virtual screening: Towards the discovery of novel JAK2 inhibitors. *RSC Advances*, 7(17), 10353–10360. https://doi.org/10.1039/c6ra24959k.
Zartler, E. R., Hanson, J., Jones, B. E., Kline, A. D., Martin, G., Mo, H., et al. (2003). RAMPED-UP NMR: Multiplexed NMR-based screening for drug discovery. *Journal of the American Chemical Society*, *125*(36), 10941–10946. https://doi.org/10.1021/ja0348593.
Zhang, J. H., Wu, X., & Sills, M. A. (2005). Probing the primary screening efficiency by multiple replicate testing: A quantitative analysis of hit confirmation and false screening results of a biochemical assay. *Journal of Biomolecular Screening*, *10*(7), 695–704. https://doi.org/10.1177/1087057105279149.
Zheng, S., Luo, X., Chen, G., Zhu, W., Shen, J., Chen, K., et al. (2005). A new rapid and effective chemistry space filter in recognizing a druglike database. *Journal of Chemical Information and Modeling*, *45*(4), 856–862. https://doi.org/10.1021/ci050031j.
Zhu, T., Cao, S., Su, P. C., Patel, R., Shah, D., Chokshi, H. B., et al. (2013). Hit identification and optimization in virtual screening: Practical recommendations based on a critical literature analysis. *Journal of Medicinal Chemistry*, *56*(17), 6560–6572. https://doi.org/10.1021/jm301916b.

CHAPTER THIRTEEN

# Knowledge-Based Approaches to Off-Target Screening

**Daniel R. McMasters**[1]
Computational Chemistry, Vertex Pharmaceuticals, Boston, MA, United States
[1]Corresponding author: e-mail address: daniel_mcmasters@vrtx.com

## Contents



## Abstract

Selective binding of a drug for its target vs other proteins is an important consideration in designing compounds with minimal risk of toxicity and therefore a key aspect of lead optimization. Screening all compounds against all known off-target proteins would be prohibitively expensive, so project teams must decide which compounds to screen in which assays. This chapter describes informatics-based methods that help prioritize testing, including screening minipanels and prioritization using predictive models (e-counterscreening).

## 1. INTRODUCTION

An important component of drug discovery involves ascertaining a molecule's selectivity for the target (or targets) of interest. If a molecule binds tightly to additional enzymes or receptors, undesired secondary or "off-target" pharmacology or toxicity may occur. Off-target binding may also

*Methods in Enzymology*, Volume 610
ISSN 0076-6879
https://doi.org/10.1016/bs.mie.2018.09.023

result in changes to the levels or activity of xenobiotic-metabolizing enzymes, resulting in the potential for drug–drug interactions or variable pharmacokinetics. Peters provided an overview of polypharmacology from the perspective of both efficacy and safety and describes some of the advantages of screening discovery compounds against panels of potential targets with known relevance to safety (Peters, 2013). It is therefore generally accepted that key compounds should be screened widely for potential secondary pharmacology. Drug discovery teams working in lead optimization are also well advised to screen representative molecules in order to obtain an early indication of possible off-target activities, while there is still time to dial them out. Given the very large number of distinct proteins and other macromolecules in an organism, however, it is impossible to screen each drug candidate against all possible off-targets, and even if possible the results would be difficult to interpret due to our incomplete understanding of the relationship between pharmacological modulation of most enzymes and receptors and the effects on an organism. It is therefore necessary to screen for potential secondary pharmacology in a judicious manner, to focus our screening resources on experiments that will provide interpretable and actionable information.

It is common for drug discovery teams to establish a tiered screening funnel, in which compounds passing each tier of the funnel proceed to be screened in the next tier of assays. This approach has the advantage that screening resources are applied only to compounds that still have the potential to become a drug based on the data collected so far. To minimize the time required for all the necessary data to be collected on a given compound and thereby expedite drug discovery, and to realize experimental efficiencies, some assays are performed in parallel at one tier. A trade-off must be made between running multiple off-target assays at the same time (which provides the team with the data as quickly as possible but at higher expense) vs running the assays sequentially (which saves assay resources by removing compounds from consideration as soon as a liability is discovered, but with the disadvantage that compounds require longer to proceed through the screening funnel).

The relationship between screening cost, time, and risk to the project can be described in the form of a project triangle (Fig. 1), in which two of the three components can be minimized simultaneously (minimize risk and time by screening all compounds and assays in parallel; minimize cost and risk by screening assays sequentially; or minimize cost and time by screening only certain compounds or assays).

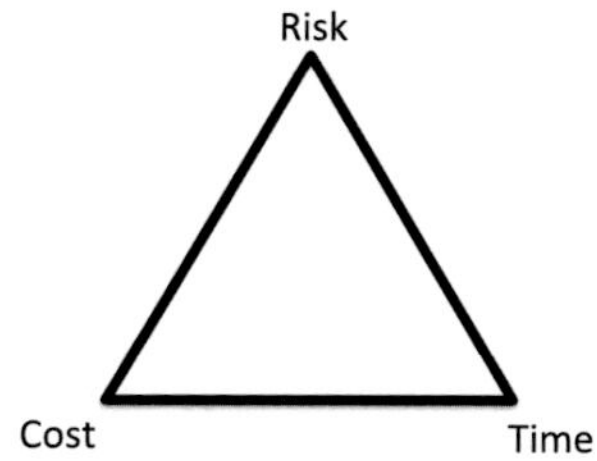


**Fig. 1** A project management triangle as applied to off-target screening. Teams can screen in a tiered manner (sacrificing time), or screen broadly without prioritizing (sacrificing cost). In order to progress rapidly without undue expense, it is necessary to accept some risk that an undesired off-target will be discovered later.

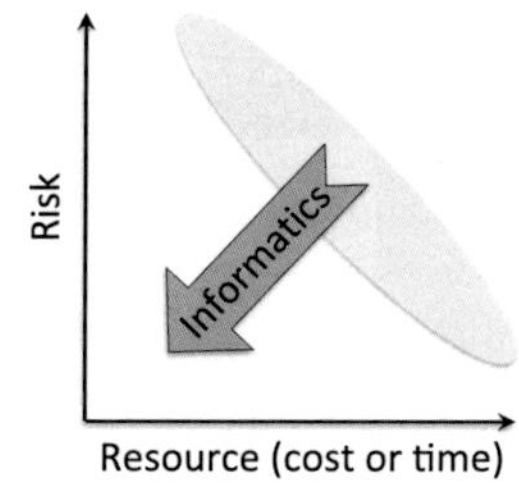


**Fig. 2** The role of informatics in off-target screening is to prioritize assays or compounds for experimental testing, thereby minimizing the risk of undiscovered polypharmacology for a given resource allocation.

Informatics can be used to decrease the risk while still minimizing cost and time, by prioritizing assays or compounds to screen based on available data or models (Fig. 2).

As a result of the tiered approach to screening, for a small number of the most commonly encountered off-targets, very many compounds are screened (broad and shallow), while for a small number of key compounds, very many targets screened (narrow and deep). Because it is not feasible to screen broadly and deeply—screening a large number of compounds against a large number of targets is cost-prohibitive—it is helpful to use knowledge-based approaches to maximize the learning from a finite screening budget (Fig. 3).

We begin with a short overview of different off-target proteins that are commonly assayed, after which we discuss three knowledge-based approaches to off-target screening: protein family-based methods to help assess selectivity among related targets, secondary pharmacology panels, and screening guided by predictive models. Off-target pharmacology that involves reactivity, such as mutagenicity or idiosyncratic toxicity, is out of scope for this discussion.

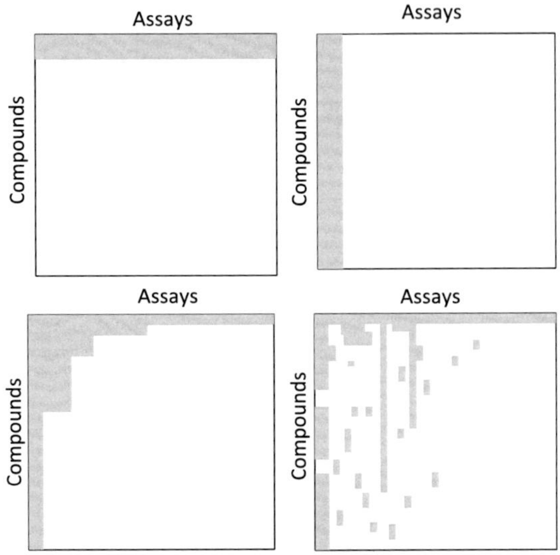


**Fig. 3** Comparison of deep (many assays but not necessarily all compounds, *upper left*) vs broad (many compounds but not all assays, *upper right*) screening. Discovery teams generally screen in a tiered manner (*lower left*), in which only compounds which meet the desired criteria at each stage of testing are submitted for the next stage. Informatics-based approaches can be used to customize the screens performed for each compound based on their predicted liabilities (*lower right*).

# 2. CLASSES OF OFF-TARGET PROTEINS

## 2.1 Proteins Related to the Target

Most drug targets are members of families of proteins that are related phylogenetically. Examples include G-protein coupled receptors (GPCRs), protein kinases, nuclear hormone receptors, serine proteases, and ion channels. The degree to which compounds that bind to the desired target also bind to these related proteins varies greatly and depends on the conservation of the protein fold and the sequence homology of the binding site.

## 2.2 Broadly Promiscuous Off-Target Proteins

A small number of off-target proteins are commonly screened very early in the screening funnel due to their promiscuity and known relevance to toxicity and drug–drug interactions. These include:

- The major human cytochrome P450 (CYP) isoforms 3A4, 2C9, and 2D6 (other isoforms are sometimes considered, such as 2C8, 2C19,

1A2), inhibition of which can cause increased exposure to drugs metabolized by these enzymes
- The pregnane X receptor (PXR), activation of which can result in induction of CYP enzymes and therefore decreased exposure to either the drug in question or other medications
- The hERG (human ether-à-go-go-related gene) cardiac potassium channel encoded by the *KCNH2* gene
- Other cardiac ion channels, such as the NaV1.5 voltage-regulated sodium channel or the CaV1.2 L-type voltage-regulated calcium channel

### 2.3 Safety Pharmacology Panels

Contract research organizations offer screening services that can be used to help elucidate the potential secondary pharmacology of a compound, such as Eurofins Panlabs. The initial screen is commonly a binding assay, which in case of activity should be followed up with a functional readout, to determine whether the binding event has functional consequences, such as agonism or antagonism. Usually only key compounds are profiled against the entire panel due to cost; a smaller subpanel can be used to obtain an initial assessment of a compound's promiscuity. A trade-off must be made between the size of the panel and the number of compounds that can be screened. For chemotypes with no known off-target liabilities, it is not necessary to screen large numbers of compounds. In this case a subpanel can be selected that contains targets that are known to result in toxicity if perturbed. For chemotypes that have been found to be active toward some of the targets, it is useful to define an even smaller panel in order to screen more compounds and obtain sufficient structure–activity relationship (SAR) data for the series and activities in question.

## 3. KNOWLEDGE-BASED APPROACHES TO OFF-TARGET SCREENING

### 3.1 Protein Family-Based Methods

If selectivity over an isoform or very closely related protein is required, it is common to screen every compound synthesized, because optimization of selectivity will drive the design of new molecules. As an example, a project to discover androgen receptor modulators may need to screen all compounds for progesterone receptor activity in order to ensure selectivity and determine the SAR differences for androgen and progesterone receptors for the chemical series of interest. Only if it has been established that a chemical series is highly selective for the desired target might the team decide to screen less broadly and merely spot check new compounds to ensure that the established selectivity is still maintained.

For profiling across larger families, such as GPCRs or especially protein kinases, it is useful to select subsets for routine screening of compounds in a resource-sparing manner. Such "minipanels" can give insight into selectivity against important members of the family as well as being predictive of a compound's selectivity profile against the entire panel. Only key compounds are then screened against the entire panel to confirm their selectivity.

Not all members of the protein family will be equally likely to share a pharmacological profile with the target of interest. There are several criteria that can be considered in determining the likelihood of a given family member binding similar compounds to the desired target. These target similarity metrics include: protein sequence similarity (full-length or active site), using bioinformatics tools like PSI-BLAST (Altschul et al., 1997); binding site shape and electrostatic similarity (Kinnings & Jackson, 2009); and pharmacological similarity (i.e., proteins that have been determined to bind the same ligands) (Sheridan, Nam, Maiorov, McMasters, & Cornell, 2009).

Additionally, it is important to decide whether the subpanel selection is meant to be representative of the entire set, so that the observed degree of selectivity in the minipanel is likely to be similar to the selectivity across the full panel; or instead "pessimistic," which is to say enriched in close relatives, such that the observed selectivity is likely to be a worst-case scenario. Choosing a set of family members that are enriched in more highly similar proteins will tend to perform better at distinguishing the most selective from moderately selective compounds. In contrast, representative sets will distinguish selective compounds from nonselective compounds.

A recent example of the use of a kinase minipanel was published by researchers from Janssen (Bembenek, Hirst, & Mirzadegan, 2018). They describe the design of a kinase minipanel to characterize the selectivity of compounds from a large number of kinase projects. A Monte Carlo approach was taken to randomly select subsets from the entire panel of 424 kinases for which 3000 compounds had assay data; for each minipanel size, 1,000,000 random selections were made. The authors found that a panel of just 20 kinases agreed with the full panel selectivity as measured by kinase hit index (ratio of kinases inhibited to total number of kinases tested, using pIC50 $\geq$ 7 or percent inhibition $\geq$85% at 10 $\mu M$ as the threshold) with $R^2$ of 0.79–0.83.

While this panel may be useful for obtaining an estimate of overall kinase activity for any compound, it may not be able to distinguish compounds that inhibit a single or just a few kinases from compounds that have a substantial number of kinase activities. When selecting chemical series for hit-to-lead activities, such as active compounds from a high-throughput screening campaign against a kinase target of interest, it may be therefore more useful to create a minipanel that is tailored to that kinase and that can distinguish highly from moderately selective compounds while testing relatively few kinases. This can be done by enriching the panel in similar kinases while still sampling diverse kinases with lower similarity. There are many ways to do this; one method is the following:

1. Decide upon a similarity metric and calculate the similarity of every kinase relative to the target kinase. It is recommended to consider the binding site similarity instead of or in addition to full-length similarity.
2. Cluster all the proteins within a relatively high similarity threshold using a high similarity criterion for clustering and take the most similar protein from each cluster.
3. Cluster all proteins within a lower similarity threshold using a somewhat lower similarity criterion and take the most similar protein from each cluster that was not sampled in step 2.
4. Continue decreasing the similarity and clustering thresholds and sampling new clusters until all protein clusters have been sampled.

The result of this procedure will be a diverse set of proteins that is enriched in off-target proteins similar to the target of interest. This will therefore serve as a "pessimistic" selectivity panel that is able to identify compounds with good selectivity for the single target of interest, as well as compounds with good overall kinome selectivity but that are not selective relative to very close homologs.

## 3.2 Correlation-Based Representative Safety Minipanels

In order to understand the potential secondary pharmacology of any potential pharmaceutical, it is typical to screen the compound against a wide variety of receptors, ion channels, and enzymes (for a review, see Peters et al., 2012). Various commercial vendors offer such screening services (such as CEREP and Eurofins). Although the total assay panel available through such services is very large, it is common to use a smaller set of assays to establish the general safety profile of compounds during lead optimization. For example, Hamon et al. describe the use of a 23-target panel at Novartis for this purpose (Hamon et al., 2009).

Data mining can be used to determine which specific assays are more or less common in a given chemotype, which can be used to devise a custom minipanel that is predictive of the entire set but due to its smaller size can be screened against more compounds for the same cost.

On a recent drug discovery project, some of the compounds synthesized during lead optimization displayed binding to multiple targets within the Eurofins panel. (A manuscript describing the project in more detail is in preparation.) There was interest in creating a smaller panel that would allow more compounds to be profiled and establish SARs for the off-target activity. In order to devise a predictive minipanel, the following informatics approach was performed.

1. Determine the correlation between each individual assay and the full panel. Determine which assay is most highly correlated and select this assay as the first member of the custom panel. It is necessary to decide on a metric for quantitating the full panel results, but it was determined that all of the metrics investigated—median percent inhibition, percent of targets active at ≥50% binding at 10 μM, percent of targets active at ≥90% binding at 10 μ*M*—were very highly correlated so gave essentially the same results.
2. Take the average of each remaining assay and the most highly correlated assay from step 1 and determine the correlation with the full panel. The highest correlation will give the assay that is most complementary with the first assay in reproducing the full panel results. Select this assay as the second member of the minipanel.
3. Continue adding assays to the growing panel, in each round selecting the assay most complementary to the existing assays. Choose enough assays to sufficiently reproduce the full panel results.

In the case of the project in question, a panel of 4 assays was found that correlated with a standard panel of 34 off-target assays with an $R^2$ of 0.90, and a 10-assay panel was found that had an $R^2$ of 0.95.

When this method was extended to the full dataset across multiple projects and chemical scaffolds, qualitatively similar were observed but with somewhat lower $R^2$ between minipanel and standard panel. Fig. 4 shows the $R^2$ between the minipanel and standard panel as a function of minipanel size. Comparison with a Monte Carlo method (100,000 random attempts) for various minipanel sizes gave similar results with the sequential addition method described above.

Unsurprisingly, the predictivity of a given minipanel to a larger panel depends on the chemical series. Creating a minipanel to predict full panel

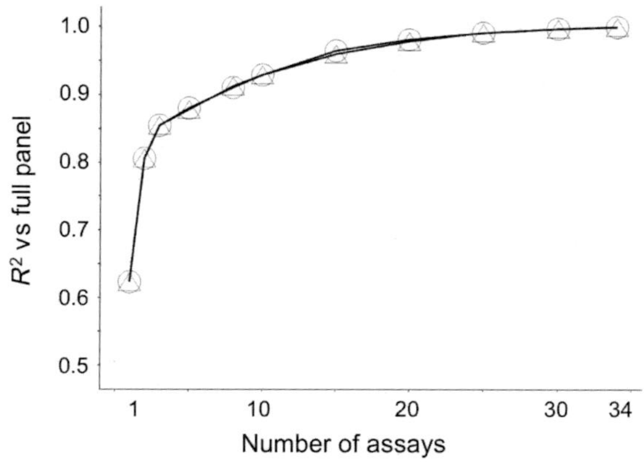


**Fig. 4** Correlation between minipanel and standard (34-assay) panel. The minipanel was selected either sequentially by maximizing correlation with each added member (*circles*) or through a Monte Carlo method in which many random sets were sampled (*triangles*).

promiscuity using data across Vertex's projects gave similar but slightly lower overall $R^2$ compared to the minipanel for a single project described above, but the performance of this global minipanel varied greatly across projects, with most projects' compounds being predicted with $R^2 > 0.8$ for minipanel size of 5–10, but some individual projects giving lower $R^2$, especially if relatively few compounds were profiled.

## 3.3 Prediction-Based Experimental Prioritization (E-Counterscreening)

During a drug discovery campaign, it is common to obtain screening data against a standard panel of promiscuous off-target proteins (CYPs, PXR, hERG, etc.). Many chemical series will have activity against one or more of these, but generally the majority of the data is negative, that is, most compounds are inactive against most of these unwanted activities, especially in the later stages of lead optimization, once the medicinal chemists have determined which chemotypes and physicochemical properties maximize the difference between the desired and undesired activities. It is tempting to reduce the screening to save resources, but this increases the risk that an undesired activity is discovered later. If an unexpected liability is introduced into a chemical series, many analogs may be synthesized with this activity, leading the optimization down the wrong path. Because of this risk, many teams continue to screen broadly even late in lead optimization when virtually all such liabilities have been eliminated.

An alternative to such broad screening is to use predictive models to determine which compounds have the greatest risk of off-target pharmacology and screen only compounds which require experimental derisking (Sheridan, McMasters, Voigt, & Wildey, 2015). This approach, referred to as "e-counterscreening," uses quantitative structure–activity relationship (QSAR) models to determine the likelihood that a given molecule will be active in a given assay. If the likelihood is below an agreed upon threshold, the compound is considered to have been sufficiently derisked; if the likelihood is above the threshold, the compound is flagged as requiring follow-up experimental confirmation of the activity.

There are a number of necessary components for implementing an "e-counterscreening" strategy:

1. QSAR models for the assay endpoints in question. The models should allow for a quantitative assessment of risk of activity for a given compound. This can be achieved by:
    a. Generating a model that outputs both the expected value and a compound-specific measure of uncertainty, such as Gaussian process regression (Burden, 2001) or conformal regression (Svensson et al., 2018).
    b. Generating predicted values and uncertainties in separate steps (Sheridan, 2013b).
    c. Building a classification model that returns the probability of membership in the "active" class. A drawback to this approach is that the threshold between active and inactive is a feature of the model and is therefore not customizable by the project team unless a separate model is built.

If none of these methods is available, the overall (not compound-specific) uncertainty of the prediction, determined from the model validation, can be used. It is important that the validation be performed in a way that simulates the actual model usage. Typical leave-one-out or other x-fold cross-validation tends to overpredict the accuracy of a predictive model, because any test molecule has near neighbors in the training set. Cluster-based cross-validation avoids this problem and comes closer to the situation encountered in drug discovery, where modifications to chemical scaffolds are continually being invented. The best method is temporal cross-validation (Sheridan, 2013a) which most closely simulates typical drug discovery.

With the quantitative probability of activity established, one must next decide what false-negative rate is acceptable. Because the predictive models are imperfect, some nonzero false-negative rate is unavoidable. The approach described by Sheridan et al. was to use the predicted activity and uncertainty to determine a "*Z*-score," that is, what multiple of the uncertainty expressed as a standard deviation away from the threshold the compound is:

$$Z\text{-score} = \frac{\text{threshold-predicted activity}}{\text{uncertainty}}$$

A compound that is predicted to have an undesired off-target activity will have a negative *Z*-score, while a compound predicted to be clean will have positive *Z*-score. The more confident the model is that the activity will be less than the threshold, the larger the *Z*-score (Fig. 5).

For a given assay and retrospective set of compounds with experimental data and predictions made prior to the compound synthesis using the same methodology as the existing model, it is possible to examine the false-negative rate (the risk of missing an undesired activity) and true negative rate (the reward in terms of savings of unnecessary screening resources) for various potential *Z*-score thresholds. If the threshold is set at a large *Z*-score, only relatively few compounds will have a sufficiently high *Z*-score and therefore be deemed inactive, resulting in low false-negative rate but also low true negative rate, and therefore little resource savings. At lower *Z*-score thresholds, the number of compounds deemed inactive will increase, but so will the false-negative rate.

This framework appropriately handles variation both in predictive model performance and hit rate. If the model has poor predictive performance on a given endpoint, the uncertainty will be higher, resulting in lower *Z*-scores and therefore fewer compounds meeting the *Z*-score threshold to avoid testing; this is the desired outcome for a low-confidence model, but still allows sufficiently robust negatives to be identified and computationally derisked. Likewise, assays with relatively high hit rates or chemical series

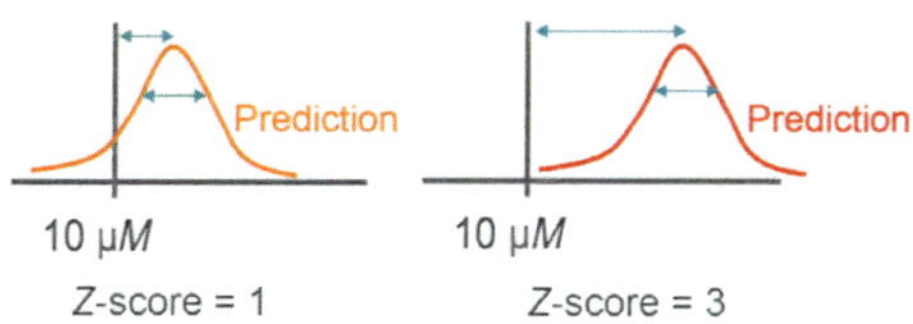


**Fig. 5** Relationship between predicted activity, prediction uncertainty, threshold, and *Z*-score. Both compounds are predicted to be clean, but according to the model the likelihood is much higher that the compound on the *left* will display the undesired off-target activity.

with a high hit rate for a given assay should also have higher predicted activities, also lowering the *Z*-scores.

Because the model returns both prediction and an uncertainty, from which a quantitative probability of activity can be estimated, in theory the chemical scaffold and compound physicochemical properties should not need to be considered. The model should be able to discern which scaffolds or properties are associated with activity and make appropriate predictions (or output higher uncertainty when unable to predict). In practice, it was observed the risk–reward plots as a function of *Z*-score varied across assays for a given scaffold and across scaffolds for a given assay. Scaffolds with a low hit rate—that is a low prior probability of activity—had a much more favorable risk–reward profile compared to scaffold/assay combinations with a higher hit rate. Thus, by monitoring the ability of the model to triage compounds for experimental testing as a function of scaffold and assay, the experimental resource can be further refined to the compounds most in need of it.

It is therefore important to benchmark the ability of the predictive models to distinguish actives from inactives, and the ability of the uncertainty estimate to assess the true uncertainty, for the chemical series of interest, and to monitor the performance on an ongoing basis. Compounds should routinely be chosen for testing regardless of the predictions to enable monitoring of the performance of the predictive models and appropriateness of the chosen threshold *Z*-score.

## REFERENCES

Altschul, S. F., Madden, T. L., Schäffer, A. A., Zhang, J., Zhang, Z., Miller, W., et al. (1997). Gapped BLAST and PSI-BLAST: A new generation of protein database search programs. *Nucleic Acids Research*, *25*, 3389–3402.

Bembenek, S. D., Hirst, G., & Mirzadegan, T. (2018). Determination of a focused mini kinase panel for early identification of selective kinase inhibitors. *Journal of Chemical Information and Modeling*, *58*, 1434–1440.

Burden, F. R. (2001). Quantitative structure – activity relationship studies using Gaussian processes. *Journal of Chemical Information and Computer Sciences*, *41*(3), 830–835.

Hamon, J., Whitebread, S., Techer-Etienne, V., Le Coq, H., Azzaoui, K., & Urban, L. (2009). In vitro safety pharmacology profiling: What else beyond hERG? *Future Medicinal Chemistry*, *1*(4), 645–665.

Kinnings, S. L., & Jackson, R. M. (2009). Binding site similarity analysis for the functional classification of the protein kinase family. *Journal of Chemical Information and Modeling*, *49*(2), 318–329.

Peters, J.-U. (2013). Polypharmacology—Foe or friend? *Journal of Medicinal Chemistry*, *56*(22), 8955–8971.

Peters, J.-U., Hert, J., Bissantz, C., Hillebrecht, A., Gerebtzoff, G., Bendels, S., et al. (2012). Can we discover pharmacological promiscuity early in the drug discovery process? *Drug Discovery Today*, *17*(7–8), 325–335.

Sheridan, R. P. (2013a). Time-split cross-validation as a method for estimating the goodness of prospective prediction. *Journal of Chemical Information and Modeling*, *53*(4), 783–790.

Sheridan, R. P. (2013b). Using random forest to model the domain applicability of another random forest model. *Journal of Chemical Information and Modeling*, *53*(11), 2837–2850.

Sheridan, R. P., McMasters, D. R., Voigt, J. H., & Wildey, M. J. (2015). eCounterscreening: Using QSAR predictions to prioritize testing for off-target activities and setting the balance between benefit and risk. *Journal of Chemical Information and Modeling*, *55*(2), 231–238.

Sheridan, R. P., Nam, K., Maiorov, V. N., McMasters, D. R., & Cornell, W. D. (2009). QSAR models for predicting the similarity in binding profiles for pairs of protein kinases and the variation of models between experimental data sets. *Journal of Chemical Information and Modeling*, *49*(8), 1974–1985.

Svensson, R., Aniceto, N., Norinder, U., Cortes-Ciriano, I., Spjuth, O., Carlsson, L., et al. (2018). Conformal regression for quantitative structure–activity relationship modeling—Quantifying prediction uncertainty. *Journal of Chemical Information and Modeling*, *58*(5), 1132–1140.

FSC
www.fsc.org
MIX
Papier | Fördert gute Waldnutzung
FSC® C083411

Kolibri 360 GmbH
Ferdinand-Jühlke-Straße 7
99095 Erfurt, Deutschland
produktsicherheit@kolibri360.de